The Human Brain during the First Trimester 15- to 18-mm Crown-Rump Lengths

This third of 15 short atlases reimagines the classic 5-volume *Atlas of Human Central Nervous System Development*. This volume presents serial sections from specimens between 15 mm and 18 mm with detailed annotations, together with 3D reconstructions. An introduction summarizes human CNS development by using high-resolution photos of methacrylate-embedded rat embryos at a similar stage of development as the human specimens in this volume. The accompanying Glossary gives definitions for all the terms used in this volume and all the others in the *Atlas*.

Key Features

- Classic anatomical atlas
- Detailed labeling of structures in the developing brain offers updated terminology and the identification of unique developmental features, such as, germinal matrices of specific neuronal populations and migratory streams of young neurons
- Appeals to neuroanatomists, developmental biologists, and clinical practitioners
- A valuable reference work on brain development that will be relevant for decades

ATLAS OF
HUMAN CENTRAL NERVOUS SYSTEM DEVELOPMENT
Series

Volume 1: The Human Brain during the First Trimester 3.5- to 4.5-mm Crown Rump Lengths

Volume 2: The Human Brain during the First Trimester 6.3- to 10.5-mm Crown Rump Lengths

Volume 3: The Human Brain during the First Trimester 15- to 18-mm Crown Rump Lengths

Volume 4: The Human Brain during the First Trimester 21- to 23-mm Crown Rump Lengths

Volume 5: The Human Brain during the First Trimester 31- to 33-mm Crown Rump Lengths

Volume 6: The Human Brain during the First Trimester 40- to 42-mm Crown Rump Lengths

Volume 7: The Human Brain during the First Trimester 57- to 60-mm Crown Rump Lengths

Volume 8: The Human Brain during the Second Trimester 96- to 150-mm Crown Rump Lengths

Volume 9: The Human Brain during the Second Trimester 160- to 170-mm Crown Rump Lengths

Volume 10: The Human Brain during the Second Trimester 190- to 210-mm Crown Rump Lengths

Volume 11: The Human Brain during the Third Trimester 225- to 235-mm Crown Rump Lengths

Volume 12: The Human Brain during the Third Trimester 260- to 270-mm Crown Rump Lengths

Volume 13: The Human Brain during the Third Trimester 310- to 350-mm Crown Rump Lengths

Volume 14: The Spinal Cord during the First Trimester

Volume 15: The Spinal Cord during the Second and Third Trimesters and the Early Postnatal Period

The Human Brain during the First Trimester 15- to 18-mm Crown-Rump Lengths

Atlas of Human Central Nervous System Development, Volume 3

Shirley A. Bayer and Joseph Altman

CRC Press
Taylor & Francis Group
Boca Raton London New York

CRC Press is an imprint of the
Taylor & Francis Group, an **informa** business

First edition published 2023
by CRC Press
6000 Broken Sound Parkway NW, Suite 300, Boca Raton, FL 33487-2742

and by CRC Press
4 Park Square, Milton Park, Abingdon, Oxon, OX14 4RN

CRC Press is an imprint of Taylor & Francis Group, LLC

LCCN no. 2022008216

ISBN: 978-1-032-18329-9 (hbk)
ISBN: 978-1-032-21928-8 (pbk)
ISBN: 978-1-003-27062-1 (ebk)

DOI: 10.1201/9781003270621

Typeset in Times Roman by KnowledgeWorks Global Ltd.

Access the support material at: https://routledge.com/9781032183299

CONTENTS

ACKNOWLEDGMENTS -- vi

AUTHORS --- viii

PART I.   **INTRODUCTION** -- 1

    Organization of the Atlas --- 1

    Specimens and Collections -- 1

    Plate Preparation --- 2

    3-Dimensional Computer Reconstructions --- 2

    Neurogenesis in Specimens (CR 15 to 18 mm) --------------------------------------- 2

    References --- 20

PART II.   **15-mm Crown Rump Length, C9247** --- 22

    Plates 1-7 A/B --- 24-37

    High Magnification Plates 8-21 A/B --- 38-65

PART III.   **15-mm Crown Rump Length, M2051** -- 66

    Plates 22-38 A/B -- 67-101

    High Magnification Plate 39 A/B -- 102-103

PART IV.   **15.8-mm Crown Rump Length, C492** -- 104

    Plates 40-56 A/B -- 106-139

PART V.   **18-mm Crown Rump Length, C1390** -- 140

    Plates 57-63 A/B -- 142-155

    High Magnification Plate 64 A/B -- 156-157

PART VI.   **17.5-mm Crown Rump Length, M2155** --- 158

    Plates 65-81 A/B -- 160-193

    High Magnification Plates 82-84 A/B -- 194-199

ACKNOWLEDGMENTS

We thank the late Dr. William DeMyer, pediatric neurologist at Indiana University Medical Center, for access to his personal library on human CNS development. We also thank the staff of the National Museum of Health and Medicine, who were at the Armed Forces Institute of Pathology, Walter Reed Hospital, Washington, D.C. when we collected data in 1995 and 1996: Dr. Adrianne Noe, Director; Archibald J. Fobbs, Curator of the Yakovlev Collection; Elizabeth C. Lockett; and William Discher. We are most grateful to the late Dr. James M. Petras at the Walter Reed Institute of Research, who made his darkroom facilities available so that we could develop all the photomicrographs on location rather than in our laboratory in Indiana. Finally, we thank Chuck Crumly, Neha Bhatt, Kara Roberts, Michele Dimont, and Rebecca Condit for expert help during production of the manuscript.

AUTHORS

Shirley A. Bayer received her PhD from Purdue University in 1974 and spent most of her scientific career working with Joseph Altman. She was a professor of biology at Indiana-Purdue University in Indianapolis for several years, where she taught courses in human anatomy and developmental neurobiology while continuing to do research in brain development. Her lengthy publication record of dozens of peer-reviewed, scientific journal articles extends back to the mid 1970s. She has co-authored several books and many articles with her late spouse, Joseph Altman. It was her research (published in Science in 1982) that proved that new neurons are added to granule cells in the dentate gyrus during adult life, a unique neuronal population that grows. That paper stimulated interest in the dormant field of adult neurogenesis.

Joseph Altman, now deceased, was born in Hungary and migrated with his family via Germany and Australia to the US. In New York, he became a graduate student in psychology in the laboratory of Hans-Lukas Teuber, earning a PhD in 1959 from New York University. He was a postdoctoral fellow at Columbia University, and later joined the faculty at the Massachusetts Institute of Technology. In 1968, he accepted a position as a professor of biology at Purdue University. During his career, he collaborated closely with Shirley A. Bayer. From the early 1960s-2016, he published many articles in peer-reviewed journals, books, monographs, and free online books that emphasized developmental processes in brain anatomy and function. His most important discovery was adult neurogenesis, the creation of new neurons in the adult brain. This discovery was made in the early 1960s while he was based at MIT, but was largely ignored in favor of the prevailing dogma that neurogenesis is limited to prenatal development. After Dr. Bayer's paper proved new neurons are added to granule cells in the hippocampus, Dr. Altman's monumental discovery became more accepted. During the 1990s, new researchers "rediscovered" and confirmed his original finding. Adult neurogenesis has recently been proven to occur in the dentate gyrus, olfactory bulb, and striatum through the measurement of Carbon-14—the levels of which changed during nuclear bomb testing throughout the 20th century—in postmortem human brains. Today, many laboratories around the world are continuing to study the importance of adult neurogenesis in brain function. In 2011, Dr. Altman was awarded the Prince of Asturias Award, an annual prize given in Spain by the Prince of Asturias Foundation to individuals, entities, or organizations globally who make notable achievements in the sciences, humanities, and public affairs. In 2012, he received the International Prize for Biology - an annual award from the Japan Society for the Promotion of Science (JSPS) for "outstanding contribution to the advancement of research in fundamental biology." This Prize is one of the most prestigious honors a scientist can receive. When Dr. Altman died in 2016, Dr. Bayer continued the work they started over 50 years ago, and in her late husband's, she created the Altman Prize, awarded each year by JSPS to an outstanding young researcher in developmental neuroscience.

INTRODUCTION

ORGANIZATION OF THE ATLAS

This is the third book in the *Atlas of Human Central Nervous System Development* series, 2nd Edition. It deals with human brain development during the middle first trimester. The five specimens in this book have crown-rump (CR) lengths from 15 to 18 mm with estimated gestation weeks (GW) from 7.4 to 7.8. To link crown-rump lengths to gestation weeks (GW), we relied on ultrasound data shown in Loughna et al. (2009). These specimens were analyzed in Volume 5 of the 1st edition (Bayer and Altman, 2008). The annotations emphasize the continuing stock-building neuroepithelium (NEP) in the cerebral cortex along the expanding shorelines of the lateral ventricles, the early shifts to a neurogenetic NEP (in the rest of the brain), the continual growth of the superarachnoid reticulum, and the interactions between the brain and peripheral structures in the head and pharyngeal arches, especially as they relate to the rhombomeres and sensory cranial nerves.

The present volume features five normal specimens. Three are cut in the frontal/horizontal plane, two in the sagittal plane. Each specimen is presented as a series of grayscale photographs of its Nissl-stained nervous system sections including the surrounding body (**Parts II** through **VI**). The photographs are shown from anterior to posterior (frontal/horizontal specimens) and medial to lateral (sagittal specimens). The dorsal part of each frontal/horizontal photo is toward the top of the page, the ventral part at the bottom, and the midline is in the vertical center. All frontal/horizontal specimens have computer-aided 3-dimensional reconstructions of their brains showing each section's location. That reconstruction clears up the ambiguity about the exact plane of sectioning through each specimen. For all sagittal specimens, the left side of each photo is facing anterior, right side posterior, top side dorsal, and bottom side ventral.

SPECIMENS AND COLLECTIONS

The specimens in this book are from the *Minot and Carnegie Collections* in the National Museum of Health and Medicine, which used to be housed at the Armed Forces Institute of Pathology (AFIP) in Walter Reed Hospital in Washington, D.C. Since the AFIP closed, the National Museum was moved to Silver Springs, MD; these collections are still available for research.

Three of the specimens are from the *Carnegie Collection* (designated by a **C** prefix), that started in the Department of Embryology of the Carnegie Institution of Washington. It was led by Franklin P. Mall (1862–1917), George L. Streeter (1873–1948), and George W. Corner (1889–1981). These specimens were collected during a 40- to 50-year time span and were histologically prepared with a variety of fixatives, embedding media, cutting planes, and histological stains. Early analyses of specimens were published in the early 1900s in *Contributions to Embryology, The Carnegie Institute of Washington* (now archived in the Smithsonian Libraries). O'Rahilly and Müller (1987, 1994) have given overviews of first trimester specimens in this collection.

Two specimens are from the *Minot Collection* (designated by an **M** prefix), which is the work of Dr. Charles Sedgwick Minot (1852-1914), an embryologist at Harvard University. Throughout his career, Minot collected about 1900 embryos from a variety of species. The 100 human embryos were probably acquired between 1900 and 1910. From our examination of these specimens and their similar appearance, we assume that they are preserved in the same way, although we could not find any records describing fixation procedures. The slides contain information on section numbers, section thickness (6 μ to 10 μ), and stain (aluminum cochineal).

PLATE PREPARATION

All sections of a given specimen were photographed at the same magnification. Sections throughout the entire specimen were photographed in serial order with Kodak technical pan black-and-white negative film (#TP442). The film was developed for 6 to 7 minutes in dilution F of Kodak HC-110 developer, stop bath for 30 seconds, Kodak fixer for 5 minutes, Kodak hypo-clearing agent for 1 minute, running water rinse for 10 minutes, and a brief rinse in Kodak photo-flo before drying.

The negatives were scanned at 2700 dots per inch (dpi) with a Nikon Coolscan-1000 35 mm negative film scanner attached to a Macintosh PowerMac G3 computer which had a plug-in driver built into Adobe Photoshop. The negatives were scanned as color positives because that brought out more subtle shades of gray. The original scans were converted to 300 dpi using the non-resampling method for image size. The powerful features of Adobe Photoshop were used to enhance contrast, correct uneven staining, and slightly darken or lighten areas of uneven exposure.

The photos chosen for annotation in **Parts II** through **VI** are presented as companion plates, designated as **A** and **B** on facing pages. **Part A** on the left page shows the full contrast photograph with labels of peripheral neural structures; **part B** on the right page shows low contrast copies of the same photograph with superimposed outlines of the labeled brain parts. The *low magnification plates* show entire sections to identify the large structures and subdivisions of the brain. The *high magnification plates* feature enlarged views of the brain to show tissue organization. This type of presentation allows a user to see the entire section as it would appear in a microscope and then consult the detailed markup in the low-contrast copy on the facing page, leaving little doubt about what is being identified. The labels themselves are not abbreviated, so the user is not constantly having to consult a list. Different fonts are used to label different classes of structures: the ventricular system is labeled in **CAPITALS**, the neuroepithelium and other germinal zones in **Helvetica bold**, transient structures in *Times bold italic*, and permanent structures in Times Roman or **Times bold**. Adobe Illustrator was used to superimpose labels and to outline structural details on the low contrast images. Plates were placed into a book layout using Adobe InDesign. Finally, high-resolution portable document files (pdf) were uploaded to CRC Press/ Taylor & Francis websites.

3-DIMENSIONAL COMPUTER RECONSTRUCTIONS

This process took five steps. *First,* image files in the series for each specimen were placed into a Photoshop stack with each image in a separate layer. *Second,* by altering the visibility and transparency of these layers, the sections were aligned to each other. After alignment, each layer was exported as a separate file. *Third,* Adobe Illustrator was used to outline the brain surface of each aligned section, and these contours were saved in separate Adobe Illustrator encapsulated postscript (eps) files. *Fourth,* the eps files were imported into 3D space (x, y, and z coordinates) using Cinema 4DXL (C4D, Maxon Computer, Inc.). For each section, points on the contours have unique x-y coordinates and the same z coordinate. By calculating the distance between sections, the entire array of contours was stretched out in the z axis. The C4D loft tool builds a spline mesh of polygons starting with the x-y points on the contour with the most anterior z coordinate and ending with the x-y points on the contour with the most posterior z coordinate. The spline meshes of the entire brain surface were rendered completely opaque at various camera angles using the C4D ray-tracing engine (**Figures 11** to **15** in **parts II** through **VI).** *Fifth,* in all frontal/horizontal-sectioned specimens, models of the brain surface posterior to a specific section were rendered with a copy of the photograph of that section texture mapped as a front cap on the model (*insets in* **Part A**).

NEUROGENESIS IN SPECIMENS (CR 15-18 mm)

The specimens in this volume are equivalent to rat embryos on embryonic days (E) 14 to E15 based on our timetables of neurogenesis using ^{3}H-thymidine dating methods (Bayer and Altman, 1995, 2012-present; Bayer et al., 1993,1995). **Table 1** lists populations being generated in the spinal cord and medulla (**Table 1A**), the pons and cerebellum (**Table 1B**), the mesencephalon (**Table 1C**), the hypothalamus and preoptic area (**Table 1D**), and the thalamus (**Table 1E**). Data for the pallidum, striatum, amygdala, and septum are in **Table 1F**. Finally, data for the neocortex, limbic cortex, olfactory cortex, hippocampal region, olfactory bulb, and anterior olfactory nucleus are in **Table 1E**. For all Tables, the left panel lists neurogenesis in E14 rats (comparable to 15-mm human specimens), the right panel in E15 rats (comparable to 16.8- to 18-mm specimens). Many of these populations are not distinguishable in the brain and spinal parenchyma, and often newly generated neurons are sequestered in the NEP before they migrate (Bayer and Altman, 2012-present). We use methacrylate-embedded rat embryos on E14 and E15 to show the fine detail of the spinal cord (**Fig. 1**, Bayer, 2013-present), medulla, pons, cerebellum, mesencephalon, diencephalon, and telencephalon (**Figs. 2-11**). All of these data are based on rats that have a similar morphological appearance to human embryos. It is assumed that similar developmental events are happening in the two species (Bayer et al., 1993, 1995; Bayer and Altman, 1995). Each table and figure set will be discussed briefly to summarize development in the spinal cord and different regions of the brain. All terms used in the annotations are defined in the comprehensive *Glossary* that accompanies the 2nd Edition Series.

Table 1A: Neurogenesis by Region

REGION and NEURAL POPULATION	CROWN RUMP LENGTH	
	15 mm	16.8-18 mm
SPINAL CORD		
Cervical somatic motor	●	
Thoracic somatic motor	●	
Thoracic visceral motor	●	
Lumbosacral somatic motor	●	
Ipsilateral sensory relay	● ●	
Dorsal horn interneurons		● ●
MEDULLA		
Ambiguus	●	● ●
Salivatory		● ●
Trigeminal (V, medullary)	● ●	● ●
Gracilis	●	●
Cuneatus	●	●
Solitary	●	●
Inferior vestibular	●	
Medial vestibular	● ●	●
Hypoglossal prepositus	● ●	●
Dorsal cochlear	●	● ●
Anteroventral cochlear	●	● ●
Posteroventral cochlear	●	● ●
Reticular formation rostral	●	●
Reticular formation caudodorsal	●	●
Reticular formation caudoventral	●	●
Raphe complex	●	●

Table 1. Neural populations in the spinal cord and medulla (**A**) that are being generated in rats on Embryonic day (E) 14 (comparable to humans at CR 15.0 mm) and on E15 (comparable to humans at CR 16.8-18 mm). *Green dots* indicate the amount of neurogenesis occurring: one dot=<15%; two dots=15-90%. Populations with 2 dots in both columns means that substantial neurogenesis occurs in both time periods. This same dot notation is used for all of the remaining parts (**B-G**) of **Table 1** on the following pages.

TABLE 1A/FIGURE 1

The somatic motor neurons complete their neurogenesis throughout the spinal cord in 15 mm specimens (**Table 1A**, *left panel*). Some presumptive somatic motor neurons are still exiting the neuroepithelium (NEP) in E14 rats (**Fig. 1A**), accumulating medial to the older somatic motor neurons (*red outline*). Some of these neurons were generated as early as E12 and are already growing cytoplasm that will extend into dendrites and long axons that enter the ventral roots of the spinal nerves. Eventually, the somatic motor neurons will form clumps that will innervate different parts of the musculature in the upper limb, neck and shoulder. There are already ventral root fibers in spinal nerves on both E14 and E15 rats (**Fig. 1AB**). The ventral spinal NEP is still active in its neurogenetic phase on E14, but there is a tremendous depletion by E15. The result is the enormous growth of the ventral horn in just 24 hours in rats (similar to human CRs of 15- to 18-mm). It is possible that many of the medial neurons in the ventral horn on E15 are microneurons that will modulate motor neuron output.

The migrating contralaterally projecting sensory relay neurons are prominent outside the spinal NEP on E14 (**Fig. 1A**) and appear as cohorts of spindle-shaped tangentially migrating cells; their neurogenesis is already complete by this day. They are extending axons into the ventral commissure. By E15 (**Fig. 1B**), the older contralaterally projecting sensory relay neurons are surrounded by younger infiltrating ipsilaterally projecting sensory relay neurons that are migrating radially from the spinal NEP (Altman and Bayer, 1984).

The dorsal spinal NEP grows dramatically between E14 and E15 in rats (*compare* **Figs 1A** and **1B**). E15 is a robust day of generation for the dorsal horn interneurons, and only a few will be generated on E16 in rats. Thus the dorsal spinal NEP is in an expansive neurogenetic phase at this stage, which can explain its enormous growth.

In the medulla, motor neurons in the hypoglossal nucleus are tentatively identified outside the medial medullary NEP on E15 (**Fig. 1D**). The same location is the site of massive migration on E14 (**Fig. 1C**). There is also adjacent robust migration just lateral to the hypoglossal nucleus that is most likely the dorsal motor nucleus of the vagus nerve; that is also tentatively identified by E15. Other neurons moving out lateral to the vagal motor nucleus may be destined for the hypoglossal prepositus nucleus. All of these nuclei have completely finished their neurogenesis and thus would be migrating at this time. On E14, a fair number of early generated reticular formation neurons are settling in the core of the medulla, and their numbers increase on E15, embedded in the criss-crossed fibrous network that is characteristic of their mature structure. The same is true of the medullary raphe complex neurons, a few settle outside the NEP on E14, many more on E15. By E15, a few neurons in the reticular formation and the raphe complex are still being produced by the medial medullary NEP (**Table 1A**, *right panel*), but it has greatly decreased.

The medullary NEP lateral and superior to the sulcus limitans is thick on both E14 and E15 as the NEP cells here are actively producing neurons in their neurogenetic phases. The most dorsal precerebellar NEP is producing neurons, and spindle-shaped cells can be traced migrating downward (*red outline*, **Fig 1C, D**) in the lateral parenchyma; this is the posterior imtramural migratory stream that contains inferior olive neurons. A few pioneer inferior olive neurons may be settling in the ventromedial medulla near the midline raphe fibers.

Between the precerebellar NEP and the NEP in the sulcus limitans, a variety of sensory neuronal populations are being generated. Groups of neurons are beginning to encircle the solitary tract, which has fibers for the first time on E15 (**Fig. 1D**). These are the afferent axons from the glossopharyngeal and vagal ganglia, which already have established entry zones, and thus it is not surprising that

THE SPINAL CORD AND MEDULLA
IN METHACRYLATE-EMBEDDED RAT EMBRYOS

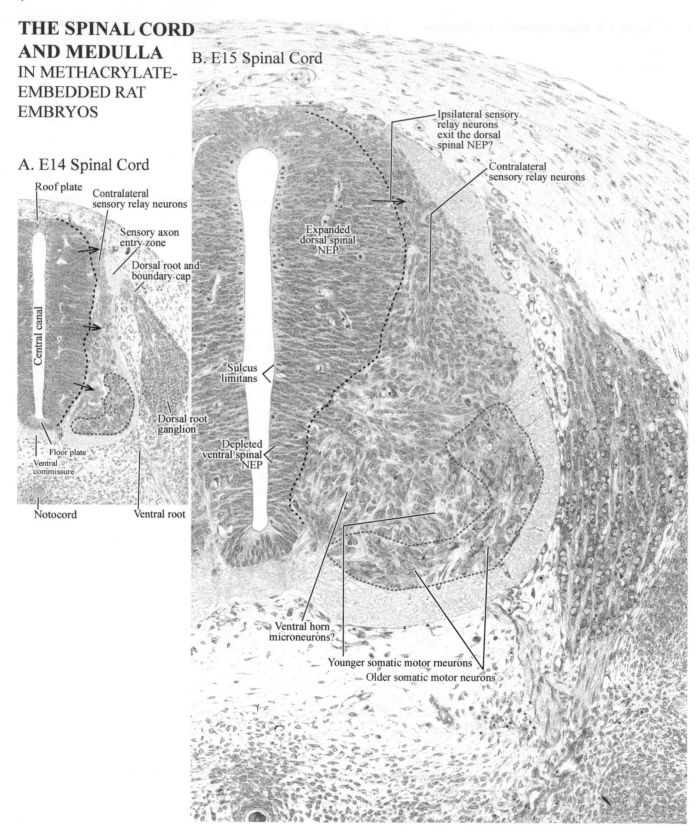

A. E14 Spinal Cord

Roof plate

Contralateral sensory relay neurons

Sensory axon entry zone

Dorsal root and boundary cap

Central canal

Dorsal root ganglion

Floor plate

Ventral commissure

Notocord

Ventral root

B. E15 Spinal Cord

Ipsilateral sensory relay neurons exit the dorsal spinal NEP?

Contralateral sensory relay neurons

Expanded dorsal spinal NEP

Sulcus limitans

Depleted ventral spinal NEP

Ventral horn microneurons?

Younger somatic motor rneurons

Older somatic motor neurons

Figure 1 (on facing pages). Frontal/horizontal sections of the spinal cord (**A, B**) and medulla (**C, D**) in rat embryos at similar stages as human specimens with CR lengths of 15-mm (**A C, E14**) and 16.8- to 18mm (**B D, E15**). The spinal cord is at the level of the cervical enlargement. The medulla is at the level of the posterior otic vesicle and the glossopharyngeal ganglion (3μ methacrylate sections, toluidine blue stain). **Source:** braindevelopment-maps.org (E14 and E15 coronal archives)

THE SPINAL CORD AND MEDULLA
IN METHACRYLATE-EMBEDDED RAT EMBRYOS

C. E14 medulla

D. E15 medulla

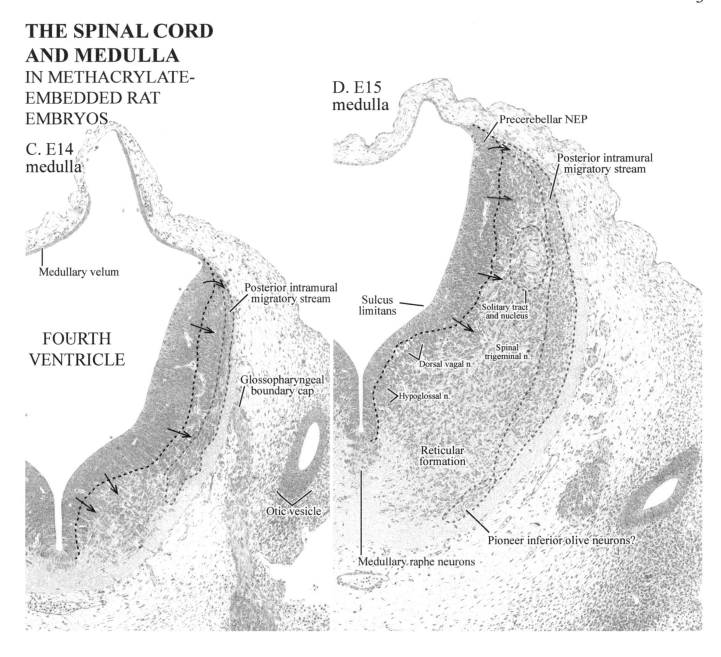

axons are already in the central solitary tract. The neurons migrating around the axonal bundle were presumptively generated by rhombomeres 4, 6, and 7. At this level, the neurons generated in rhombomere 6 are the most likely suspects for those accumulating around the tract. But neurons may migrate anteriorly and posteriorly within the solitary nucleus.

This level of the medulla probably contains NEPs that are generating the spinal nucleus of the trigeminal because it extends all the way into the spinal cord to blend with the substantia gelatinosa. Both E14 and E15 are robust days for neuron generation (**Table 1A**). However, a nuclear group is not definite by E15, but some pioneer spinal trigeminal neurons may be settling in the vicinity of the label placement in **Fig. 1D**.

This level of the medulla also may have the posterior remnants of the NEP that is generating the vestibular nuclei. If so, the neurons exiting the NEP in the vicinity of the sulcus limitans may be migrating vestibular neurons, since most vestibular nuclear neurons were generated prior to E14, and the only nuclei left to be generated are the medial and inferior (**Table 1A**).

Other neurons that are actively being generated are outside this section plane of the medulla, including the gracilis and cuneatus nuclei (they would be in more posterior levels), and the cochlear nuclei (they would be in more anterior levels).

THE CEREBELLUM AND PONS
IN METHACRYLATE-EMBEDDED RAT EMBRYOS

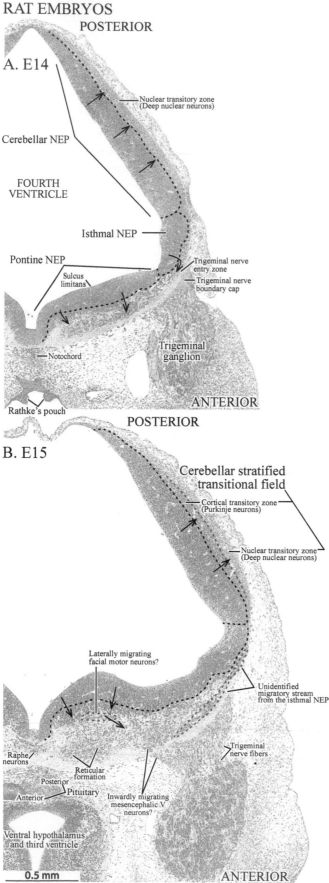

A. E14

POSTERIOR

Nuclear transitory zone (Deep nuclear neurons)

Cerebellar NEP

FOURTH VENTRICLE

Isthmal NEP

Pontine NEP

Sulcus limitans

Trigeminal nerve entry zone

Trigeminal nerve boundary cap

Notochord

Trigeminal ganglion

Rathke's pouch

ANTERIOR

B. E15

POSTERIOR

Cerebellar stratified transitional field

Cortical transitory zone (Purkinje neurons)

Nuclear transitory zone (Deep nuclear neurons)

Laterally migrating facial motor neurons?

Unidentified migratory stream from the isthmal NEP

Raphe neurons

Reticular formation

Posterior

Anterior

Pituitary

Inwardly migrating mesencephalic V neurons?

Trigeminal nerve fibers

Ventral hypothalamus and third ventricle

0.5 mm

ANTERIOR

Table 1B: Neurogenesis by Region		
REGION and NEURAL POPULATION	CROWN RUMP LENGTH 15 mm	16.8-18 mm
PONS		
Trigeminal (V) principal sensory	••	•
Infratrigeminal	•	••
Facial (VII) nucleus	•	
Parabrachial	•	
Dorsal tegmental	••	••
Ventral tegmental	•	••
Raphe complex	••	••
Reticular formation	•	•
Dorsal n. lateral lemniscus	•	
Ventral n. lateral lemniscus	•	•
Trapezoid (medial)		••
Lateral superior olive	•	
CEREBELLUM and PRECEREBELLAR NUCLEI		
Deep nuclei	••	
Purkinje neurons	••	••
Pontine reticular nucleus		•
Inferior olive	•	
External Cuneate	•	••
Lateral reticular	••	•

TABLES 1B-C/FIGURES 2-3

The pontine neuroepithelium (NEP) on E14 (**Fig. 2A**) is active and in the neurogenetic phase; motor structures have already been generated (mainly by the NEP medial to the sulcus limitans) except for a few neurons in the facial motor nucleus (**Table 1B**, *left column*). The NEP lateral to the sulcus limitans is also thick and active. Both parts of the NEP show massive outward migration (*arrows*). On E15 (**Fig. 2B**), reticular formation neurons and raphe neurons are tentatively identified in the pontine parenchyma, including a group of neurons migrating laterally just outside the NEP surrounded by fibers. They may be headed for the facial motor nucleus which will settle in the ventrolateral pons. By E15 some large neurons appear to be migrating into the brain with the most medial trigeminal nerve fibers. These are very likely the neurons of the mesencephalic nucleus of V that will go to the midbrain and settle around the central gray. The laterally migrating neurons along the superficial border of the trigeminal nerve fibers may be destined to settle in the trigeminal nuclei in the pons.

The cerebellar NEP is very thick and active on both E14 (**Fig. 2A**) and E15 (**Fig. 2B**). Deep neurons are finishing neurogenesis on E15, and Purkinje neurons are generated robustly on both E14 and E15 (**Table 1B**). There are some early generated deep nuclear neurons sojourning outside the NEP on E14 in a nuclear transitory zone. By E15, some Purkinje cells have moved out of the cerebellar NEP into

Figure 2. Horizontal sections of the anterior pons and cerebellum in rat embryos on E14 (A, similar to human CRs of 15-mm) and E15 (B, similar to human CRs from 16.8- to 18-mm) at the level of the trigeminal nerve entry zone. (3µ methacrylate section, toluidine blue stain)
braindevelopmentmaps.org (E14-E15 horizontal archive)

Table 1C: Neurogenesis by Region

REGION and NEURAL POPULATION	CROWN RUMP LENGTH 15 mm	CROWN RUMP LENGTH 16.8-18 mm
MESENCEPHALIC TEGMENTUM/ISTHMUS		
Edinger Westphal III nucleus	●	●
Nucleus of Darkschewitsch	●	●
Parabigeminal nucleus	●●	●
Red nucleus (magnocellular)	●	
Red nucleus (parvocellular)	●	
Substantia nigra compacta	●●	●●
Substantia nigra reticulata	●●	●●
Ventral tegmental area (lateral)	●●	●●
Ventral tegmental area (medial)		●
Dorsal interpeduncular nucleus	●●	●●
Ventral interpeduncular nucleus	●	●
Raphe complex	●●	●
Dorsal central gray	●●	●
Lateral central gray	●	●●
Ventral central gray	●●	●
SUPERIOR COLLICULUS		
stratum album	●	●●
stratum griseum profundum	●	●●
stratum lemnisci	●	●
magnocellular zone	●	●
stratum griseum intermediate	●	●
stratum opticum	●	●●
stratum griseum superficial	●	●
stratum zonale	●	●●
INFERIOR COLLICULUS		
Anterolateral		●●
Posterolateral		●●

their own sojourn zone, the cortical transitory zone. Both zones are part of the cerebellar stratified transitional field.

The midbrain tegmental and the tectal NEPs in horizontal sections (**Fig. 3**) are very thick; all are in the neurogenetic phase. The only neuronal populations that can be identified outside the NEP are pioneer neurons in the oculomotor complex and possibly some neurons in the ventral tegmental area/substantia nigra. Fewer neurons are outside the NEP lateral to the sulcus limitans. No doubt the tegmental NEP contains many sequestered postmitotic neurons because many populations have robust neurogenesis at both times (**Table 1C**). The superior collicular NEP generates a variety of neurons destined to settle in different layers. A few large neurons are just visible outside the NEP on E14 (**Fig 3A**), a few more on E15 (**Fig. 3B**); the remaining neurons are actively being generated on E15 (**Table 1C**, *right column*), which accounts for the few neurons in the superior collicular parenchyma.

Figure 3. Horizontal sections of the midbrain in rat embryos on E14 (A, similar to human CRs of 15- mm) and E15 (B, similar to human CRs from 16.8- to 18-mm) at the level of the trigeminal nerve entry zone. (3μ methacrylate section, toluidine blue stain)
braindevelopmentmaps.org (E14-E15 horizontal archive)

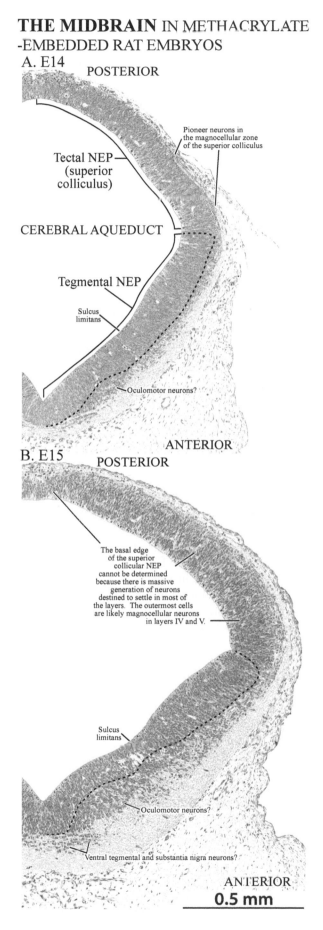

THE MIDBRAIN IN METHACRYLATE -EMBEDDED RAT EMBRYOS

A. E14

POSTERIOR

Pioneer neurons in the magnocellular zone of the superior colliculus

Tectal NEP (superior colliculus)

CEREBRAL AQUEDUCT

Tegmental NEP

Sulcus limitans

Oculomotor neurons?

ANTERIOR

B. E15

POSTERIOR

The basal edge of the superior collicular NEP cannot be determined because there is massive generation of neurons destined to settle in most of the layers. The outermost cells are likely magnocellular neurons in layers IV and V.

Sulcus limitans

Oculomotor neurons?

Ventral tegmental and substantia nigra neurons?

ANTERIOR

0.5 mm

Table 1D: Neurogenesis by Region		
REGION and NEURAL POPULATION	CROWN RUMP LENGTH	
	15 mm	16.8-18 mm
PREOPTIC A./HYPOTHALAMUS		
Lateral preoptic area	● ●	●
Medial preoptic area	● ●	● ●
Medial preoptic nucleus	●	● ●
Sexually dimorphic nucleus		●
Periventricular preoptic nucleus		●
Median preoptic nucleus	●	● ●
Lateral hypothalamic area	●	●
Paraventricular nucleus	● ●	● ●
Supraoptic nucleus	● ●	● ●
Ventromedial nucleus	● ●	● ●
Dorsomedial nucleus	●	● ●
Anterobasal nucleus	● ●	●
Arcuate nucleus		● ●
Suprachiasmatic nucleus	● ●	● ●
Premammillary nucleus	● ●	● ●
Suprammillary nucleus	●	●
Tuberommillary nucleus		●
Lateral mammillary nucleus	●	
Medial mammillary n. (dorsal)	●	● ●

TABLES 1D-E/FIGURES 4-7

Neurogenesis throughout the preoptic area and hypothalamus (**Table 1D**) is robust and many of its neuronal populations are being generated. In the thalamus (**Table 1E**), the anterior complex is just getting started (the neuroepithelium [NEP] is in stockbuilding stage), while posterior complex neurons are well on their way to being completed; for example, both geniculate bodies complete neurogenesis on E14 (**Table 1E**, *left column*, comparable to human CRs of 15 mm). As a result, all the NEPs in the diencephalon are quite thick except for some parts of the subthalamic and lateral hypothalamic NEPs.

We have chosen to illustrate diencephalic development in a series of four photos of methacrylate embedded rat embryos on E15 (comparable to human CRs between 16.8-18 mm) in frontal sections from anterior (**Fig. 4**) to progressively more posterior levels (**Figs. 5-7**). At the anterior level the diamond-shaped third ventricle has the preoptic NEP in the upper part of the diamond, with lateral preoptic area neurons settling outside it. The thalamus and subthalamus are not in this section plane. The lateral points in the diamond are the NEPs extending into the optic recess. The lower point of the diamond is taken up by the anterior hypothalamic NEP, where the lateral hypothalamic area and the anterobasal nucleus are settling. All of the NEPs at this level are well into their neurogenetic phases.

The section in **Figure 5** cuts through the pituitary gland below the infundibular recess. This part of the hypothalamus is the presumptive location of the neuroepithelia that are generating neurosecretory neurons in the supraoptic, paraventricular, and arcuate nuclei. It is possible that part of the medial hypothalamic NEP may also be generating the dorsomedial and ventromedial nuclei. The lateral hypothalamic NEP is superior (dorsal) to the medial NEP, and its neuronal progeny are migrating away in a curved ventrolateral course (*all large arrows indicate the direction of postmitotic neuronal migration*). Note that the lateral hypothalamic NEP is thinner than the medial NEP, indicating that its neurogenetic phase is coming to a close.

The subthalamic NEP is also thinner and a dense wave of radially migrating cells is adjacent to its basal edge. Farther laterally, neurons are less dense and are beginning to differentiate. This part of the diencephalon is adjacent to the basal telencephalon and it is possible that there is an exchange between neurons generated by diencephalic NEPs and telencephalic NEPs (you can see a fusion of the two NEPs in **Fig. 10**). This section shows the medial basal telencephalic NEP that is the presumed source of the bed nucleus of the stria terminalis. But the neurons settling outside this NEP may be part of the medial segment of the globus pallidus (entopeduncular nucleus). Some may also be neurons in the bed nucleus of the stria terminalis, but

Table 1E: Neurogenesis by Region		
REGION and NEURAL POPULATION	CROWN RUMP LENGTH	
	15 mm	16.8-18 mm
EPITHALAMUS/THALAMUS		
Anterodorsal		●
Anteroventral		●
Anteromedial		●
Anterolateral	●	● ●
Ventral complex	●	● ●
Ventroposterior (lateral)	●	● ●
Ventroposterior (medial)	●	● ●
Dorsal lateral geniculate	● ●	
Ventral lateral geniculate	● ●	
Pulvinar		● ●
Medial geniculate	● ●	
Reticular	● ●	
Medial Dorsal		●
Paratenial		●
Parafascicular	●	● ●
Reuniens		●
Rhomboid		●
Lateral habenula	● ●	● ●
Medial habenula		●
Subthalamic nucleus	● ●	● ●

the only way to determine the border is to note that the anterior lobule lacks an SVZ. The NEPs in this lobule will generate the medial dorsal nucleus and the anterior thalamic nuclear complex. All of these nuclei are just beginning to be generated (**Table 1E**, *right column*), so there are very few neurons outside this NEP. The presumptive habenular NEP lies just above the anterior lobule NEP and blends with the glioepithelia in the pineal recess.

The sections shown in **Figures 6** and **7** slice through NEPs in the superior lobule that are generating neurons in the lateral and medial geniculate bodies. The lateral geniculate neurons are generated just inferior to the pineal recess and move out of the NEP to migrate ventrally (within the section plane) in the superficial parenchyma. Eventually, these neurons will settle in the lateral thalamus to meet up with the optic tract. The tentative identification of the habenular NEP splits the superior lobule. As yet, habenular neurons are probably sequestered in the NEP because so few are outside it. Medial geniculate neurons also migrate outside their NEP and go posteriorly (perpendicular to the section plane) to eventually settle in a clump in the most posterior thalamus, adjacent to the midbrain where they will be near the auditory fibers coming from the inferior colliculus. The intermediate lobule NEP is in both posterior sections, because there is a subventricular zone there. Also, there is massive neuronal migration outside the intermediate lobule NEP, presumably to settle in the VPA/VPL nuclei. The reticular protuberance disappears in the most posterior diencephalic section (*compare* **Figs 6** and **7**), but there is a larger accumulation of presumptive reticular neurons migrating dorsolaterally (*curved arrows*).

The subthalamic NEP appears thick in **Figure 6** (sequestered postmitotic neurons cause the basal edge to be indistinct), but has thinned out in **Figure 7**. Indeed, neurogenesis in the subthalamus is coming to a close. The most notable feature here is that a lateral accumulation of neurons may be those settling in the subthalamic nucleus. Their neurons are generated in the hypothalamic mammillary NEP and migrate dorsolaterally. Indeed, a few spindle-shaped cells can be identified in a migratory stream heading upward from the mammillary NEP in both **Figures 6** and **7**.

The lateral hypothalamic NEP is thin in **Figure 6** and even more depleted **Figure 7**. That thinning indicates that neurogenesis is nearly complete at these posterior levels. Below that lies the premammillary and mammillary NEPs that are well into their neurogenetic phases. Postmitotic neurons outside these NEPs may be destined to settle in the premammillary and supramammillary nuclei. The lateral mammillary nucleus neurons are quite prominent outside the NEP in the most posterior section (**Fig. 7**).

the thickness of the NEP could be that most bed nucleus neurons are sequestered in basal parts of the NEP. The stria terminalis has a fair number of fibers that are probably from limbic areas of the brainstem heading toward the amygdala.

The thalamic NEP is superior to the subthalamic NEP. This section cuts through the anterior part of the reticular protuberance just above the reticular NEP. The reticular protuberance is the source of the oldest neurons in the reticular nucleus that settle closer to the midline than the remainder of the reticular nucleus. Some reticular neurons are outside the NEP. The neuroepithelium in the intermediate lobule is directly superior to the reticular NEP. Note that there are a few mitotic cells just outside the basal NEP in a subventricular zone (SVZ). The presence of an SVZ is diagnostic for the identification of the intermediate lobule. Neurons generated here will project to cortical areas around the central sulcus (the paracentral lobule). The ventral complex (ventral anterior and ventral lateral nuclei, VA/VL) is generated by more anterior parts of the intermediate lobule, while the posteroventral complex (ventral posterolateral and ventral posteromedial nuclei, VPL/VPM) are generated by more posterior parts of the intermediate lobule. We can find no differences between the VA/VL and the VPA/VPL NEPs. The anterior lobule NEP is directly above the intermediate lobule NEP. Again,

10

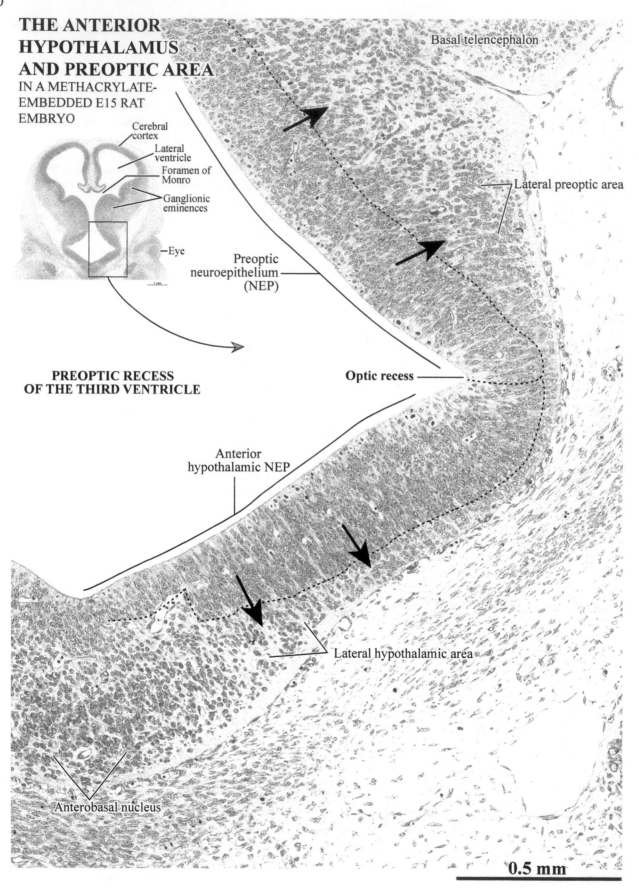

THE ANTERIOR HYPOTHALAMUS AND PREOPTIC AREA
IN A METHACRYLATE-EMBEDDED E15 RAT EMBRYO

Cerebral cortex

Lateral ventricle

Foramen of Monro

Ganglionic eminences

Eye

Preoptic neuroepithelium (NEP)

PREOPTIC RECESS OF THE THIRD VENTRICLE

Anterior hypothalamic NEP

Basal telencephalon

Lateral preoptic area

Optic recess

Lateral hypothalamic area

Anterobasal nucleus

0.5 mm

Figure 4. Frontal (coronal) section through the anterior forebrain that slices the preoptic area and anterior hypothalamic neuroepithelia in a rat embryo at a similar stage of development to human specimens with crown rump lengths of 16.8- to 18-mm. (3 μ methacrylate section, toluidine blue stain) Source: braindevelopmentmaps.org (E15 coronal archive)

THE DIENCEPHALON
IN A METHACRYLATE-
EMBEDDED E13 RAT EMBRYO

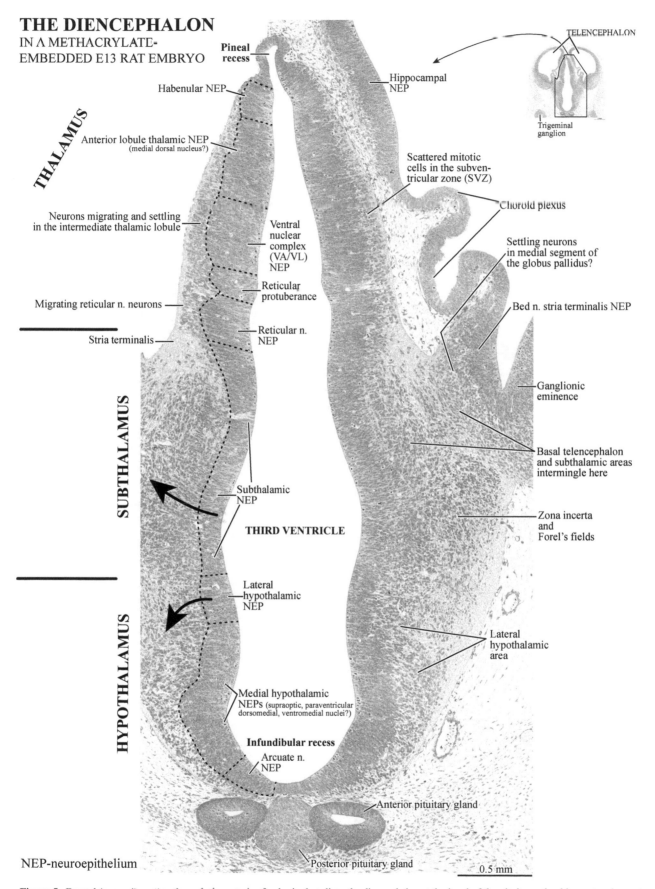

Pineal recess

TELENCEPHALON

Trigeminal ganglion

Habenular NEP

THALAMUS

Anterior lobule thalamic NEP
(medial dorsal nucleus?)

Hippocampal NEP

Scattered mitotic cells in the subventricular zone (SVZ)

Choroid plexus

Neurons migrating and settling in the intermediate thalamic lobule

Ventral nuclear complex (VA/VL) NEP

Settling neurons in medial segment of the globus pallidus?

Reticular protuberance

Bed n. stria terminalis NEP

Migrating reticular n. neurons

Reticular n. NEP

Stria terminalis

Ganglionic eminence

SUBTHALAMUS

Basal telencephalon and subthalamic areas intermingle here

Subthalamic NEP

THIRD VENTRICLE

Zona incerta and Forel's fields

Lateral hypothalamic NEP

HYPOTHALAMUS

Lateral hypothalamic area

Medial hypothalamic NEPs (supraoptic, paraventricular dorsomedial, ventromedial nuclei?)

Infundibular recess

Arcuate n. NEP

Anterior pituitary gland

NEP-neuroepithelium

Posterior pituitary gland

0.5 mm

Figure 5. Frontal (coronal) section through the anterior forebrain that slices the diencephalon at the level of the pituitary gland in a rat embryo at a similar stage of development to human specimens with crown rump lengths of 16.8- to 18-mm. (3 μ methacrylate section, toluidine blue stain) Source: braindevelopmentmaps.org (E15 coronal archive)

12

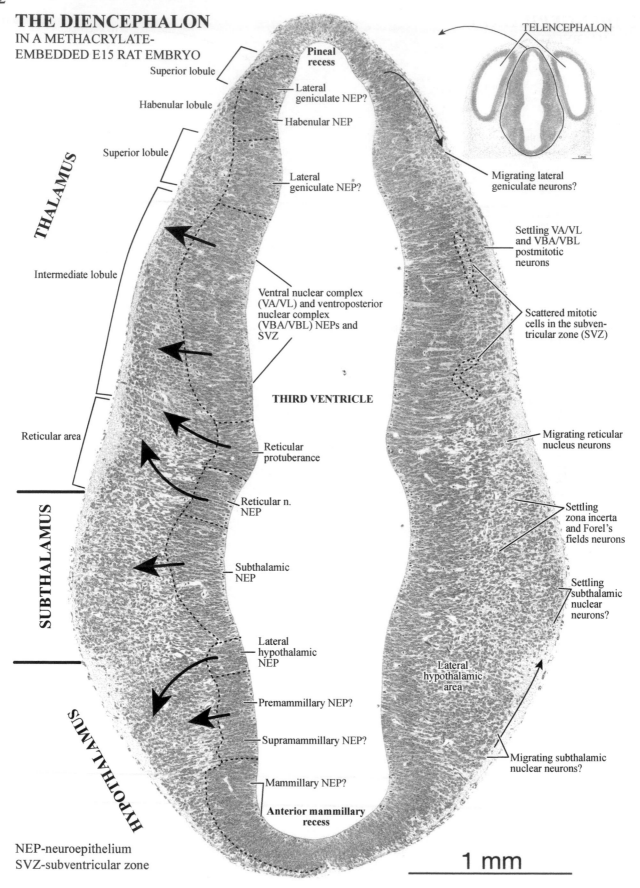

THE DIENCEPHALON
IN A METHACRYLATE-
EMBEDDED E15 RAT EMBRYO

Superior lobule

Habenular lobule

Superior lobule

THALAMUS

Intermediate lobule

Reticular area

SUBTHALAMUS

HYPOTHALAMUS

Pineal recess

Lateral geniculate NEP?

Habenular NEP

Lateral geniculate NEP?

Ventral nuclear complex (VA/VL) and ventroposterior nuclear complex (VBA/VBL) NEPs and SVZ

THIRD VENTRICLE

Reticular protuberance

Reticular n. NEP

Subthalamic NEP

Lateral hypothalamic NEP

Premammillary NEP?

Supramammillary NEP?

Mammillary NEP?

Anterior mammillary recess

TELENCEPHALON

Migrating lateral geniculate neurons?

Settling VA/VL and VBA/VBL postmitotic neurons

Scattered mitotic cells in the subventricular zone (SVZ)

Migrating reticular nucleus neurons

Settling zona incerta and Forel's fields neurons

Settling subthalamic nuclear neurons?

Lateral hypothalamic area

Migrating subthalamic nuclear neurons?

NEP-neuroepithelium
SVZ-subventricular zone

1 mm

Figure 6. Frontal (coronal) section through the anterior forebrain that slices the diencephalon through anterior parts of the mammillary body in a rat embryo at a similar stage of development to human specimens with crown rump lengths of 16.8- to 18-mm. (3 μ methacrylate section, toluidine blue stain) Source: braindevelopmentmaps.org (E15 coronal archive)

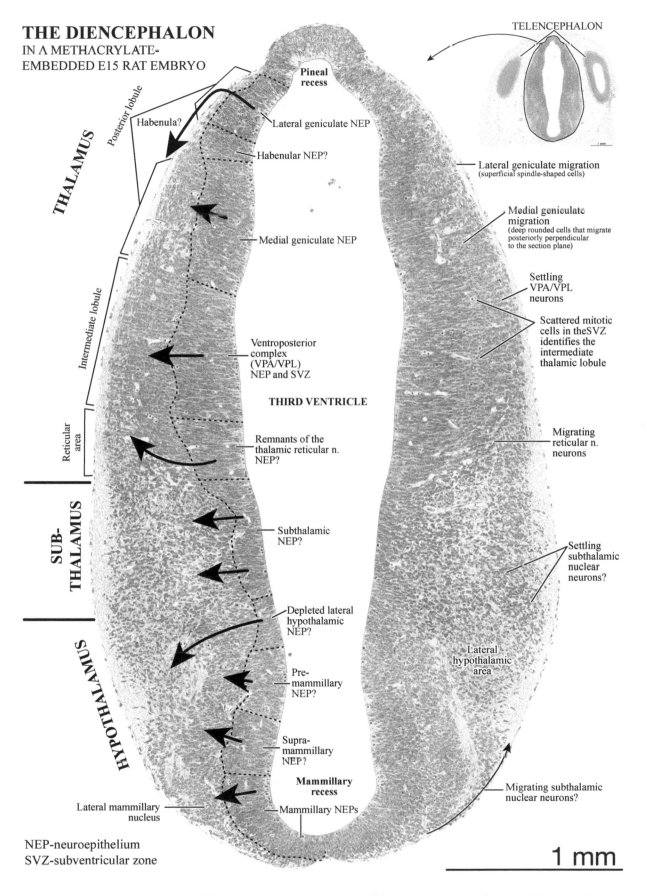

THE DIENCEPHALON
IN A METHACRYLATE-
EMBEDDED E15 RAT EMBRYO

TELENCEPHALON

THALAMUS

Posterior lobule

Habenula?

Pineal recess

Lateral geniculate NEP

Habenular NEP?

Medial geniculate NEP

Intermediate lobule

Reticular area

Ventroposterior complex (VPA/VPL) NEP and SVZ

THIRD VENTRICLE

Remnants of the thalamic reticular n. NEP?

SUB-THALAMUS

Subthalamic NEP?

Depleted lateral hypothalamic NEP?

Pre-mammillary NEP?

Supra-mammillary NEP?

Mammillary recess

HYPOTHALAMUS

Lateral mammillary nucleus

Mammillary NEPs

Lateral geniculate migration
(superficial spindle-shaped cells)

Medial geniculate migration
(deep rounded cells that migrate posteriorly perpendicular to the section plane)

Settling VPA/VPL neurons

Scattered mitotic cells in theSVZ identifies the intermediate thalamic lobule

Migrating reticular n. neurons

Settling subthalamic nuclear neurons?

Lateral hypothalamic area

Migrating subthalamic nuclear neurons?

NEP-neuroepithelium
SVZ-subventricular zone

1 mm

Figure 7. Frontal (coronal) section through the anterior forebrain that slices the diencephalon through posterior parts of the mammillary body in a rat embryo at a similar stage of development to human specimens with crown rump lengths of 16.8- to 18-mm. (3 μ methacrylate section, toluidine blue stain) Source: braindevelopmentmaps.org (E15 coronal archive)

Table 1F: Neurogenesis by Region

REGION and NEURAL POPULATION	CROWN RUMP LENGTH	
	15 mm	16.8-18 mm
PALLIDUM AND STRIATUM		
Entopeduncular nucleus	•	
Globus pallidus (external segment)	• •	•
Substantia innominata	•	• •
Basal nucleus of Meynert	•	• •
Olfactory tubercle (large neurons)	• •	• •
Olfactory tubercle (small neurons)	•	•
Caudate and putamen		•
Nucleus accumbens		•
AMYGDALA		
Anterior amygdaloid area	• •	•
Nucleus of the accessory olfactory tract	•	•
Nucleus of the lateral olfactory tract	•	• •
Central nucleus	•	•
Intercalated masses		•
Medial nucleus	• •	• •
Anterior cortical nucleus	• •	• •
Posterior cortical nucleus	•	• •
Basomedial nucleus	•	• •
Basolateral nucleus	•	• •
Lateral nucleus	•	• •
Bed n. stria terminalis (anterior)	•	• •
Bed n. stria terminalis (preoptic continuation)	•	• •
SEPTUM		
Medial nucleus	• •	• •
Diagonal band (vertical limb)	•	• •
Triangular nucleus	•	• •
Lateral nucleus	•	• •
Bed nucleus of the anterior commissure	•	• •

TABLES 1F-G/FIGURES 8-11

Neurogenesis throughout the basal telencephalon (**Table 1F**) is continuing to be robust in rat embryos on E14 (similar to human embryos with CR 15 mm, *left column*); for example, the entopeduncular nucleus finishes and many other populations are being generated. Neurogenesis is at a high level in rat embryos on E15 (similar to human embryos with CRs from 16.8-18 mm, *right column*). Note that the substantia innominata, basal nucleus of Meynert, and large neurons in the olfactory tubercle are being generated in large numbers. All amygdaloid neuroepithelia (NEPs) are in neurogenetic stage and neurons are being generated in every nuclear group. The same is true for all septal NEPs. In contrast, neurogenesis in the striatum is just getting started because the subventricular zone (SVZ) that will generate the medium spiny neurons is just being set up.

Neurogenesis in the cortical regions (**Table 1G**) of the telencephalon is either completely finished (Cajal-Retzius neurons) or nearly finished (subplate neurons) in neocortical areas, while neurons in layers VI and V are just getting started. The deep layers of the olfactory cortex started neurogenesis in E13 rats (similar to 10-10-5 mm human embryos (*see* Table 1E in Bayer and Altman, 2022,) and continue in the specimens in this Volume, but now the superficial layers are also being generated. The entorhinal cortex and subiculum begin their neurogenetic phases in the hippocampal region, so their NEPs leave stockbuilding stages to enter neurogenetic stages. In contrast, the NEPs producing the hippocampus proper have not even started, so they are not listed in **Table 1G**. The olfactory neurons in the accessory bulb and main bulb are produced in the basal telencephalic neuroepithelia (NEPs).

As for the diencephalon, we have chosen to illustrate telencephalic development with three frontal/coronal sections from anterior (**Fig. 8**) to progressively more posterior levels (**Figs. 9-10**). **Figure 11** shows a horizontal section from the frontal to the occipital poles. Many of these images show presumptive migrating neurons if you increase the magnification in the electronic versions. It is highly encouraged for readers to download the original images available in braindevelopmentmaps.org where you can zoom to high magnification and clearly see individual cells, fibers and migratory streams of postmitotic neurons.

Figure 8 is a section just behind the frontal telencephalic pole cutting through the olfactory invagination that contains both the vomeronasal nerve and the olfactory nerve. Some spindle-shaped neurons outside the medial ganglionic NEP are tentatively identified as migrating neurons. The accessory bulb output neurons may be the ones migrating and settling near the spot where the vomeronasal nerve is approaching the basomedial telencephalon; that nerve is their main input in mature brains. Farther ventrally, some main olfactory bulb mitral cells may be migrating from the medial ganglionic NEP (*curved arrows*) toward the area where the olfactory nerve is approaching the telencephalon. There is no olfactory bulb yet, but the contact zones for these nerves probably induce the NEP to evaginate into an olfactory bulb. If you want to learn more about olfactory bulb development, there is a book available on braindevelopmentmaps.org; you can find it by clicking on item 6 in the "Annotated and labeled zoomified image sets" at the bottom of the navigation page (Bayer, 2017).

The cortical NEPs are very thick at this level because they are being cut tangentially. That allows the different parts of the primordial plexiform layer to be more easily identified. Cohorts of spindle-shaped neurons are in a lateral migratory stream outside the neocortical and limbic cortical NEPs (*curved arrows*); some of these may be migrating ventrally to eventually settle in the primary olfactory cortex and claustrum. All areas of the cortical NEP are also the source of radially migrating cells (*straight arrows*).

Figure 9 cuts through the telencephalon at the level of the eye and shows the prominent anterolateral and medi-

Table 1G: Neurogenesis by Region

REGION and NEURAL POPULATION	CROWN RUMP LENGTH 15 mm	16.8-18 mm
NEOCORTEX and LIMBIC CORTEX		
Cajal-Retzius neurons	●●	●
Layer V	●	●
Layer VI	●	●●
Subplate VII	●●	●●
OLFACTORY CORTEX		
Layer II (anterior)	●	●
Layer II (posterior)	●	●●
Layers III-IV (anterior)	●●	●●
Layers III-IV (posterior)	●	●●
HIPPOCAMPAL REGION		
Entorhinal cortex Layer II		●●
Entorhinal cortex Layer III		●
Entorhinal cortex Layer IV	●	●●
Entorhinal cortex Layers V-VI	●	●●
Subiculum		●
OLFACTORY BULB		
Output neurons (accessory bulb)	●	●
Mitral cells (main bulb)	●	●●
Internal tufted cells (main bulb)	●	●
ANTERIOR OLFACTORY NUCLEUS		
Pars externa		●
AON proper		●

al ganglionic eminences as well as the septum. All NEPs producing neurons for these areas are well into their neurogenetic stages but are still thick because many postmitotic neurons are sequestered in their basal parts. A few mitotic cells are outside the dense cellular regions adjacent to the NEP in the ganglionic eminences. These are the pioneer members of the subventricular zone (SVZ) that will generate the medium spiny neurons in the striatum and nucleus accumbens. Some spindle-shaped cells, initially seen outside the medial ganglionic eminence NEP, can be followed in a migratory stream that travels through the SVZ directed toward the cortex (deep curved arrows) The lateral migratory stream (superficial arrows) from the cortex can be tentatively followed heading toward the primary olfactory cortex, where both superficial and deep layers can be distinguished. The cortical NEPs in this slice take up most of the cortical wall, but a maturation gradient can be discerned. Note that the primordial plexiform layer is thicker outside the lateral limbic and lateral neocortical NEPs than it is outside the medial neocortical and medial limbic (cingulate) cortical NEPs. This maturation gradient will be present through all stages of cortical maturation: lateral areas are always ahead of dorsomedial areas. The

parenchyma expands outside the medial prefrontal limbic cortical NEP. Neurons in this area of the cortex are also generated earlier and differentiate earlier than dorsal neo- and limbic cortex.

The section in **Figure 10** slices the telencephalon at the level of the amygdala and the anterior thalamus. The basal telencephalic NEP is on the medial side of the deep cleft in the lateral ventricle where it is fused with the subthalamic NEP on the diencephalic side. This medial NEP probably contains the germinal site of parts of the entopeduncular nucleus and the external segment of the globus pallidus. Farther back (*see* **Fig. 5**) the germinal source of the bed nucleus of the stria terminalis is probably present at this location. Just lateral to the slight indentation of the amygdaloid NEP is a remnant of the posterolateral ganglionic eminence, recognizable because it contains mitotic cells in an SVZ. Posterior parts of the primary olfactory cortex are settling ventrolateral to the amygdala, and spindle-shaped cells from the lateral migratory stream of the cortex can be followed into the lateral parts of the presumptive amygdala. Possibly, these are cortical neurons invading the basolateral nuclear complex, where many cells look like modified pyramidal cells. The parts of the cortex in this section are characterized by a thick NEP that takes up most of the brain wall. Still, there is a thicker primordial plexiform layer outside the lateral limbic and neocortical areas than the medial neocortical and limbic cortical areas. The parenchyma expands again adjacent to the subicular/hippocampal NEP. That does not mean that this part of the cortex is generated earlier or maturing earlier, rather it is due to the curved structure of the future hippocampus that is just beginning to develop adjacent to the fimbria/fornix glioepithelium that blends with the choroid plexus.

The section in **Figure 11** gives a different perspective of the cerebral cortex, cutting through it horizontally from the frontal to the occipital poles above the basal telencephalon (the ganglionic eminences are ventral to this section). The horizontal plane is slicing most of the cortical NEPs perpendicularly, so one can easily see maturation gradients as judged by the thickness of the primordial plexiform layer. It is thickest on the mid-lateral side where the insular part of the cortex will develop and thins out as one goes anterior and posterior (curved arrows). Another feature of this section plane is the lack of any spindle-shaped cells. The migrating cells are still there as indicated in the lateral migratory stream, but they are positioned perpendicular to the section plane and appear as small, rounded cells rather than long, spindle-shaped cells.

THE TELENCEPHALON
IN A METHACRYLATE-
EMBEDDED E15 RAT EMBRYO
(OLFACTORY LEVEL)

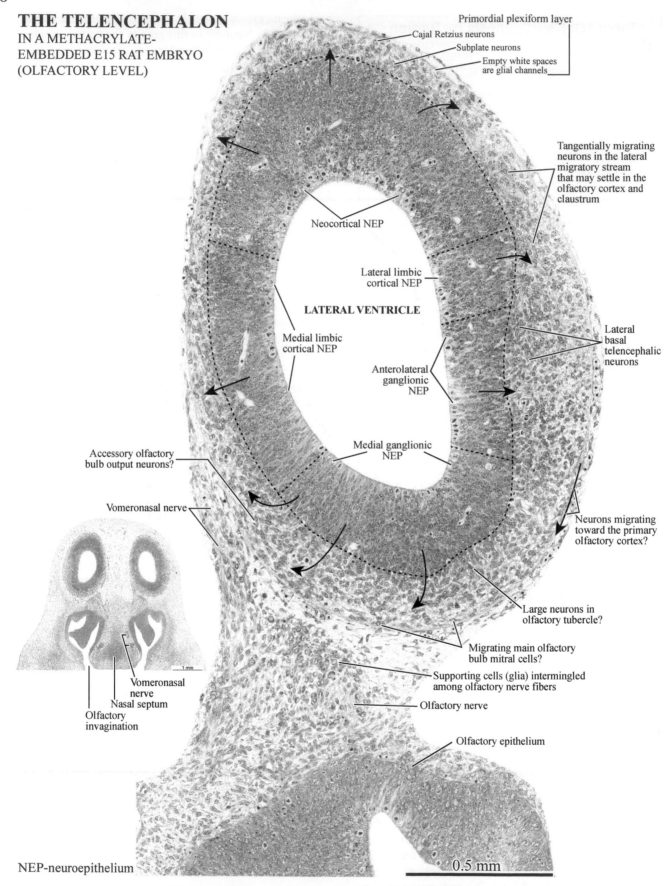

Primordial plexiform layer

Cajal Retzius neurons

Subplate neurons

Empty white spaces
are glial channels

Tangentially migrating
neurons in the lateral
migratory stream
that may settle in the
olfactory cortex and
claustrum

Neocortical NEP

Lateral limbic
cortical NEP

LATERAL VENTRICLE

Lateral
basal
telencephalic
neurons

Medial limbic
cortical NEP

Anterolateral
ganglionic
NEP

Accessory olfactory
bulb output neurons?

Medial ganglionic
NEP

Vomeronasal nerve

Neurons migrating
toward the primary
olfactory cortex?

Vomeronasal
nerve

Nasal septum

Olfactory
invagination

Large neurons in
olfactory tubercle?

Migrating main olfactory
bulb mitral cells?

Supporting cells (glia) intermingled
among olfactory nerve fibers

Olfactory nerve

Olfactory epithelium

NEP-neuroepithelium

0.5 mm

Figure 8. Frontal (coronal) section through the anterior telencephalon that slices the olfactory invagination, the olfactory nerve, and the vomeronasal nerve in a rat embryo at a similar stage of development to human specimens with crown rump lengths of 16.8- to 18-mm. (3 μ methacrylate section, toluidine blue stain) Source: braindevelopmentmaps.org (E15 coronal archive)

THE TELENCEPHALON
IN A METHACRYLATE-
EMBEDDED
E15 RAT EMBRYO
(EYE LEVEL)

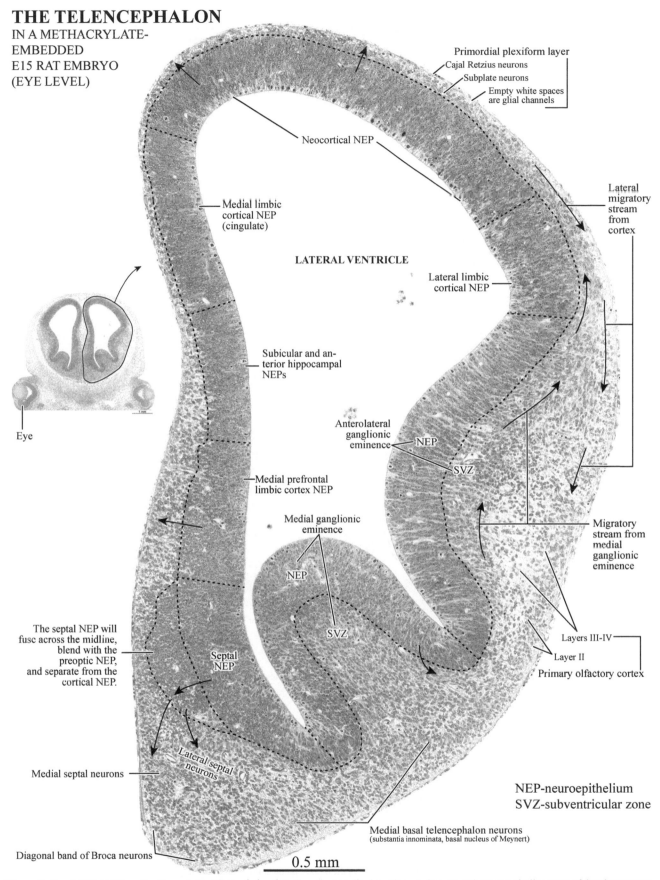

Primordial plexiform layer
Cajal Retzius neurons
Subplate neurons
Empty white spaces
are glial channels

Neocortical NEP

Lateral
migratory
stream
from
cortex

Medial limbic
cortical NEP
(cingulate)

LATERAL VENTRICLE

Lateral limbic
cortical NEP

Eye

Subicular and an-
terior hippocampal
NEPs

Anterolateral
ganglionic
eminence — NEP

SVZ

Migratory
stream from
medial
ganglionic
eminence

Medial prefrontal
limbic cortex NEP

Medial ganglionic
eminence

NEP

SVZ

The septal NEP will
fuse across the midline,
blend with the
preoptic NEP,
and separate from the
cortical NEP.

Septal
NEP

Layers III-IV

Layer II

Primary olfactory cortex

Medial septal neurons

Lateral septal
neurons

NEP-neuroepithelium
SVZ-subventricular zone

Diagonal band of Broca neurons

Medial basal telencephalon neurons
(substantia innominata, basal nucleus of Meynert)

0.5 mm

Figure 9. Frontal (coronal) section through the telencephalon that slices the ganglionic eminences in a rat embryo at a similar stage of development as human specimens of crown rump lengths 16.8- to 18-mm. (3 μ methacrylate section, toluidine blue stain) Source: braindevelopmentmaps.org (E15 coronal archive)

18

THE TELENCEPHALON
IN A METHACRYLATE-
EMBEDDED
E15 RAT EMBRYO
(ANTERIOR
AMYGDALA
LEVEL)

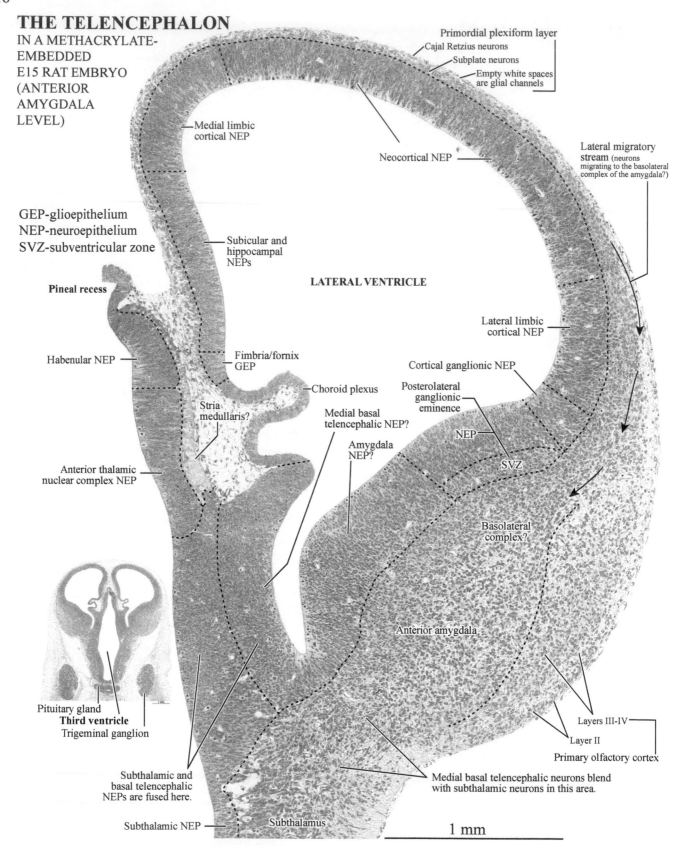

Primordial plexiform layer

Cajal Retzius neurons

Subplate neurons

Empty white spaces
are glial channels

Medial limbic
cortical NEP

Neocortical NEP

Lateral migratory
stream (neurons
migrating to the basolateral
complex of the amygdala?)

GEP-glioepithelium
NEP-neuroepithelium
SVZ-subventricular zone

Subicular and
hippocampal
NEPs

LATERAL VENTRICLE

Lateral limbic
cortical NEP

Pineal recess

Habenular NEP

Fimbria/fornix
GEP

Choroid plexus

Cortical ganglionic NEP

Posterolateral
ganglionic
eminence

Stria
medullaris?

Medial basal
telencephalic NEP?

NEP

Amygdala
NEP?

SVZ

Anterior thalamic
nuclear complex NEP

Basolateral
complex?

Pituitary gland
Third ventricle
Trigeminal ganglion

Anterior amygdala

Layers III-IV

Layer II

Primary olfactory cortex

Subthalamic and
basal telencephalic
NEPs are fused here.

Medial basal telencephalic neurons blend
with subthalamic neurons in this area.

Subthalamic NEP

Subthalamus

1 mm

Figure 10. Frontal (coronal) section through the telencephalon that slices the diencephalon at the level of the trigeminal ganglion in a rat embryo at a similar stage of development to human specimens of crown rump lengths of 16.8- to 18-mm. (3 μ methacrylate section, toluidine blue stain) Source: braindevelopmentmaps.org (E15 coronal archive)

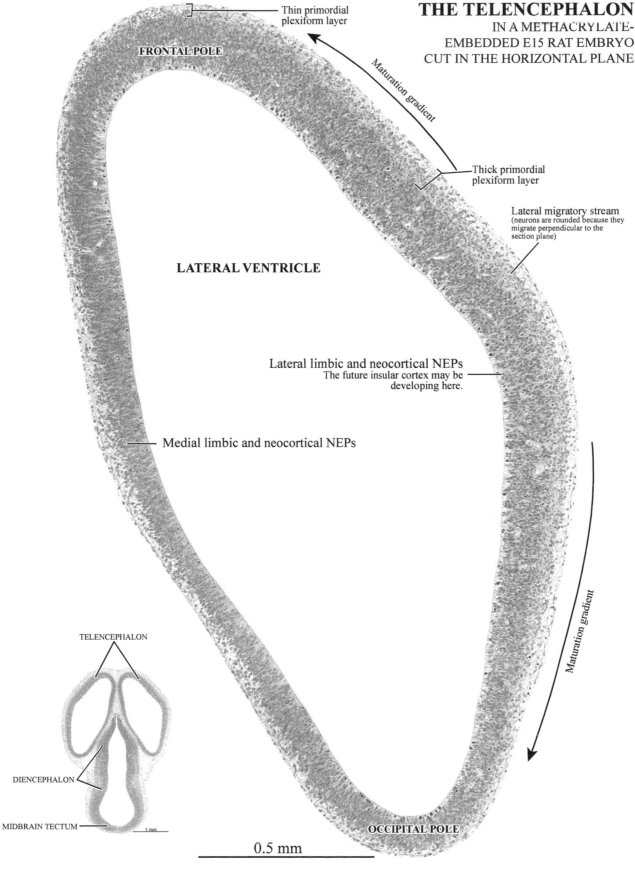

Thin primordial
plexiform layer

THE TELENCEPHALON
IN A METHACRYLATE-
EMBEDDED E15 RAT EMBRYO
CUT IN THE HORIZONTAL PLANE

FRONTAL POLE

Maturation gradient

Thick primordial
plexiform layer

Lateral migratory stream
(neurons are rounded because they
migrate perpendicular to the
section plane)

LATERAL VENTRICLE

Lateral limbic and neocortical NEPs
The future insular cortex may be
developing here.

Medial limbic and neocortical NEPs

Maturation gradient

TELENCEPHALON

DIENCEPHALON

MIDBRAIN TECTUM

1 mm

OCCIPITAL POLE

0.5 mm

Figure 11. Horizontal section through the telencephalic cerebral cortex that slices it from the frontal to occipital poles in a rat embryo at a similar stage of development to human specimens of crown rump lengths of 16.8- to 18-mm. (3 μ methacrylate section, toluidine blue stain) Source: braindevelopmentmaps.org (E15 coronal archive)

REFERENCES

Altman J, Bayer SA (1984) *The Development of the Rat Spinal Cord. Advances in Anatomy Embryology and Cell Biology*, Vol. 85, Berlin, Springer -Verlag.

Bayer, SA (2013-present) www.braindevelopmentmaps. org (This website is an image database of methacrylate-embedded normal rat embryos and paraffin-embedded rat embryos exposed to ^{3}H-Thymidine.)

Bayer SA (2017) *Development of the Olfactory Bulb.* braindevelopmentmaps.org. Laboratory of Developmental Neurobiology, Ocala, FL.

Bayer, SA (2022) *Glossary to Accompany Atlas of Human Central Nervous System Development (Second Edition)* Laboratory of Developmental Neurobiology, Ocala, FL.

Bayer SA, Altman J, Russo RJ, Zhang X (1993) Timetables of neurogenesis in the human brain based on experimentally determined patterns in the rat. *Neurotoxicology* **14**: 83-144.

Bayer SA, Altman J, Russo RJ, Zhang X (1995) Embryology. In: *Pediatric Neuropathology*, Serge Duckett, Ed. Williams and Wilkins, pp. 54-107.

Bayer SA, Altman J (1995) Development: Some principles of neurogenesis, neuronal migration and neural circuit formation. In: *The Rat Nervous System*, 2nd Edition, George Paxinos, Ed. Academic Press, Orlando, Florida., pp. 1079-1098.

Bayer SA, Altman J (2008) *Atlas of Human Central Nervous System Development* (First Edition), Volume 5, CRC Press.

Bayer SA, Altman J (2012-present) www.neurondevelopment.org (This website has downloadable pdf files of our scientific papers on rat brain development grouped by subject.)

Bayer SA, Altman J (2022) *The Human Brain During the First Trimester 6.3- to 10.5-mm Crown-Rump Lengths, Atlas of Human Central Nervous System Development* (Second Edition), Volume 2, CRC Press/Taylor and Francis.

Loughna P, Citty L, Evans T, Chudleigh T (2009) Fetal size and dating: Charts recommended for clinical obstetric practice, *Ultrasound*, 17:161-167.

O'Rahilly R, Müller F. (1987) *Developmental Stages in Human Embryos, Carnegie Institution of Washington*, Publication 637.

O'Rahilly R; Müller F. (1994) *The Embryonic Human Brain*, Wiley-Liss, New York.

Van Hartesveldt C, Moore B, Hartman, BK (1986) Transient midline raphe glial structure in the developing rat, *J Comp Neurol* 253:174-184.

PART II: C9247
CR 15.0 mm (GW 7.4)
Sagittal

Carnegie Collection specimen #9247 (designated here as C9247) was collected in 1954 from a tubal pregnancy. The crown-rump length (CR) is 15 mm estimated to be at gestational week (GW) 7.4. C9247 was fixed in formalin, embedded in a celloidin/paraffin mix, and was cut in 8-μm sagittal sections that were stained with azan. Various orientations of the computer-aided 3-D reconstruction of M2051's brain are used to show the gross external features of a GW7.4 brain (**Figure 1**). C9247's sections are perfectly aligned in the sagittal plane. Considering all of the specimens in every volume of the *Atlas*, this is one of the best for quality of histological preservation and adherence to a section plane. Indeed, nearly an entire volume could be dedicated to the analysis of this brain. We photographed 64 sections at low magnification from the left to the right sides of the brain. Seven sections from the left side of the brain are illustrated in **Plates 1 to 7A/B**. Each illustrated section shows the brain with all surrounding tissues. Labels in **A Plates** (normal-contrast images) identify non-neural structures, peripheral neural structures, and brain ventricular divisions; labels in **B Plates** (low-contrast images) identify central neural structures. **Plates 8 to 21 A/B** show high-magnification views of all parts of the developing brain. The sagittal plane is ideal to show the relative sizes of major brain subdivisions and the entry zones of sensory nerve fibers.

The telencephalon is the smallest overall brain structure, composed mainly of a "stockbuilding" neuroepithelium (NEP) in the cortex surrounding an expanding telencephalic superventricle. The primordial plexiform layer adjacent to the cortical NEP has only scattered Cajal-Retzius cells, as illustrated in the dorsomedial less mature cortex. Photos of the ventrolateral cortex have more Cajal-Retzius cells in the layer. Many migrating neurons are adjacent to the basal ganglionic and basal telencephalic neuroepithelia forming mounds in the floor of the telencephalon. The evaginating olfactory NEP is just becoming visible (**Plate 3**), and a few pioneer olfactory nerve fibers get near the brain surface, perhaps contacting it.

The diencephalon is the larger forebrain structure. The neurogenetic neuroepithelium surrounds a dorsally expanding superventricle in the future thalamic area. The subthalamic and hypothalamic NEPs are shrinking because stockbuilding stem cells are depleted as they generate neurons. Migrating and settling young neurons accumulate in a thick band outside the NEP in the ventral diencephalic parenchyma adjacent to a thin subpial fibrous band.

The mesencephalon is a prominent arch between the pontine and diencephalic flexures. The roof (tectum and pretectum) contains neurogenetic NEPs adjacent to a very thin layer of pioneer migrating neurons. There are a some fiber bundles in the posterior commissure. The tegmental and isthmal NEPs are rapidly unloading their neuronal progeny in dense bands in the adjacent parenchyma. The outermost clumps of young neurons appear to interact with axons in the thick subpial fiber band.

The rhombencephalon is the largest brain structure. Both the pons and medulla have depleting NEPs that are unloading neuronal and glial progeny into an expanding parenchyma. The pons and medulla contain longitudinal bands of migrating cells, denser just outside the neuroepithelium, less dense in the core, and again denser adjacent to the subpial fiber band. The genu of the facial motor nerve forms fascicles adjacent to the NEP in both medial and lateral sections; these fascicles never reach the pial surface. What is presumed to be the solitary tract is a prominent core fiber tract in the medulla. Lateral sections show large peripheral sensory nerves contacting the brain. The mesencephalic nuclear neurons associated with the trigeminal nerve are migrating into the brain. The subpial fiber band is thicker where the axons from sensory ganglia enter the brain and appear to mingle with migrating neurons at the entry zones. Peripheral nerves have dense glia (Schwann cells), while central fiber tracts are clear. The cerebellum stands out as the most immature part of the rhombencephalon. All parts of the cerebellar NEP are generating deep nuclear neurons, Purkinje cells, and glial stem cells. It appears to be very thick, but many Purkinje cells and deep nuclear neurons are sequestered in the basal part prior to entry into the cerebellar transitional fields, which has more distinct layers, especially laterally (**Plate 13**).

The midline raphe glial structure is best observed in the sagittal plane and **Plates 19 to 21 A/B** are exclusively devoted to it. What is most mysterious is a lens-shaped group of cells that are migrating within the section plane just below the NEP proper and above the layer of morphocytes. This cell group has been observed in many specimens (**Plate 20**).

EXTERNAL FEATURES OF THE GW7.4 BRAIN

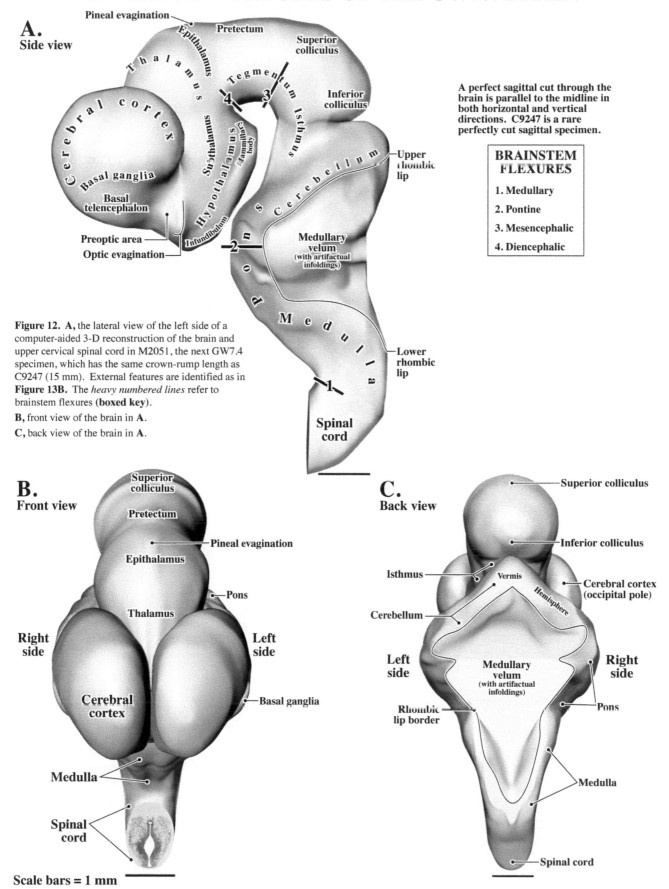

A.
Side view

Pineal evagination
Epithalamus
Pretectum
Superior colliculus
Thalamus
Tegmentum
Cerebral cortex
Inferior colliculus
4
3
Subthalamus
Isthmus
Basal ganglia
Mammillary body
Hypothalamus
Basal telencephalon
Upper rhombic lip
Preoptic area
Infundibulum
Optic evagination
2
Pons
Cerebellum
Medullary velum
(with artifactual infoldings)
Medulla
Lower rhombic lip
1
Spinal cord

A perfect sagittal cut through the brain is parallel to the midline in both horizontal and vertical directions. C9247 is a rare perfectly cut sagittal specimen.

BRAINSTEM FLEXURES

1. Medullary
2. Pontine
3. Mesencephalic
4. Diencephalic

Figure 12. **A,** the lateral view of the left side of a computer-aided 3-D reconstruction of the brain and upper cervical spinal cord in M2051, the next GW7.4 specimen, which has the same crown-rump length as C9247 (15 mm). External features are identified as in **Figure 13B.** The *heavy numbered lines* refer to brainstem flexures (**boxed key**).

B, front view of the brain in **A**.

C, back view of the brain in **A**.

B.
Front view

Superior colliculus
Pretectum
Pineal evagination
Epithalamus
Pons
Thalamus
Right side
Left side
Cerebral cortex
Basal ganglia
Medulla
Spinal cord

Scale bars = 1 mm

C.
Back view

Superior colliculus
Inferior colliculus
Isthmus
Vermis
Cerebral cortex (occipital pole)
Hemisphere
Cerebellum
Left side
Right side
Medullary velum (with artifactual infoldings)
Rhombic lip border
Pons
Medulla
Spinal cord

24

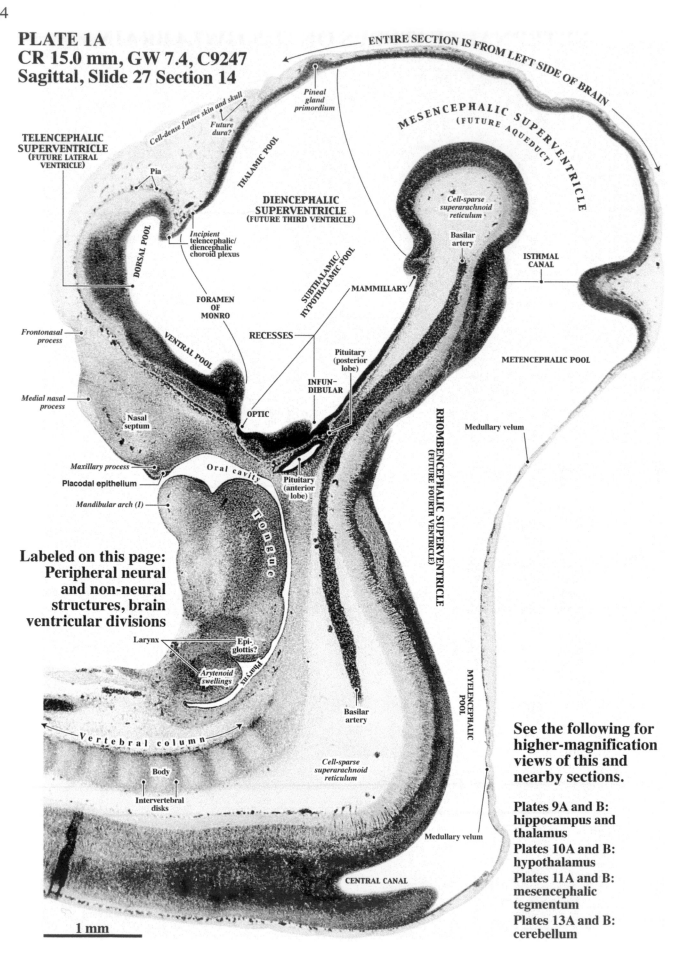

PLATE 1A
CR 15.0 mm, GW 7.4, C9247
Sagittal, Slide 27 Section 14

ENTIRE SECTION IS FROM LEFT SIDE OF BRAIN

Pineal gland primordium

MESENCEPHALIC SUPERVENTRICLE
(FUTURE AQUEDUCT)

Cell-dense future skin and skull

Future dura?

TELENCEPHALIC
SUPERVENTRICLE
(FUTURE LATERAL
VENTRICLE)

Pia

THALAMIC POOL

DIENCEPHALIC
SUPERVENTRICLE
(FUTURE THIRD VENTRICLE)

Cell-sparse superarachnoid reticulum

Basilar artery

ISTHMAL
CANAL

DORSAL POOL

Incipient telencephalic/ diencephalic choroid plexus

SUBTHALAMIC/ HYPOTHALAMIC POOL

MAMMILLARY

Frontonasal process

FORAMEN
OF
MONRO

VENTRAL POOL

RECESSES

METENCEPHALIC POOL

Pituitary (posterior lobe)

Medial nasal process

INFUN-
DIBULAR

Nasal septum

OPTIC

Medullary velum

Maxillary process

Oral cavity

Pituitary (anterior lobe)

RHOMBENCEPHALIC SUPERVENTRICLE
(FUTURE FOURTH VENTRICLE)

Placodal epithelium

Mandibular arch (I)

Tongue

**Labeled on this page:
Peripheral neural
and non-neural
structures, brain
ventricular divisions**

MYELENCEPHALIC
POOL

Larynx

Epi-
glottis?

Arytenoid swellings

Pharynx

Basilar artery

Vertebral column

**See the following for
higher-magnification
views of this and
nearby sections.**

Cell-sparse superarachnoid reticulum

Body

Intervertebral disks

**Plates 9A and B:
hippocampus and
thalamus**

Medullary velum

**Plates 10A and B:
hypothalamus**

**Plates 11A and B:
mesencephalic
tegmentum**

CENTRAL CANAL

**Plates 13A and B:
cerebellum**

1 mm

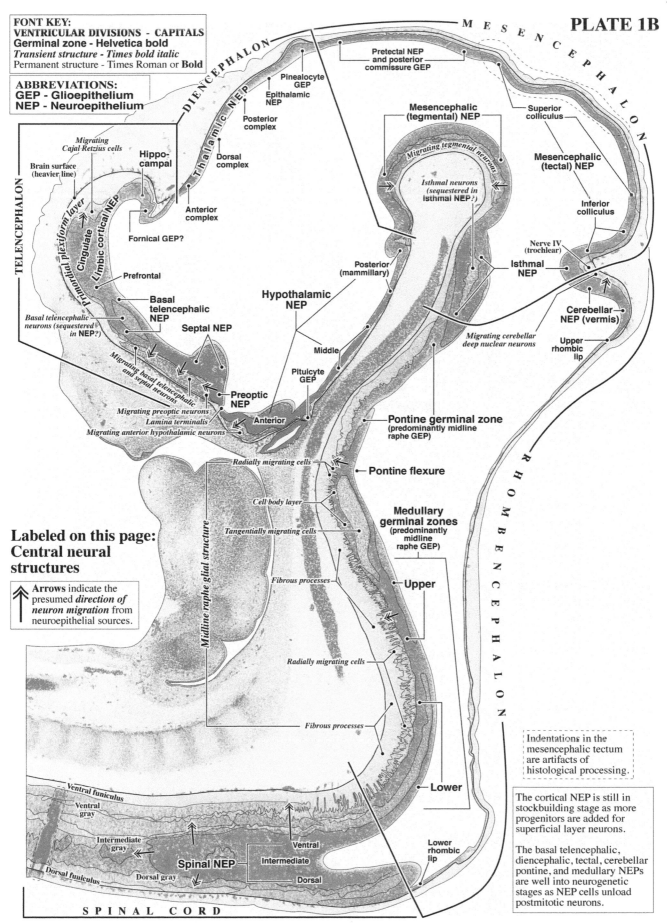

FONT KEY:
VENTRICULAR DIVISIONS - CAPITALS
Germinal zone - Helvetica bold
Transient structure - Times bold italic
Permanent structure - Times Roman or **Bold**

ABBREVIATIONS:
GEP - Glioepithelium
NEP - Neuroepithelium

M E S E N C E P H A L O N

Pretectal NEP
and posterior
commissure GEP

D I E N C E P H A L O N

Pinealocyte
GEP

Epithalamic
NEP

*Migrating
Cajal-Retzius cells*

Posterior
complex

Hippo-
campal

Superior
colliculus

Mesencephalic
(tegmental) NEP

Migrating tegmental neurons

Mesencephalic
(tectal) NEP

Brain surface
(heavier line)

Dorsal
complex

Thalamic NEP

*Isthmal neurons
(sequestered in
Isthmal NEP?)*

Inferior
colliculus

Anterior
complex

Primordial plexiform layer

Cingulate

Limbic cortical NEP

Fornical GEP?

Nerve IV
(trochlear)

Isthmal
NEP

Prefrontal

Posterior
(mammillary)

TELENCEPHALON

Basal
telencephalic
NEP

Hypothalamic
NEP

Cerebellar
NEP (vermis)

*Basal telencephalic
neurons (sequestered
in NEP?)*

Septal NEP

Middle

*Migrating cerebellar
deep nuclear neurons*

Upper
rhombic
lip

*Migrating basal telencephalic
and septal neurons*

Pituicyte
GEP

Pontine germinal zone
(predominantly midline
raphe GEP)

Preoptic
NEP

Migrating preoptic neurons

Lamina terminalis

Anterior

Migrating anterior hypothalamic neurons

Radially migrating cells

Pontine flexure

R H O M B E N C E P H A L O N

Cell body layer

Medullary
germinal zones
(predominantly
midline
raphe GEP)

**Labeled on this page:
Central neural
structures**

Midline raphe glial structure

Tangentially migrating cells

Fibrous processes

Upper

Arrows indicate the
presumed *direction of
neuron migration* from
neuroepithelial sources.

Radially migrating cells

Fibrous processes

Lower

Indentations in the
mesencephalic tectum
are artifacts of
histological processing.

Ventral funiculus

Ventral
gray

Intermediate
gray

Ventral

Lower
rhombic
lip

The cortical NEP is still in
stockbuilding stage as more
progenitors are added for
superficial layer neurons.

Spinal NEP

Intermediate

Dorsal funiculus

Dorsal gray

Dorsal

The basal telencephalic,
diencephalic, tectal, cerebellar
pontine, and medullary NEPs
are well into neurogenetic
stages as NEP cells unload
postmitotic neurons.

S P I N A L   C O R D

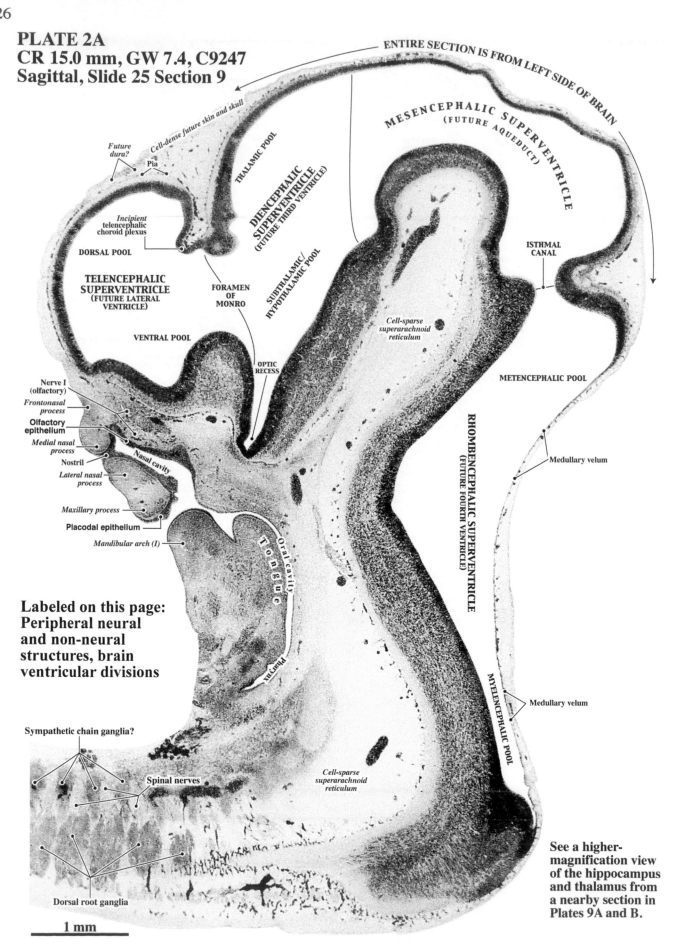

26

PLATE 2A
CR 15.0 mm, GW 7.4, C9247
Sagittal, Slide 25 Section 9

ENTIRE SECTION IS FROM LEFT SIDE OF BRAIN

MESENCEPHALIC SUPERVENTRICLE
(FUTURE AQUEDUCT)

Future dura?

Cell-dense future skin and skull

Pia

THALAMIC POOL

DIENCEPHALIC SUPERVENTRICLE
(FUTURE THIRD VENTRICLE)

Incipient telencephalic choroid plexus

DORSAL POOL

ISTHMAL CANAL

Cell-sparse superarachnoid reticulum

TELENCEPHALIC SUPERVENTRICLE
(FUTURE LATERAL VENTRICLE)

FORAMEN OF MONRO

SUBTHALAMIC/ HYPOTHALAMIC POOL

METENCEPHALIC POOL

VENTRAL POOL

OPTIC RECESS

Nerve I (olfactory)

Frontonasal process

Olfactory epithelium

Medial nasal process

Nostril

Nasal cavity

Lateral nasal process

RHOMBENCEPHALIC SUPERVENTRICLE
(FUTURE FOURTH VENTRICLE)

Maxillary process

Placodal epithelium

Mandibular arch (I)

Oral cavity

Tongue

Medullary velum

Labeled on this page: Peripheral neural and non-neural structures, brain ventricular divisions

Pharynx

MYELENCEPHALIC POOL

Medullary velum

Sympathetic chain ganglia?

Spinal nerves

Cell-sparse superarachnoid reticulum

See a higher-magnification view of the hippocampus and thalamus from a nearby section in Plates 9A and B.

Dorsal root ganglia

1 mm

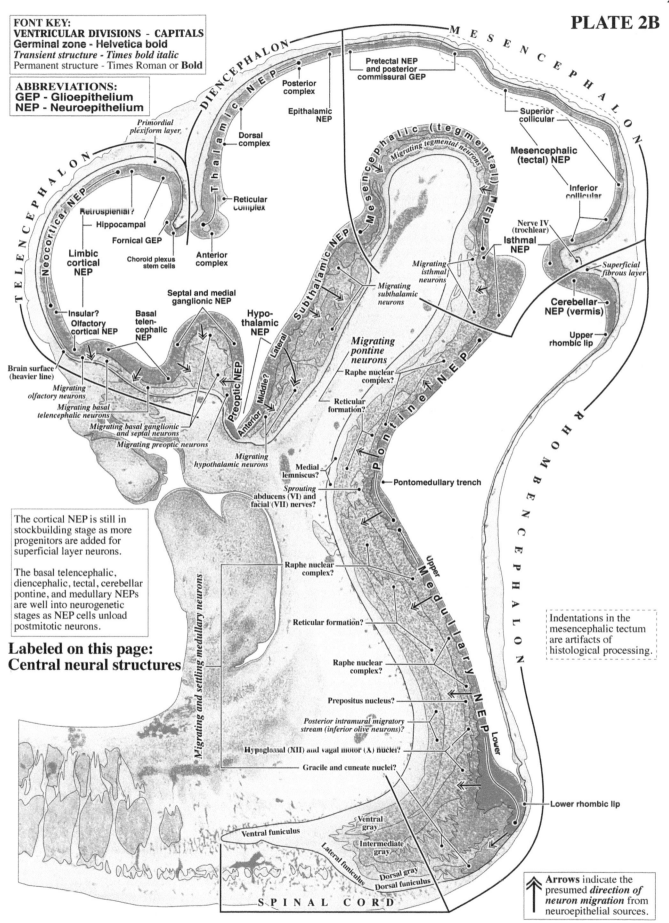

FONT KEY:
VENTRICULAR DIVISIONS - CAPITALS
Germinal zone - Helvetica bold
Transient structure - Times bold italic
Permanent structure - Times Roman or **Bold**

ABBREVIATIONS:
GEP - Glioepithelium
NEP - Neuroepithelium

Primordial plexiform layer

DIENCEPHALON

Thalamic NEP

Posterior complex

Epithalamic NEP

Pretectal NEP and posterior commissural GEP

MESENCEPHALON

Dorsal complex

Reticular complex

TELENCEPHALON

Neocortical NEP

Retrosplenial?

Hippocampal

Fornical GEP

Choroid plexus stem cells

Limbic cortical NEP

Anterior complex

Septal and medial ganglionic NEP

Hypothalamic NEP

Basal telencephalic NEP

Olfactory cortical NEP

Insular?

Brain surface (heavier line)

Migrating olfactory neurons

Migrating basal telencephalic neurons

Migrating basal ganglionic and septal neurons

Migrating preoptic neurons

Preoptic NEP

Anterior Middle?

Migrating hypothalamic neurons

Lateral

Subthalamic NEP

Mesencephalic (tegmental) NEP

Migrating tegmental neurons

Mesencephalic (tectal) NEP

Superior collicular

Inferior collicular

Nerve IV (trochlear)

Isthmal NEP

Superficial fibrous layer

Cerebellar NEP (vermis)

Upper rhombic lip

Migrating isthmal neurons

Migrating subthalamic neurons

Migrating pontine neurons

Raphe nuclear complex?

Reticular formation?

Pontine NEP

Medial lemniscus?

Sprouting abducens (VI) and facial (VII) nerves?

Pontomedullary trench

Upper

Medullary NEP

RHOMBENCEPHALON

Lower

The cortical NEP is still in stockbuilding stage as more progenitors are added for superficial layer neurons.

The basal telencephalic, diencephalic, tectal, cerebellar pontine, and medullary NEPs are well into neurogenetic stages as NEP cells unload postmitotic neurons.

Labeled on this page:
Central neural structures

Raphe nuclear complex?

Reticular formation?

Raphe nuclear complex?

Prepositus nucleus?

Posterior intramural migratory stream (inferior olive neurons)?

Hypoglossal (XII) and vagal motor (X) nuclei?

Gracile and cuneate nuclei?

Migrating and settling medullary neurons

Lower rhombic lip

Indentations in the mesencephalic tectum are artifacts of histological processing.

Ventral gray

Intermediate gray

Ventral funiculus

Lateral funiculus

Dorsal gray

Dorsal funiculus

SPINAL CORD

Arrows indicate the presumed *direction of neuron migration* from neuroepithelial sources.

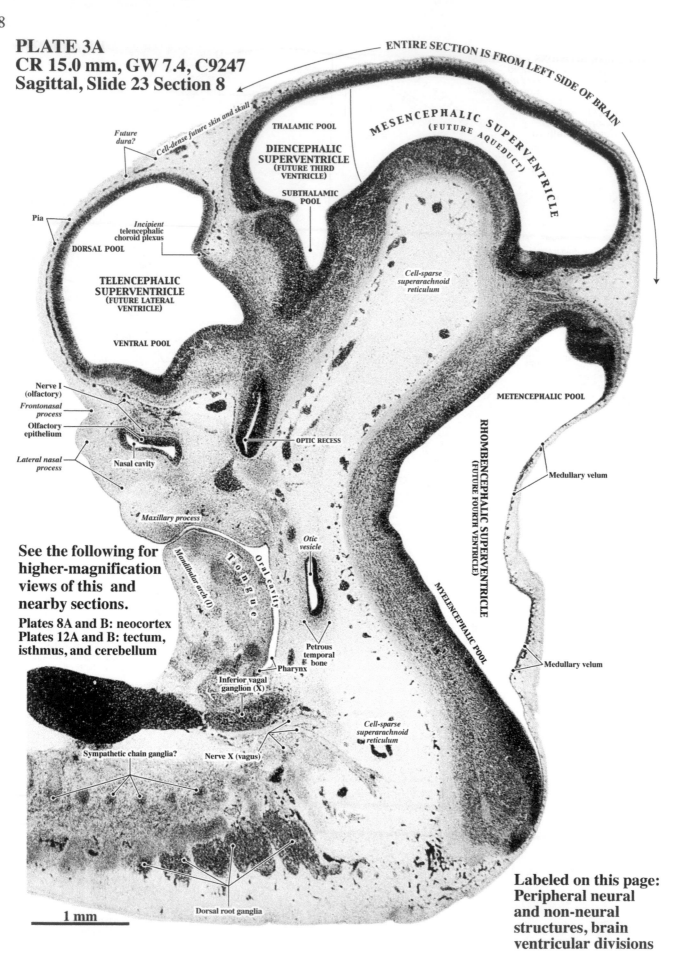

28

PLATE 3A
CR 15.0 mm, GW 7.4, C9247
Sagittal, Slide 23 Section 8

ENTIRE SECTION IS FROM LEFT SIDE OF BRAIN

Future dura?

Cell-dense future skin and skull

THALAMIC POOL

MESENCEPHALIC SUPERVENTRICLE
(FUTURE AQUEDUCT)

DIENCEPHALIC
SUPERVENTRICLE
(FUTURE THIRD
VENTRICLE)

SUBTHALAMIC
POOL

Pia

*Incipient
telencephalic
choroid plexus*

DORSAL POOL

*Cell-sparse
superarachnoid
reticulum*

**TELENCEPHALIC
SUPERVENTRICLE**
(FUTURE LATERAL
VENTRICLE)

VENTRAL POOL

Nerve I
(olfactory)

*Frontonasal
process*

METENCEPHALIC POOL

Olfactory
epithelium

OPTIC RECESS

RHOMBENCEPHALIC SUPERVENTRICLE
(FUTURE FOURTH VENTRICLE)

*Lateral nasal
process*

Nasal cavity

Medullary velum

Maxillary process

**See the following for
higher-magnification
views of this and
nearby sections.**

*Otic
vesicle*

MYELENCEPHALIC POOL

Plates 8A and B: neocortex
**Plates 12A and B: tectum,
isthmus, and cerebellum**

Mandibular arch (I)

Tongue

Oral cavity

Petrous
temporal
bone

Medullary velum

Pharynx

Inferior vagal
ganglion (X)

*Cell-sparse
superarachnoid
reticulum*

Sympathetic chain ganglia?

Nerve X (vagus)

**Labeled on this page:
Peripheral neural
and non-neural
structures, brain
ventricular divisions**

Dorsal root ganglia

1 mm

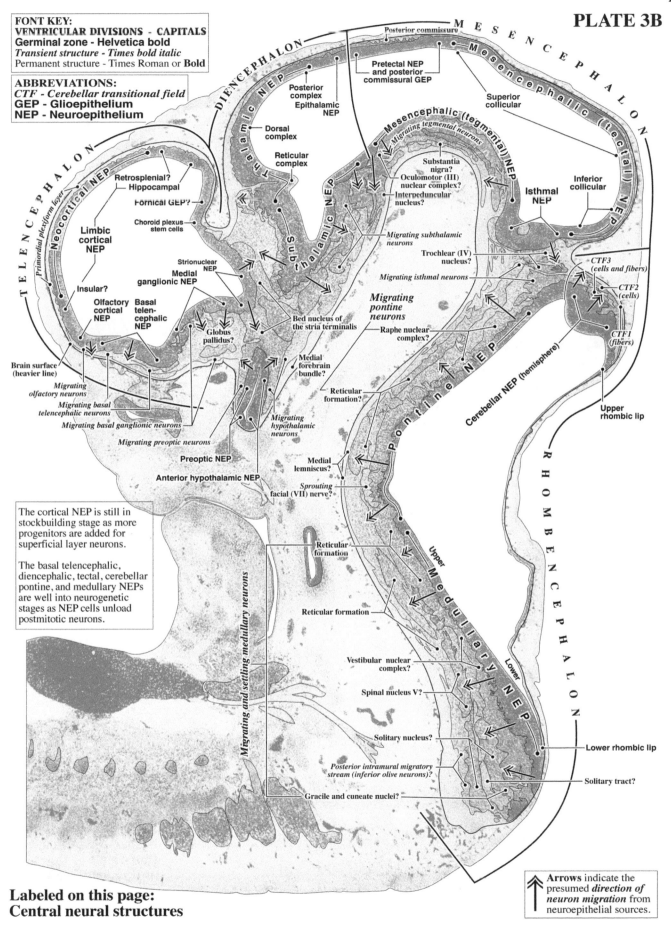

FONT KEY:
VENTRICULAR DIVISIONS - CAPITALS
Germinal zone - Helvetica bold
Transient structure - Times bold italic
Permanent structure - Times Roman or **Bold**

ABBREVIATIONS:
CTF - Cerebellar transitional field
GEP - Glioepithelium
NEP - Neuroepithelium

MESENCEPHALON

DIENCEPHALON

Posterior commissure

Mesencephalic (tectal) NEP

TELENCEPHALON

Thalamic NEP

Posterior complex Epithalamic NEP

Dorsal complex

Reticular complex

Pretectal NEP and posterior commissural GEP

Superior collicular

Mesencephalic (tegmental) NEP

Migrating tegmental neurons

Substantia nigra?
Oculomotor (III) nuclear complex?
Interpeduncular nucleus?

Inferior collicular

Isthmal NEP

Neocortical NEP

Primordial plexiform layer

Retrosplenial? Hippocampal

Fornical GEP?

Choroid plexus stem cells

Subthalamic NEP

Migrating subthalamic neurons

Trochlear (IV) nucleus?

Migrating isthmal neurons

CTF3 *(cells and fibers)*

CTF2 *(cells)*

Limbic cortical NEP

Strionuclear NEP

Medial ganglionic NEP

Insular?

Olfactory cortical NEP

Basal telencephalic NEP

Globus pallidus?

Bed nucleus of the stria terminalis

Migrating pontine neurons

Raphe nuclear complex?

CTF1 *(fibers)*

Brain surface (heavier line)

Migrating olfactory neurons

Migrating basal telencephalic neurons

Migrating basal ganglionic neurons

Migrating preoptic neurons

Preoptic NEP

Anterior hypothalamic NEP

Medial forebrain bundle?

Migrating hypothalamic neurons

Reticular formation?

Pontine NEP

Cerebellar NEP (hemisphere)

Upper rhombic lip

Medial lemniscus?

Sprouting facial (VII) nerve?

The cortical NEP is still in stockbuilding stage as more progenitors are added for superficial layer neurons.

The basal telencephalic, diencephalic, tectal, cerebellar pontine, and medullary NEPs are well into neurogenetic stages as NEP cells unload postmitotic neurons.

Migrating and settling medullary neurons

Reticular formation

Reticular formation

Upper Medullary NEP

RHOMBENCEPHALON

Lower

Vestibular nuclear complex?

Spinal nucleus V?

Solitary nucleus?

Posterior intramural migratory stream (inferior olive neurons)?

Gracile and cuneate nuclei?

Lower rhombic lip

Solitary tract?

Labeled on this page:
Central neural structures

Arrows indicate the presumed *direction of neuron migration* from neuroepithelial sources.

30

PLATE 4A
CR 15 mm, GW 7.4, C9247
Sagittal, Slide 21 Section 8

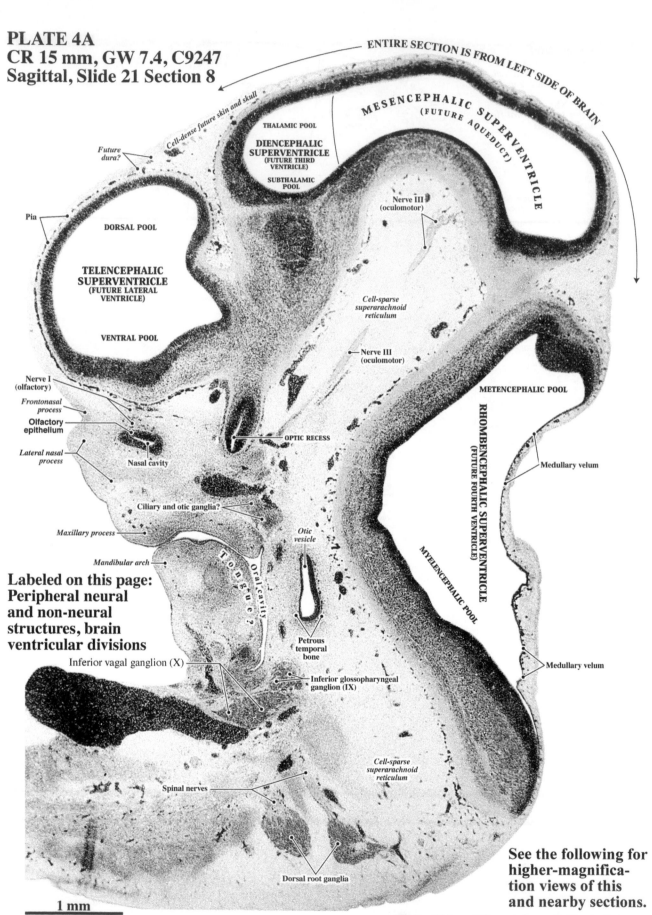

Cell-dense future skin and skull

ENTIRE SECTION IS FROM LEFT SIDE OF BRAIN

MESENCEPHALIC SUPERVENTRICLE
(FUTURE AQUEDUCT)

THALAMIC POOL

DIENCEPHALIC
SUPERVENTRICLE
(FUTURE THIRD
VENTRICLE)

SUBTHALAMIC
POOL

Future dura?

Pia

DORSAL POOL

TELENCEPHALIC
SUPERVENTRICLE
(FUTURE LATERAL
VENTRICLE)

VENTRAL POOL

Nerve III
(oculomotor)

*Cell-sparse
superarachnoid
reticulum*

Nerve III
(oculomotor)

METENCEPHALIC POOL

RHOMBENCEPHALIC SUPERVENTRICLE
(FUTURE FOURTH VENTRICLE)

Nerve I
(olfactory)

*Frontonasal
process*

Olfactory
epithelium

*Lateral nasal
process*

Nasal cavity

OPTIC RECESS

Medullary velum

Ciliary and otic ganglia?

Maxillary process

Mandibular arch

T o n g u e ?

Oral cavity

*Otic
vesicle*

MYELENCEPHALIC POOL

**Labeled on this page:
Peripheral neural
and non-neural
structures, brain
ventricular divisions**

Petrous
temporal
bone

Medullary velum

Inferior vagal ganglion (X)

Inferior glossopharyngeal
ganglion (IX)

*Cell-sparse
superarachnoid
reticulum*

Spinal nerves

Dorsal root ganglia

**See the following for
higher-magnifica-
tion views of this
and nearby sections.**

1 mm

**Plates 11A and B: mesencephalic tegmentum
Plates 18A and B: pons and upper medulla**

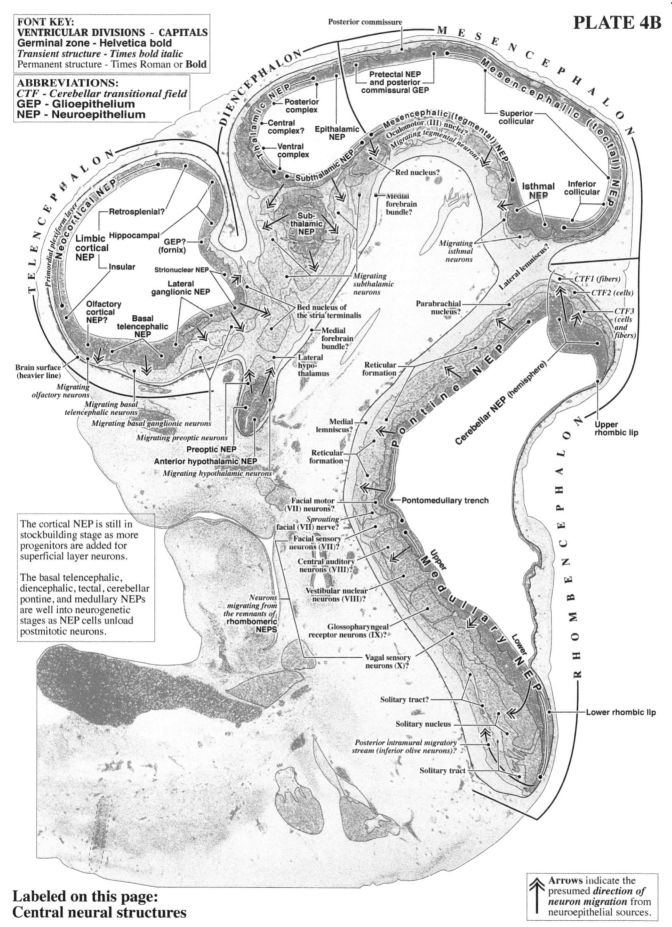

FONT KEY:
VENTRICULAR DIVISIONS - CAPITALS
Germinal zone - Helvetica bold
Transient structure - Times bold italic
Permanent structure - Times Roman or **Bold**

ABBREVIATIONS:
CTF - Cerebellar transitional field
GEP - Glioepithelium
NEP - Neuroepithelium

Posterior commissure

M E S E N C E P H A L O N

D I E N C E P H A L O N

Pretectal NEP
and posterior
commissural GEP

Mesencephalic (tectial) NEP

Thalamic NEP

Posterior
complex

Central
complex?

Ventral
complex

Epithalamic
NEP

Mesencephalic (tegmental) NEP

Oculomotor (III) nuclei?

Migrating tegmental neurons

Superior
collicular

**Isthmal
NEP**

Inferior
collicular

T E L E N C E P H A L O N

Subthalamic NEP

Red nucleus?

Primordial plexiform layer

Neocortical NEP

Retrosplenial?

Hippocampal

**Limbic
cortical
NEP**

Insular

**GEP?
(fornix)**

Sub-
thalamic
NEP

Medial
forebrain
bundle?

*Migrating
isthmal
neurons*

Lateral lemniscus?

CTF1 (fibers)

CTF2 (cells)

**Olfactory
cortical
NEP?**

**Lateral
ganglionic NEP**

Strionuclear NEP

Parabrachial
nucleus?

*CTF3
(cells
and
fibers)*

**Basal
telencephalic
NEP**

Bed nucleus of
the stria terminalis

Medial
forebrain
bundle?

*Migrating
subthalamic
neurons*

Reticular
formation

Pontine NEP

Brain surface
(heavier line)

Lateral
hypo-
thalamus

*Migrating
olfactory neurons*

Cerebellar NEP (hemisphere)

Upper
rhombic lip

*Migrating basal
telencephalic neurons*

Migrating basal ganglionic neurons

Medial
lemniscus?

R H O M B E N C E P H A L O N

Migrating preoptic neurons

Preoptic NEP

Reticular
formation

Anterior hypothalamic NEP

Migrating hypothalamic neurons

Upper Medullary NEP

Facial motor
(VII) neurons?

Pontomedullary trench

*Sprouting
facial (VII) nerve?*

The cortical NEP is still in
stockbuilding stage as more
progenitors are added for
superficial layer neurons.

The basal telencephalic,
diencephalic, tectal, cerebellar
pontine, and medullary NEPs
are well into neurogenetic
stages as NEP cells unload
postmitotic neurons.

Facial sensory
neurons (VII)?

Central auditory
neurons (VIII)?

Vestibular nuclear
neurons (VIII)?

Lower

*Neurons
migrating from
the remnants of
rhombomeric
NEPS*

Glossopharyngeal
receptor neurons (IX)?

Vagal sensory
neurons (X)?

Lower rhombic lip

Solitary tract?

Solitary nucleus

*Posterior intramural migratory
stream (inferior olive neurons)?*

Solitary tract

**Labeled on this page:
Central neural structures**

Arrows indicate the
presumed *direction of
neuron migration* from
neuroepithelial sources.

PLATE 5A
CR 15 mm, GW 7.4, C9247
Sagittal, Slide 19 Section 8

Labeled on this page:
Peripheral neural
and non-neural
structures, brain
ventricular divisions

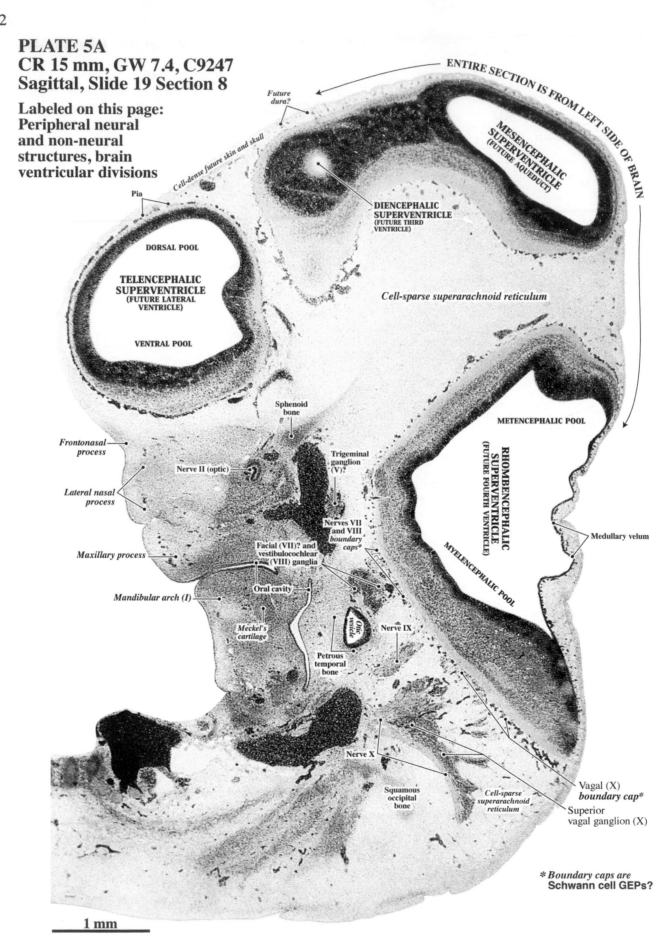

ENTIRE SECTION IS FROM LEFT SIDE OF BRAIN

Future dura?

Cell-dense future skin and skull

MESENCEPHALIC
SUPERVENTRICLE
(FUTURE AQUEDUCT)

DIENCEPHALIC
SUPERVENTRICLE
(FUTURE THIRD
VENTRICLE)

Pia

DORSAL POOL

Cell-sparse superarachnoid reticulum

TELENCEPHALIC
SUPERVENTRICLE
(FUTURE LATERAL
VENTRICLE)

VENTRAL POOL

METENCEPHALIC POOL

Sphenoid
bone

*Frontonasal
process*

Trigeminal
ganglion
(V)?

RHOMBENCEPHALIC
SUPERVENTRICLE
(FUTURE FOURTH VENTRICLE)

Nerve II (optic)

*Lateral nasal
process*

Nerves VII
and VIII
*boundary
caps**

Facial (VII)? and
vestibulocochlear
(VIII) ganglia

MYELENCEPHALIC POOL

Medullary velum

Maxillary process

Oral cavity

Mandibular arch (I)

*Otic
vesicle*

Nerve IX

*Meckel's
cartilage*

Petrous
temporal
bone

Nerve X

Squamous
occipital
bone

*Cell-sparse
superarachnoid
reticulum*

Vagal (X)
*boundary cap**

Superior
vagal ganglion (X)

*** Boundary caps are
Schwann cell GEPs?*

1 mm

PLATE 5B

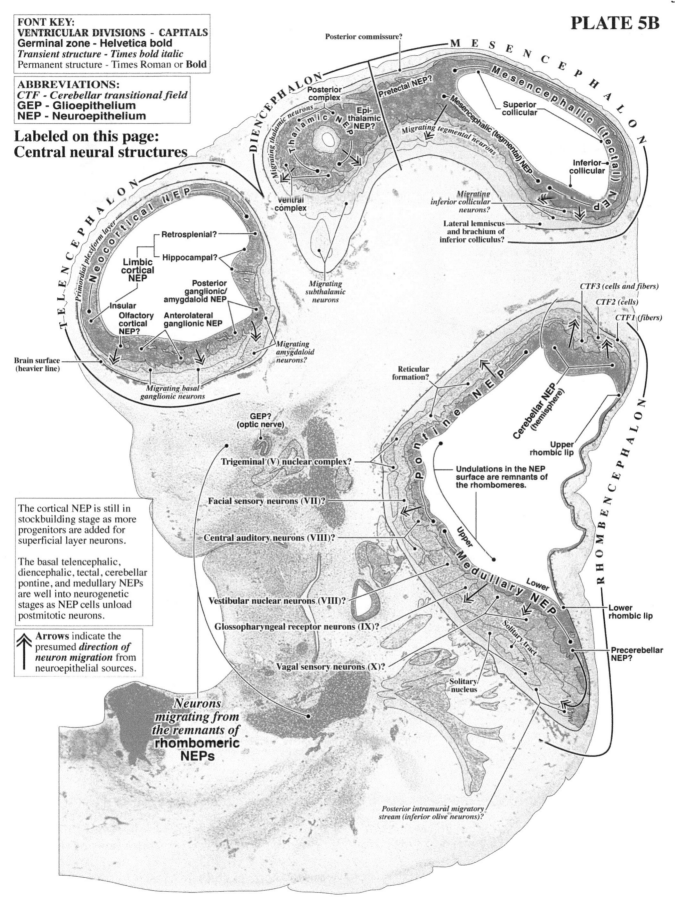

FONT KEY:
VENTRICULAR DIVISIONS - CAPITALS
Germinal zone - Helvetica bold
Transient structure - Times bold italic
Permanent structure - Times Roman or **Bold**

ABBREVIATIONS:
CTF - Cerebellar transitional field
GEP - Glioepithelium
NEP - Neuroepithelium

Labeled on this page:
Central neural structures

Posterior commissure?

MESENCEPHALON

DIENCEPHALON

Posterior complex

Epi-thalamic NEP?

Pretectal NEP?

Mesencephalic (tectal) NEP?

Superior collicular

Thalamic NEP

Migrating thalamic neurons

Mesencephalic (tegmental) NEP

Migrating tegmental neurons

Inferior collicular

TELENCEPHALON

Neocortical NEP

Primordial plexiform layer

ventral complex

Migrating inferior collicular neurons?

Lateral lemniscus and brachium of inferior colliculus?

Retrosplenial?

Limbic cortical NEP

Hippocampal?

Posterior ganglionic/amygdaloid NEP

Insular

Olfactory cortical NEP?

Anterolateral ganglionic NEP

Migrating subthalamic neurons

CTF3 (cells and fibers)
CTF2 (cells)
CTF1 (fibers)

Brain surface (heavier line)

Migrating amygdaloid neurons?

Migrating basal ganglionic neurons

Reticular formation?

Pontine NEP

Cerebellar NEP (hemisphere)

RHOMBENCEPHALON

GEP? (optic nerve)

Upper rhombic lip

Undulations in the NEP surface are remnants of the rhombomeres.

Trigeminal (V) nuclear complex?

The cortical NEP is still in stockbuilding stage as more progenitors are added for superficial layer neurons.

The basal telencephalic, diencephalic, tectal, cerebellar pontine, and medullary NEPs are well into neurogenetic stages as NEP cells unload postmitotic neurons.

Facial sensory neurons (VII)?

Upper

Medullary NEP

Central auditory neurons (VIII)?

Lower

Lower rhombic lip

Vestibular nuclear neurons (VIII)?

Solitary tract

Glossopharyngeal receptor neurons (IX)?

Precerebellar NEP?

Arrows indicate the presumed *direction of neuron migration* from neuroepithelial sources.

Vagal sensory neurons (X)?

Solitary nucleus

Neurons migrating from the remnants of **rhombomeric NEPs**

Posterior intramural migratory stream (inferior olive neurons)?

34

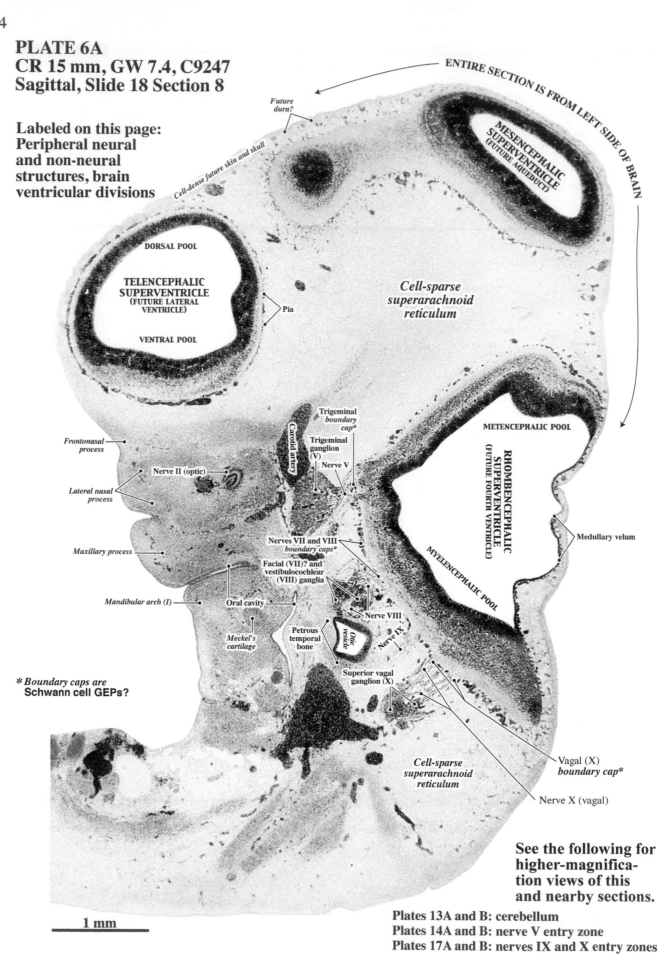

PLATE 6A
CR 15 mm, GW 7.4, C9247
Sagittal, Slide 18 Section 8

Labeled on this page:
Peripheral neural
and non-neural
structures, brain
ventricular divisions

ENTIRE SECTION IS FROM LEFT SIDE OF BRAIN

Future dura?

Cell-dense future skin and skull

MESENCEPHALIC SUPERVENTRICLE (FUTURE AQUEDUCT)

DORSAL POOL

TELENCEPHALIC SUPERVENTRICLE (FUTURE LATERAL VENTRICLE)

VENTRAL POOL

Pia

Cell-sparse superarachnoid reticulum

METENCEPHALIC POOL

RHOMBENCEPHALIC SUPERVENTRICLE (FUTURE FOURTH VENTRICLE)

Frontonasal process

Carotid artery

Trigeminal boundary cap*

Trigeminal ganglion (V)

Nerve II (optic)

Nerve V

Lateral nasal process

Maxillary process

Nerves VII and VIII boundary caps*

Facial (VII)? and vestibulocochlear (VIII) ganglia

Medullary velum

MYELENCEPHALIC POOL

Mandibular arch (I)

Oral cavity

Nerve VIII

Meckel's cartilage

Petrous temporal bone

Otic vesicle

Nerve IX

***Boundary caps are Schwann cell GEPs?**

Superior vagal ganglion (X)

Vagal (X) *boundary cap***

Cell-sparse superarachnoid reticulum

Nerve X (vagal)

1 mm

See the following for higher-magnification views of this and nearby sections.

Plates 13A and B: cerebellum
Plates 14A and B: nerve V entry zone
Plates 17A and B: nerves IX and X entry zones

PLATE 6B

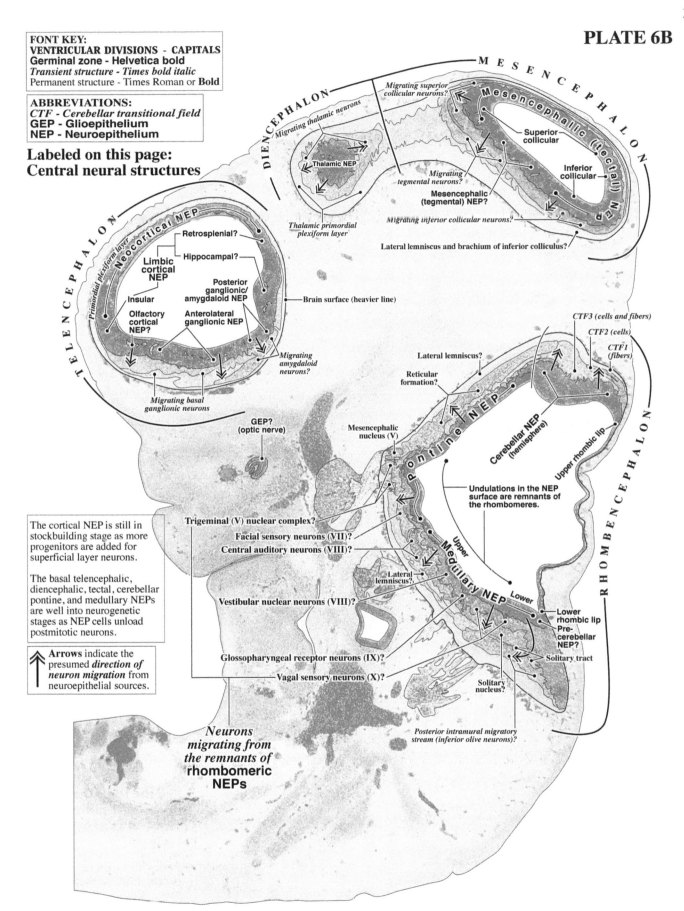

FONT KEY:
VENTRICULAR DIVISIONS - CAPITALS
Germinal zone - Helvetica bold
Transient structure - Times bold italic
Permanent structure - Times Roman or **Bold**

ABBREVIATIONS:
CTF - Cerebellar transitional field
GEP - Glioepithelium
NEP - Neuroepithelium

Labeled on this page:
Central neural structures

MESENCEPHALON

DIENCEPHALON

Migrating superior collicular neurons?

Mesencephalic (tectal) NEP

Superior collicular

Inferior collicular

Migrating thalamic neurons

Thalamic NEP

Migrating tegmental neurons?

Mesencephalic (tegmental) NEP?

Migrating inferior collicular neurons?

Thalamic primordial plexiform layer

Lateral lemniscus and brachium of inferior colliculus?

TELENCEPHALON

Primordial plexiform layer

Neocortical NEP

Retrosplenial?

Limbic cortical NEP

Hippocampal?

Insular

Posterior ganglionic/amygdaloid NEP

Olfactory cortical NEP?

Anterolateral ganglionic NEP

Brain surface (heavier line)

CTF3 (cells and fibers)

CTF2 (cells)

CTF1 (fibers)

Lateral lemniscus?

Reticular formation?

Migrating amygdaloid neurons?

Cerebellar NEP (hemisphere)

Upper rhombic lip

Migrating basal ganglionic neurons

GEP? (optic nerve)

Mesencephalic nucleus (V)

Pontine NEP

Undulations in the NEP surface are remnants of the rhombomeres.

RHOMBENCEPHALON

The cortical NEP is still in stockbuilding stage as more progenitors are added for superficial layer neurons.

The basal telencephalic, diencephalic, tectal, cerebellar pontine, and medullary NEPs are well into neurogenetic stages as NEP cells unload postmitotic neurons.

Trigeminal (V) nuclear complex?

Facial sensory neurons (VII)?

Central auditory neurons (VIII)?

Upper

Medullary NEP

Lateral lemniscus?

Lower

Lower rhombic lip Pre-cerebellar NEP?

Vestibular nuclear neurons (VIII)?

Glossopharyngeal receptor neurons (IX)?

Solitary tract

Solitary nucleus?

Arrows indicate the presumed *direction of neuron migration* from neuroepithelial sources.

Vagal sensory neurons (X)?

Posterior intramural migratory stream (inferior olive neurons)?

Neurons migrating from the remnants of **rhombomeric NEPs**

PLATE 7A
CR 15 mm, GW 7.4, C9247
Sagittal, Slide 16 Section 8

**Labeled on this page:
Peripheral neural
and non-neural
structures, brain
ventricular divisions**

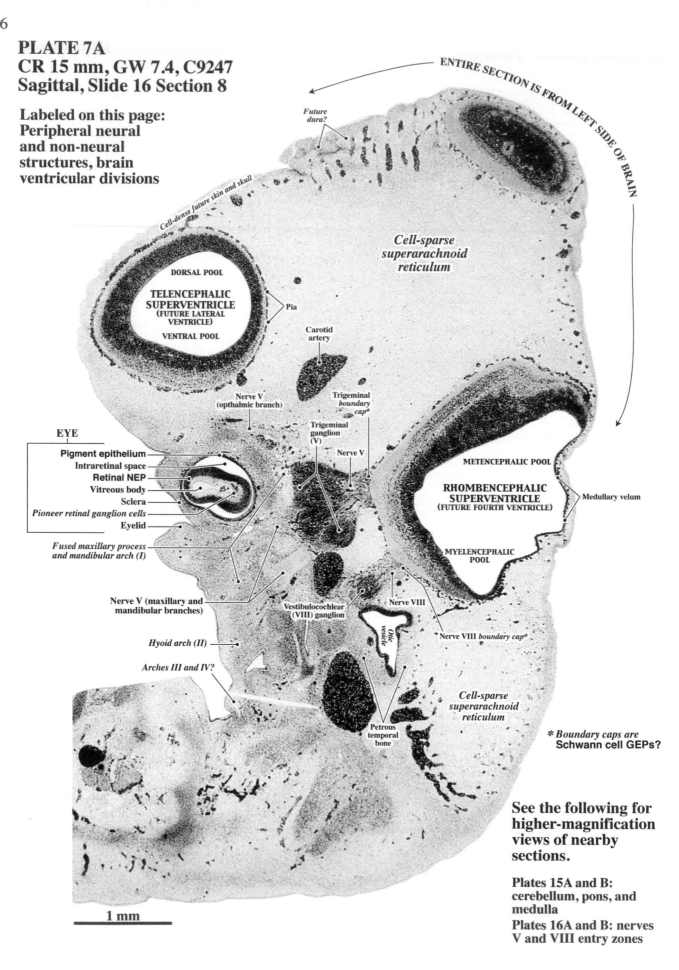

ENTIRE SECTION IS FROM LEFT SIDE OF BRAIN

Future dura?

Cell-dense future skin and skull

*Cell-sparse
superarachnoid
reticulum*

DORSAL POOL

**TELENCEPHALIC
SUPERVENTRICLE
(FUTURE LATERAL
VENTRICLE)**

VENTRAL POOL

Pia

Carotid
artery

Nerve V
(opthalmic branch)

Trigeminal
*boundary
cap**

Trigeminal
ganglion
(V)

Nerve V

METENCEPHALIC POOL

**RHOMBENCEPHALIC
SUPERVENTRICLE
(FUTURE FOURTH VENTRICLE)**

Medullary velum

EYE

Pigment epithelium
Intraretinal space
Retinal NEP
Vitreous body
Sclera
Pioneer retinal ganglion cells
Eyelid

**MYELENCEPHALIC
POOL**

*Fused maxillary process
and mandibular arch (I)*

Nerve V (maxillary and
mandibular branches)

Vestibulocochlear
(VIII) ganglion

Nerve VIII

Nerve VIII *boundary cap**

*Otic
vesicle*

Hyoid arch (II)

Arches III and IV?

*Cell-sparse
superarachnoid
reticulum*

Petrous
temporal
bone

*** Boundary caps are
Schwann cell GEPs?**

1 mm

**See the following for
higher-magnification
views of nearby
sections.**

**Plates 15A and B:
cerebellum, pons, and
medulla**

**Plates 16A and B: nerves
V and VIII entry zones**

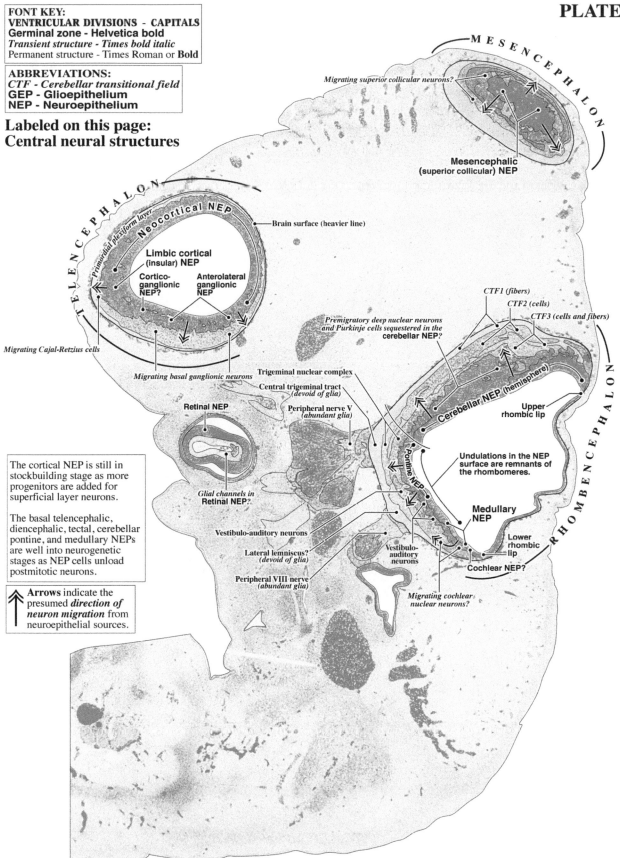

FONT KEY:
VENTRICULAR DIVISIONS - CAPITALS
Germinal zone - Helvetica bold
Transient structure - Times bold italic
Permanent structure - Times Roman or **Bold**

ABBREVIATIONS:
CTF - Cerebellar transitional field
GEP - Glioepithelium
NEP - Neuroepithelium

Labeled on this page:
Central neural structures

The cortical NEP is still in stockbuilding stage as more progenitors are added for superficial layer neurons.

The basal telencephalic, diencephalic, tectal, cerebellar pontine, and medullary NEPs are well into neurogenetic stages as NEP cells unload postmitotic neurons.

Arrows indicate the presumed *direction of neuron migration* from neuroepithelial sources.

M E S E N C E P H A L O N

Migrating superior collicular neurons?

Mesencephalic (superior collicular) NEP

T E L E N C E P H A L O N

Primordial plexiform layer

Neocortical NEP

Brain surface (heavier line)

Limbic cortical (insular) NEP

Cortico-ganglionic NEP?

Anterolateral ganglionic NEP

Migrating Cajal-Retzius cells

Migrating basal ganglionic neurons

Trigeminal nuclear complex

Central trigeminal tract (devoid of glia)

Premigratory deep nuclear neurons and Purkinje cells sequestered in the cerebellar NEP?

CTF1 (fibers)

CTF2 (cells)

CTF3 (cells and fibers)

Cerebellar NEP (hemisphere)

Upper rhombic lip

Retinal NEP

Peripheral nerve V (abundant glia)

Glial channels in **Retinal NEP?**

Pontine NEP

Undulations in the NEP surface are remnants of the rhombomeres.

Medullary NEP

Vestibulo-auditory neurons

Lateral lemniscus? (devoid of glia)

Vestibulo-auditory neurons

Lower rhombic lip

Peripheral VIII nerve (abundant glia)

Migrating cochlear nuclear neurons?

Cochlear NEP?

R H O M B E N C E P H A L O N

DORSAL NEOCORTEX

PLATE 8A

CR 15 mm, GW 7.4, C9247
Sagittal, Slide 23 Section 8

See the entire section in Plates 3A and B.

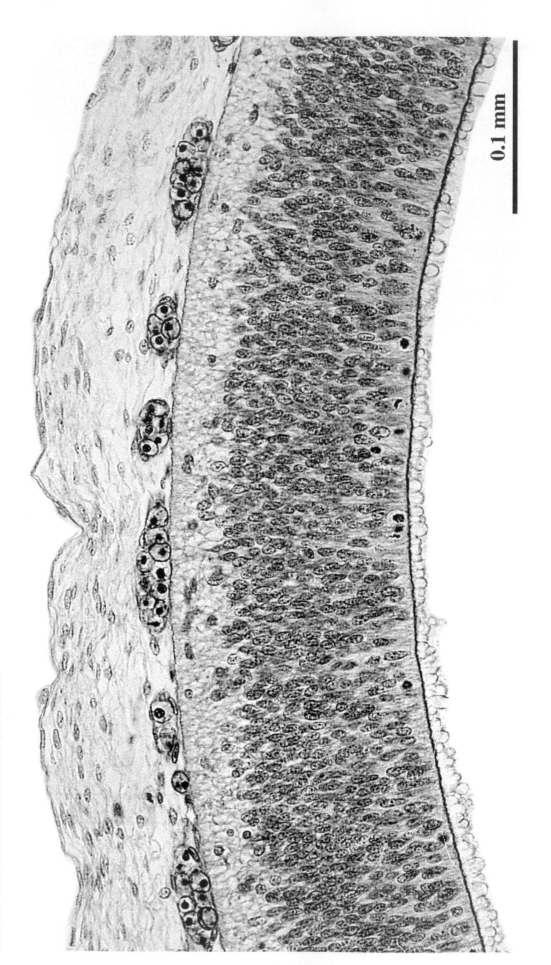

0.1 mm

PLATE 8B

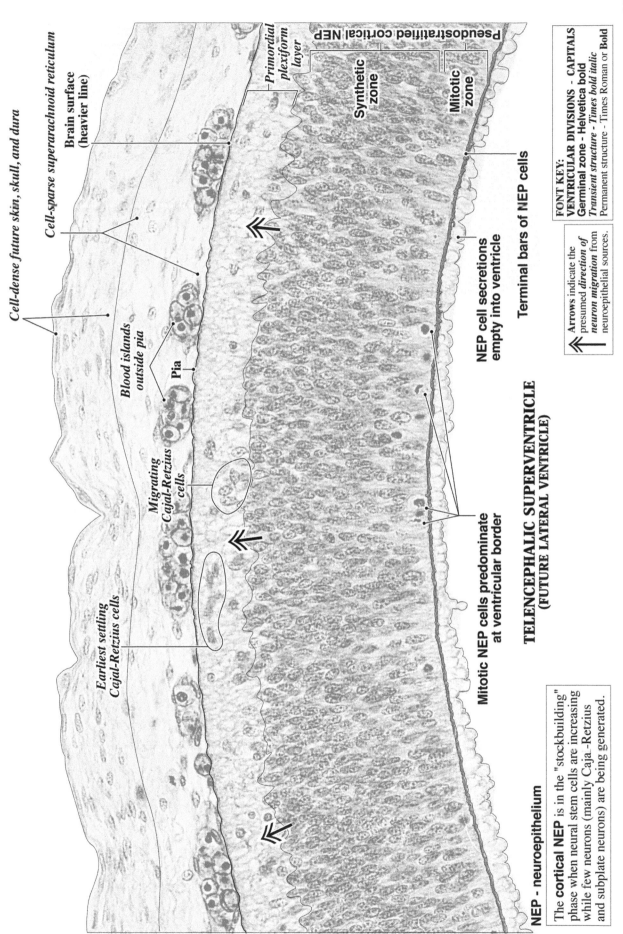

NEP - neuroepithelium

The **cortical NEP** is in the "stockbuilding" phase when neural stem cells are increasing while few neurons (mainly Caja - Retzius and subplate neurons) are being generated.

Cell-dense future skin, skull, and dura

Cell-sparse superarachnoid reticulum

Brain surface (heavier line)

Primordial plexiform layer

Pseudostratified cortical NEP

Synthetic zone

Mitotic zone

Terminal bars of NEP cells

NEP cell secretions empty into ventricle

Blood islands outside pia

Pia

Migrating Cajal-Retzius cells

Earliest settling Cajal-Retzius cells

Mitotic NEP cells predominate at ventricular border

TELENCEPHALIC SUPERVENTRICLE
(FUTURE LATERAL VENTRICLE)

39

FONT KEY:
VENTRICULAR DIVISIONS - CAPITALS
Germinal zone - Helvetica bold
Transient structure - Times bold italic
Permanent structure - Times Roman or **Bold**

Arrows indicate the presumed *direction of neuron migration* from neuroepithelial sources.

HIPPOCAMPUS AND THALAMUS

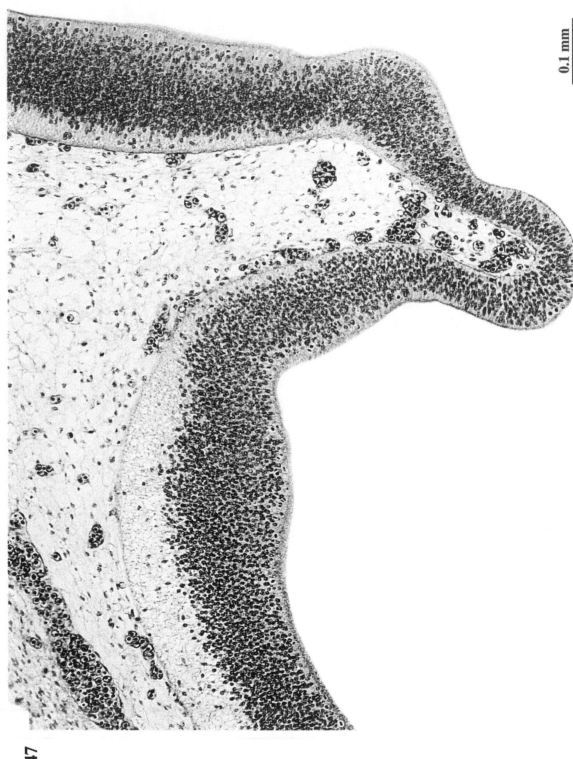

0.1 mm

PLATE 9A

CR 15 mm
GW 7.4, C9247
Sagittal
Slide 26
Section 9

See nearby
sections in
Plates 1-2A/B.

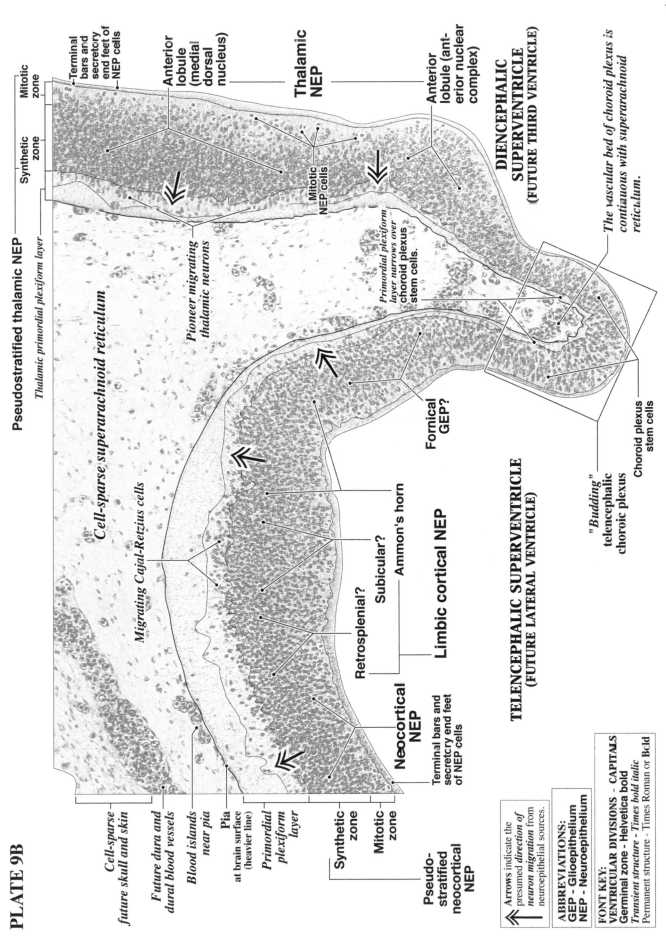

41

PLATE 9B

Pseudostratified thalamic NEP

Thalamic primordial plexiform layer

Mitotic zone

Synthetic zone

Terminal bars and secretory end feet of NEP cells

Anterior lobule (medial dorsal nucleus)

Thalamic NEP

Mitotic NEP cells

Anterior lobule (anterior nuclear complex)

DIENCEPHALIC SUPERVENTRICLE (FUTURE THIRD VENTRICLE)

The vascular bed of choroid plexus is continuous with superarachnoid reticulum.

Cell-sparse superarachnoid reticulum

Pioneer migrating thalamic neurons

Primordial plexiform layer narrows over choroid plexus stem cells.

Migrating Cajal-Retzius cells

Fornical GEP?

Choroid plexus stem cells

"Budding" telencephalic choroic plexus

Choroid plexus stem cells

Ammon's horn

Subicular?

Retrosplenial?

Limbic cortical NEP

TELENCEPHALIC SUPERVENTRICLE (FUTURE LATERAL VENTRICLE)

Neocortical NEP

Terminal bars and secretcry end feet of NEP cells

Cell-sparse future skull and skin

Future dura and dural blood vessels

Blood islands near pia

Pia at brain surface (heavier line)

Primordial plexiform layer

Synthetic zone

Mitotic zone

Pseudo-stratified neocortical NEP

Arrows indicate the presumed *direction of neuron migration* from neuroepithelial sources.

ABBREVIATIONS:
GEP - Glioepithelium
NEP - Neuroepithelium

FONT KEY:
VENTRICULAR DIVISIONS - CAPITALS
Germinal zone - Helvetica bold
Transient structure - Times bold italic
Permanent structure - Times Roman or Bold

PLATE 10A

HYPOTHALAMUS
See the entire section
in Plates 1A/B.

CR 15 mm, GW 7.4, C9247
Sagittal, Slide 27 Section 14

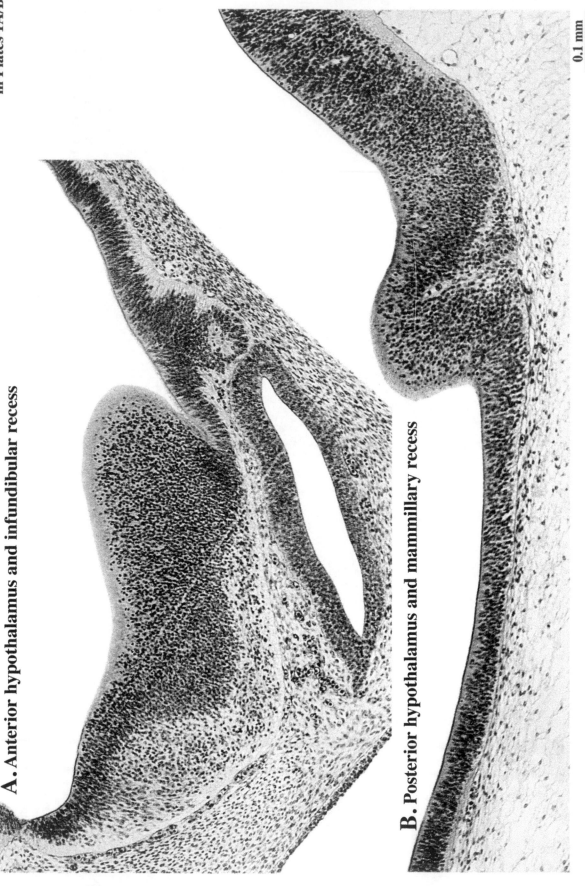

A. Anterior hypothalamus and infundibular recess

B. Posterior hypothalamus and mammillary recess

0.1 mm

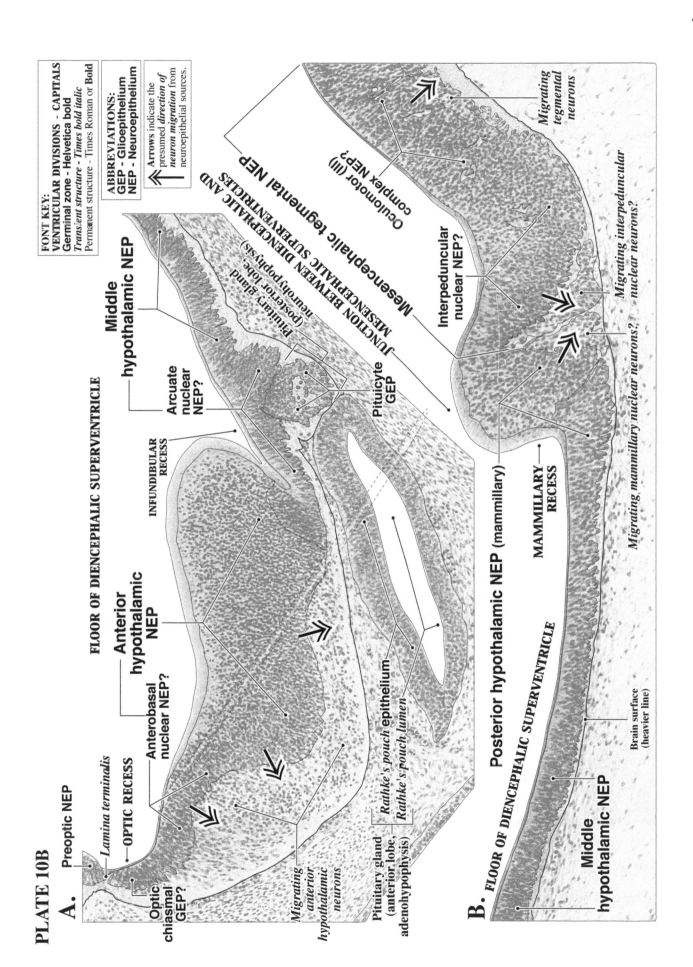

PLATE 10B

43

FONT KEY:
VENTRICULAR DIVISIONS - CAPITALS
Germinal zone - Helvetica bold
Transient structure - *Times bold italic*
Permanent structure - Times Roman or Bold

ABBREVIATIONS:
GEP - Glioepithelium
NEP - Neuroepithelium

⇐ Arrows indicate the presumed *direction of neuron migration* from neuroepithelial sources.

A.

Preoptic NEP

Lamina terminalis

OPTIC RECESS

Optic chiasmal GEP?

FLOOR OF DIENCEPHALIC SUPERVENTRICLE

Anterobasal nuclear NEP?

Anterior hypothalamic NEP

Middle hypothalamic NEP

INFUNDIBULAR RECESS

Arcuate nuclear NEP?

Migrating anterior hypothalamic neurons

Pituitary gland (anterior lobe, adenohypophysis)

Rathke's pouch epithelium
Rathke's pouch lumen

Pituitary gland (posterior lobe, neurohypophysis)

Pituicyte GEP

JUNCTION BETWEEN DIENCEPHALIC SUPERVENTRICLES AND MESENCEPHALIC tegmental NEP

Oculomotor (III) complex NEP?

Interpeduncular nuclear NEP?

Migrating tegmental neurons

Migrating interpeduncular nuclear neurons?

Migrating mammillary nuclear neurons?; nuclear neurons?

B. *FLOOR OF DIENCEPHALIC SUPERVENTRICLE*

Posterior hypothalamic NEP (mammillary)

MAMMILLARY RECESS

Brain surface (heavier line)

Middle hypothalamic NEP

44

PLATE 11A
CR 15 mm, GW 7.4, C9247
Sagittal

A.
Near Plate 1
Slide 27
Section 7

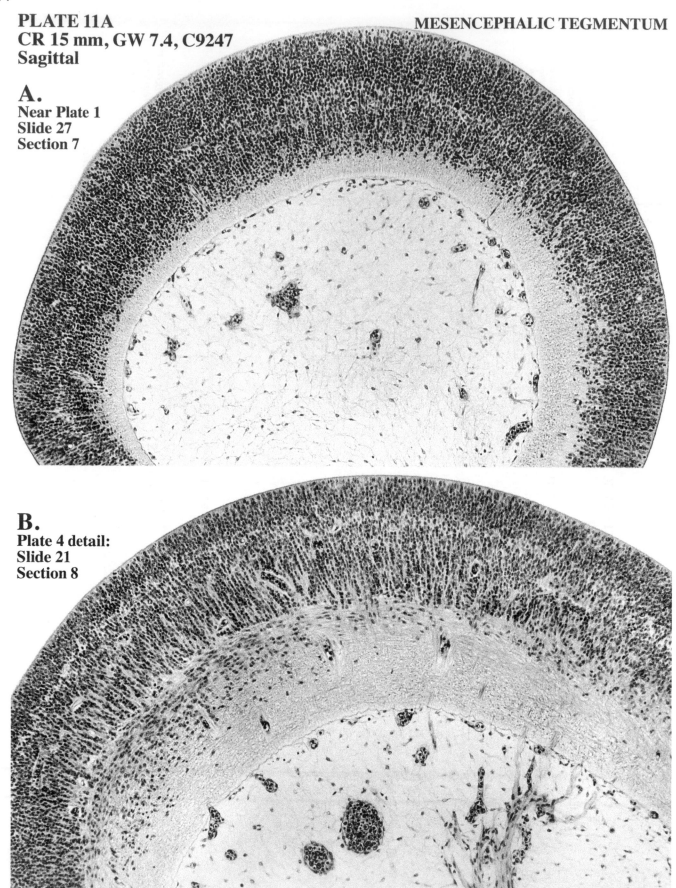

B.
Plate 4 detail:
Slide 21
Section 8

See Plates 1A/B and 4A/B for complete sections.

0.1 mm

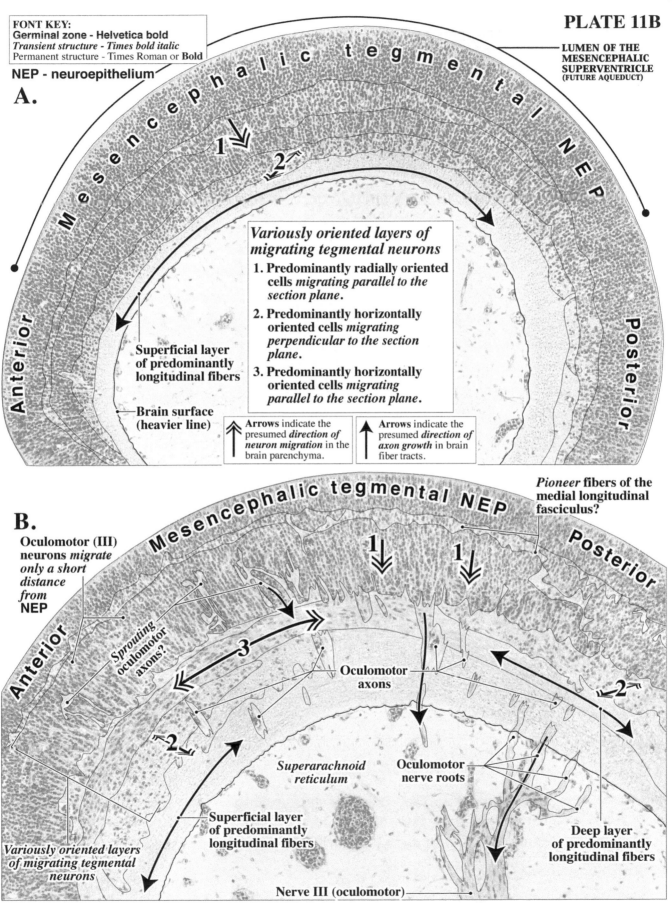

A.

FONT KEY:
Germinal zone - Helvetica bold
Transient structure - Times bold italic
Permanent structure - Times Roman or **Bold**

NEP - neuroepithelium

Mesencephalic tegmental NEP

LUMEN OF THE
MESENCEPHALIC
SUPERVENTRICLE
(FUTURE AQUEDUCT)

Anterior

Posterior

1

2

Variously oriented layers of migrating tegmental neurons

1. Predominantly radially oriented cells *migrating parallel to the section plane.*

2. Predominantly horizontally oriented cells *migrating perpendicular to the section plane.*

3. Predominantly horizontally oriented cells *migrating parallel to the section plane.*

Superficial layer of predominantly longitudinal fibers

Brain surface (heavier line)

Arrows indicate the presumed *direction of neuron migration* in the brain parenchyma.

Arrows indicate the presumed *direction of axon growth* in brain fiber tracts.

B.

Mesencephalic tegmental NEP

Pioneer fibers of the medial longitudinal fasciculus?

Oculomotor (III) neurons *migrate only a short distance from* **NEP**

Posterior

Anterior

1

1

Sprouting oculomotor axons?

3

2

Oculomotor axons

2

Superarachnoid reticulum

Oculomotor nerve roots

Superficial layer of predominantly longitudinal fibers

Deep layer of predominantly longitudinal fibers

Variously oriented layers of migrating tegmental neurons

Nerve III (oculomotor)

PLATE 12A

CR 15 mm
GW 7.4
C9247
Sagittal
Slide 24, Section 8

See a nearby complete
section in Plates 3A/B.

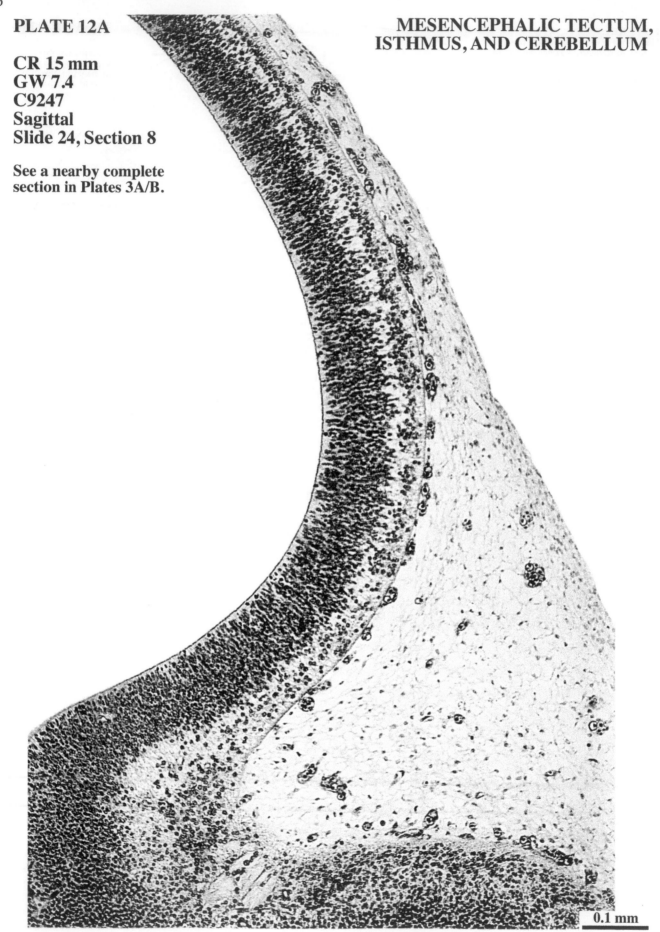

0.1 mm

PLATE 12B

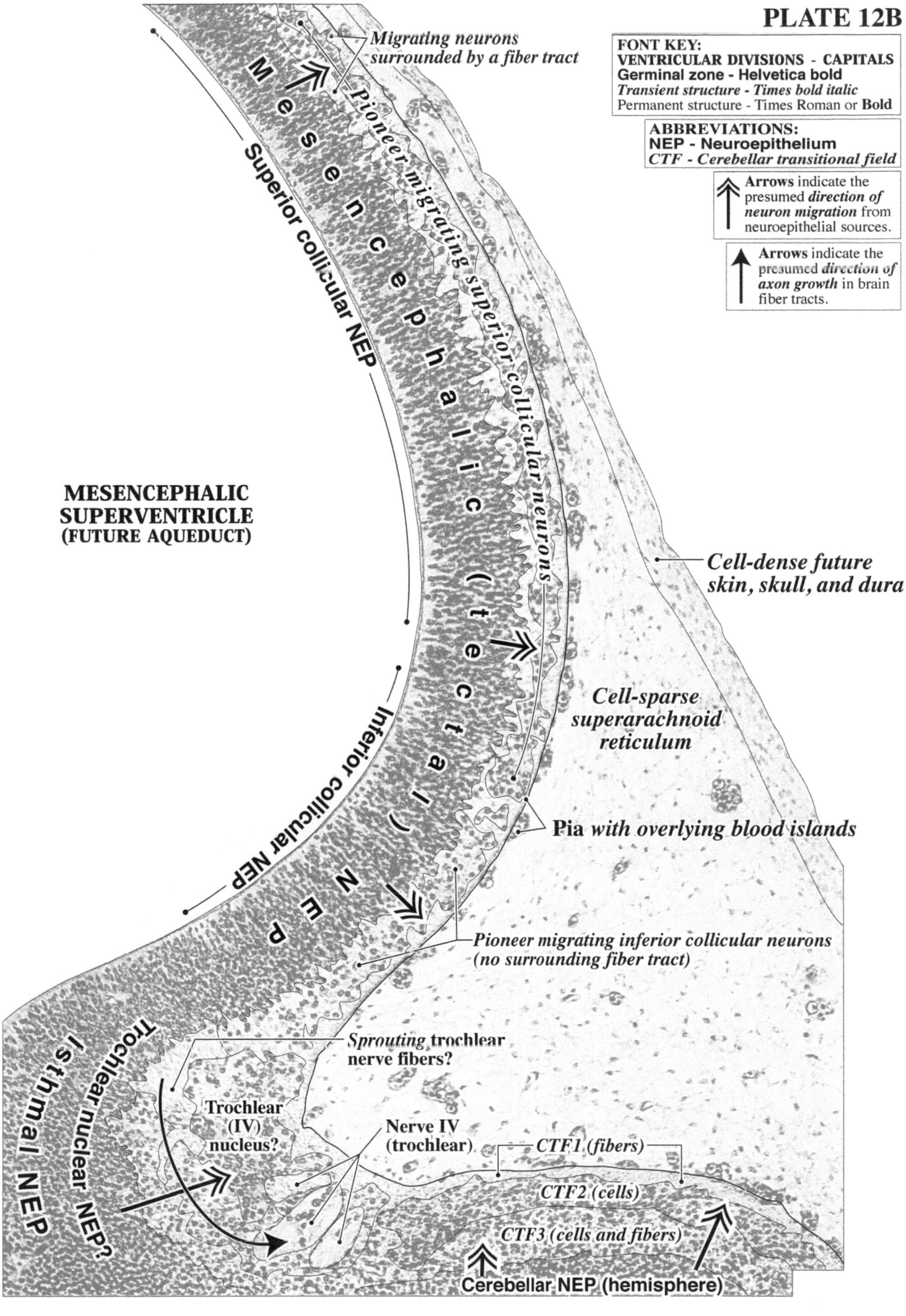

FONT KEY:
VENTRICULAR DIVISIONS - CAPITALS
Germinal zone - Helvetica bold
Transient structure - Times bold italic
Permanent structure - Times Roman or **Bold**

ABBREVIATIONS:
NEP - Neuroepithelium
CTF - Cerebellar transitional field

Arrows indicate the presumed *direction of neuron migration* from neuroepithelial sources.

Arrows indicate the presumed *direction of axon growth* in brain fiber tracts.

Migrating neurons surrounded by a fiber tract

M e s e n c e p h a l i c

Pioneer migrating superior collicular neurons

Superior collicular NEP

MESENCEPHALIC SUPERVENTRICLE (FUTURE AQUEDUCT)

Cell-dense future skin, skull, and dura

Inferior collicular NEP

N E P   (t e c t a l)

Cell-sparse superarachnoid reticulum

Pia *with overlying blood islands*

Pioneer migrating inferior collicular neurons (no surrounding fiber tract)

Isthmal NEP

Trochlear nuclear NEP?

Sprouting trochlear nerve fibers?

Trochlear (IV) nucleus?

Nerve IV (trochlear)

CTF1 (fibers)

CTF2 (cells)

CTF3 (cells and fibers)

Cerebellar NEP (hemisphere)

PLATE 13A

A.

Slide 27, Section 14
See the complete section in Plates 1A/B.

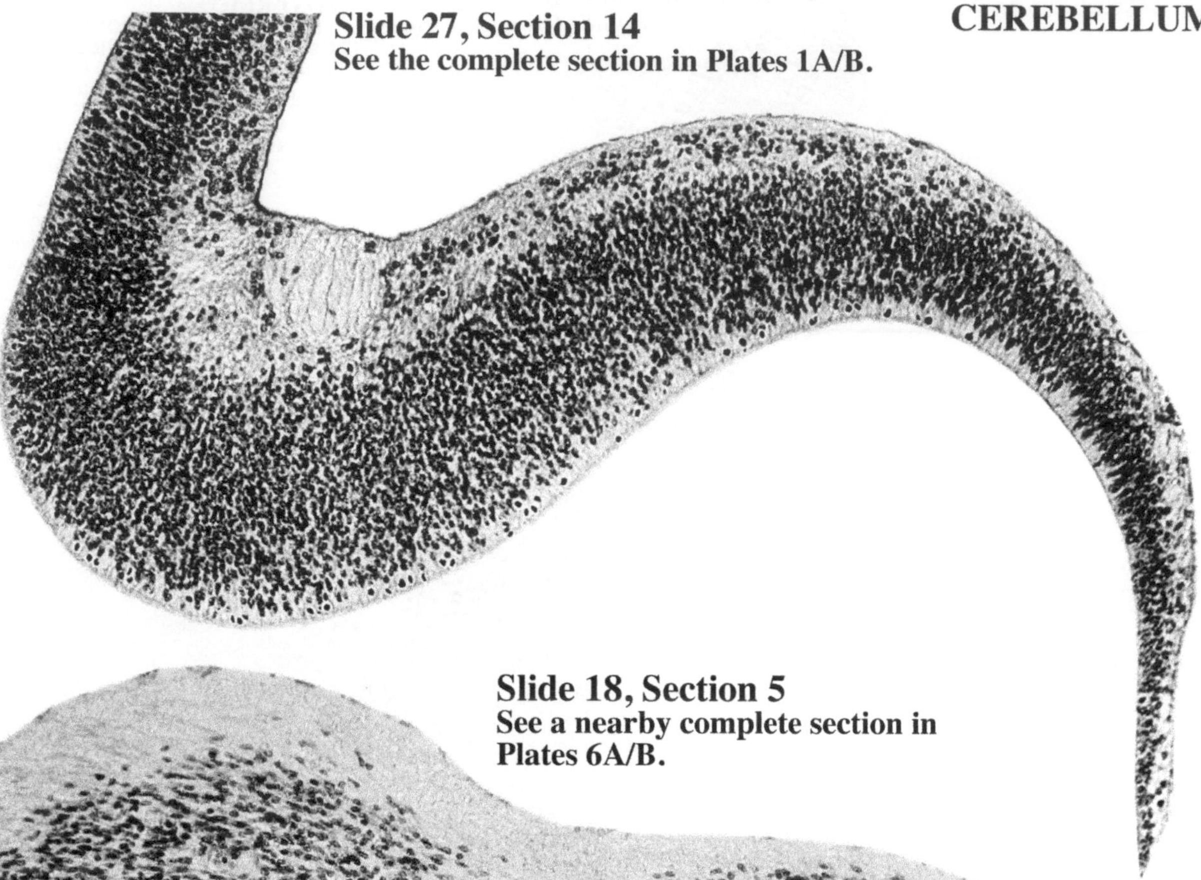

B.

Slide 18, Section 5
See a nearby complete section in
Plates 6A/B.

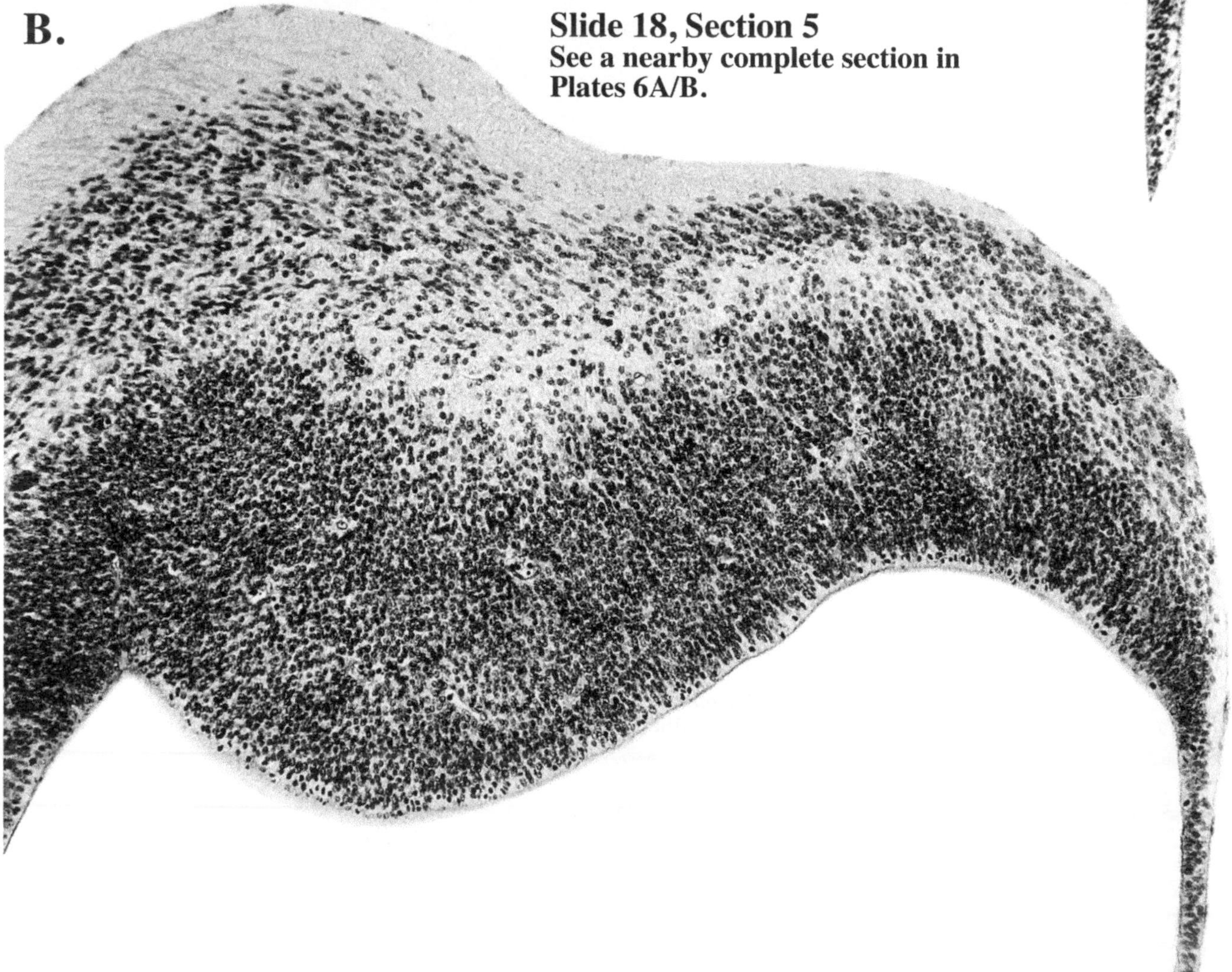

0.1 mm

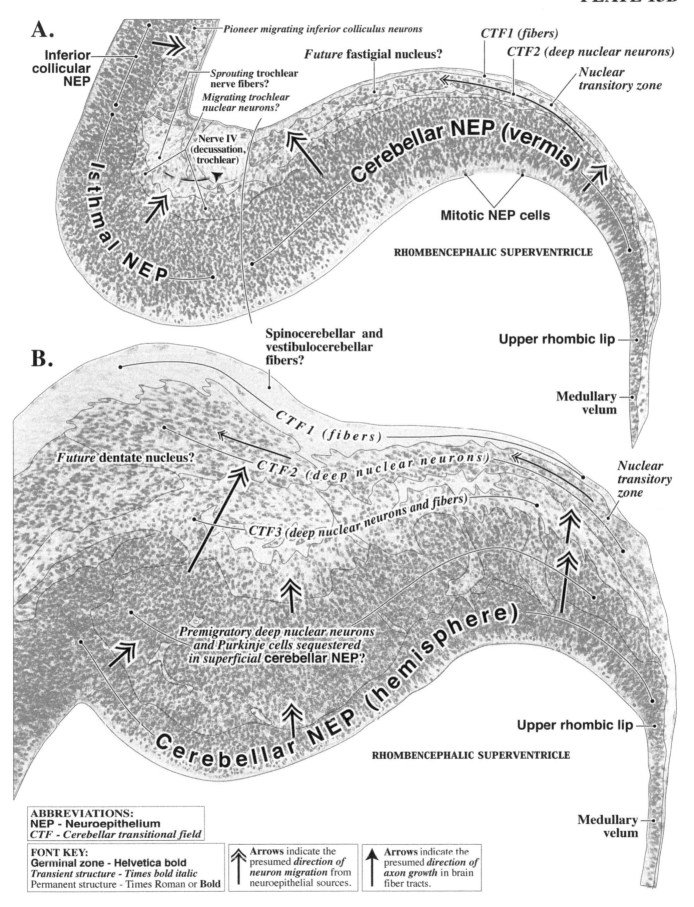

A.

Pioneer migrating inferior colliculus neurons

Inferior collicular NEP

Sprouting trochlear nerve fibers?

Migrating trochlear nuclear neurons?

Future fastigial nucleus?

CTF1 (fibers)

CTF2 (deep nuclear neurons)

Nuclear transitory zone

Nerve IV (decussation, trochlear)

Isthmal NEP

Cerebellar NEP (vermis)

Mitotic NEP cells

RHOMBENCEPHALIC SUPERVENTRICLE

Upper rhombic lip

Medullary velum

B.

Spinocerebellar and vestibulocerebellar fibers?

CTF1 (fibers)

Future dentate nucleus?

CTF2 (deep nuclear neurons)

CTF3 (deep nuclear neurons and fibers)

Nuclear transitory zone

Premigratory deep nuclear neurons and Purkinje cells sequestered in superficial cerebellar NEP?

Cerebellar NEP (hemisphere)

RHOMBENCEPHALIC SUPERVENTRICLE

Upper rhombic lip

Medullary velum

ABBREVIATIONS:
NEP - Neuroepithelium
CTF - Cerebellar transitional field

FONT KEY:
Germinal zone - Helvetica bold
Transient structure - Times bold italic
Permanent structure - Times Roman or **Bold**

Arrows indicate the presumed *direction of neuron migration* from neuroepithelial sources.

Arrows indicate the presumed *direction of axon growth* in brain fiber tracts.

PLATE 14A

CR 15 mm, GW 7.4, C9247, Sagittal
Slide 19,
Section 3

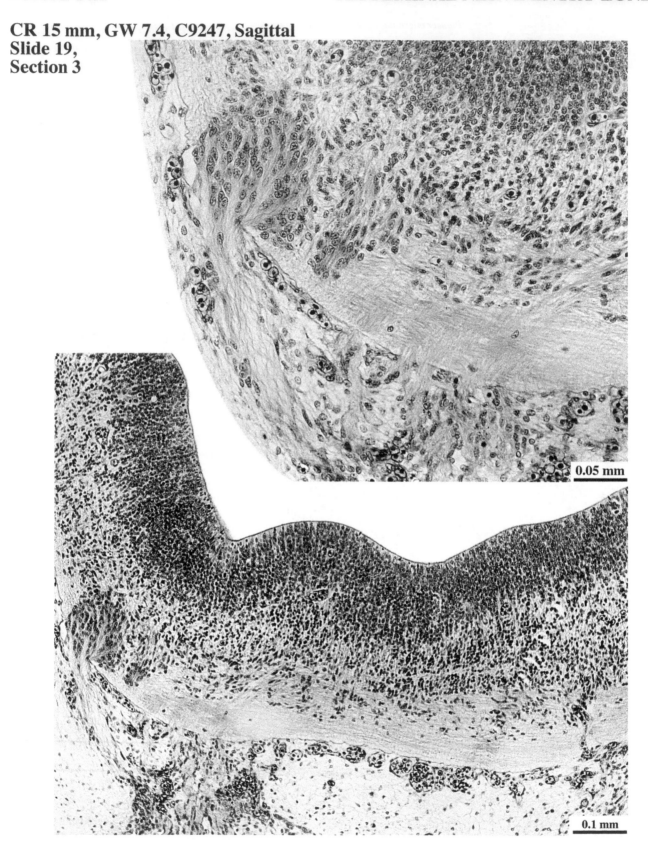

0.05 mm

0.1 mm

See complete sections in Plates 6-7A/B.

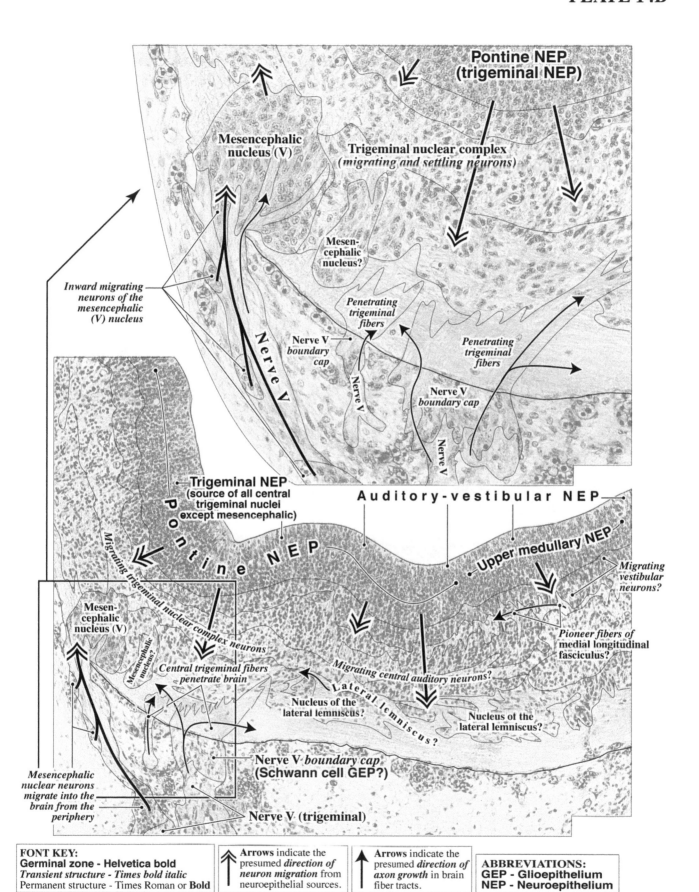

Pontine NEP
(trigeminal NEP)

Mesencephalic
nucleus (V)

Trigeminal nuclear complex
(migrating and settling neurons)

Mesen-
cephalic
nucleus?

*Penetrating
trigeminal
fibers*

*Inward migrating
neurons of the
mesencephalic
(V) nucleus*

Nerve V
*boundary
cap*

*Penetrating
trigeminal
fibers*

Nerve V
boundary cap

Nerve V

Nerve V

Trigeminal NEP
(source of all central
trigeminal nuclei
except mesencephalic)

A u d i t o r y - v e s t i b u l a r   N E P

P o n t i n e   N E P

Upper medullary NEP

Migrating trigeminal nuclear complex neurons

*Migrating
vestibular
neurons?*

Mesen-
cephalic
nucleus (V)

*Mesencephalic
nucleus?*

*Central trigeminal fibers
penetrate brain*

Migrating central auditory neurons?

L a t e r a l   l e m n i s c u s ?

*Nucleus of the
lateral lemniscus?*

*Pioneer fibers of
medial longitudinal
fasciculus?*

*Nucleus of the
lateral lemniscus?*

*Mesencephalic
nuclear neurons
migrate into the
brain from the
periphery*

Nerve V *boundary cap*
(Schwann cell GEP?)

Nerve V (trigeminal)

FONT KEY:
Germinal zone - Helvetica bold
Transient structure - Times bold italic
Permanent structure - Times Roman or **Bold**

Arrows indicate the
presumed *direction of
neuron migration* from
neuroepithelial sources.

Arrows indicate the
presumed *direction of
axon growth* in brain
fiber tracts.

ABBREVIATIONS:
GEP - Glioepithelium
NEP - Neuroepithelium

PLATE 15A

CR 15 mm, GW 7.4, C9247, Sagittal
Slide 16, Section 3

LATERAL CEREBELLUM,
PONS, AND MEDULLA

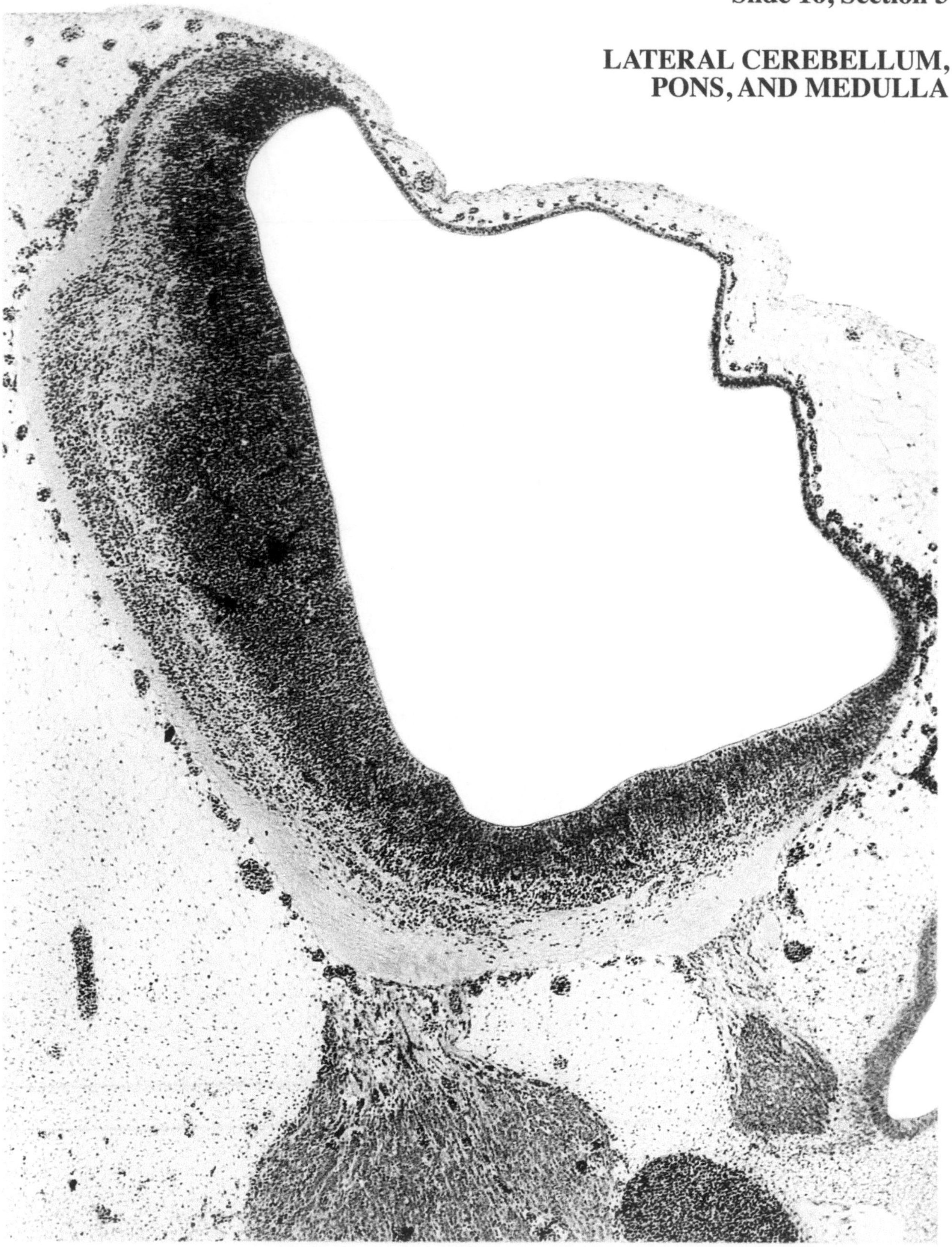

0.5 mm

See a nearby complete section in Plate 7A/B.

PLATE 15B

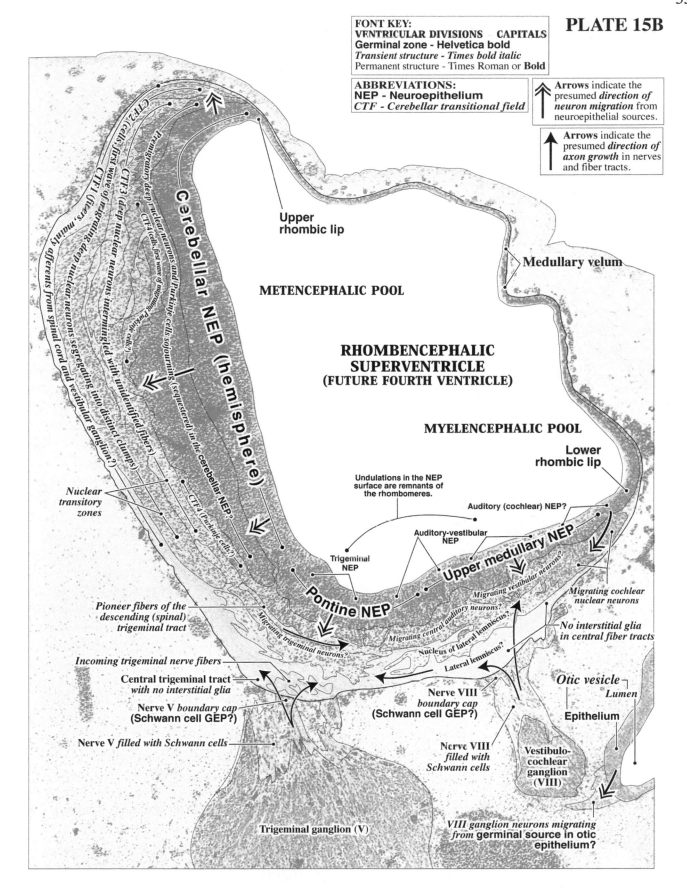

FONT KEY:
VENTRICULAR DIVISIONS   CAPITALS
Germinal zone - Helvetica bold
Transient structure - Times bold italic
Permanent structure - Times Roman or **Bold**

ABBREVIATIONS:
NEP - Neuroepithelium
CTF - Cerebellar transitional field

Arrows indicate the presumed *direction of neuron migration* from neuroepithelial sources.

Arrows indicate the presumed *direction of axon growth* in nerves and fiber tracts.

Upper rhombic lip

Medullary velum

METENCEPHALIC POOL

RHOMBENCEPHALIC SUPERVENTRICLE (FUTURE FOURTH VENTRICLE)

MYELENCEPHALIC POOL

Lower rhombic lip

CTF2 (cells: first wave of migrating Purkinje cells?)
CTF1 (layers: mainly afferents from spinal cord and vestibular ganglion?)
Premigratory deep nuclear neurons and Purkinje cells sojourning (sequestered in the cerebellar NEP?)
CTF3 (deep nuclear neurons intermingled with unidentified fibers)
CTF4 (cells: first wave of migrating deep nuclear neurons segregating into distinct clumps?)

Cerebellar NEP (hemisphere)

Nuclear transitory zones

CTF4 (Purkinje cells?)

Undulations in the NEP surface are remnants of the rhombomeres.

Auditory (cochlear) NEP?

Auditory-vestibular NEP

Trigeminal NEP

Upper medullary NEP

Migrating vestibular neurons?

Migrating cochlear nuclear neurons

Pontine NEP

Migrating trigeminal neurons?

Migrating central auditory neurons?

No interstitial glia in central fiber tracts

Nucleus of lateral lemniscus?

Lateral lemniscus?

Otic vesicle

Lumen

Pioneer fibers of the descending (spinal) trigeminal tract

Incoming trigeminal nerve fibers

Central trigeminal tract *with no interstitial glia*

Nerve V *boundary cap* (Schwann cell GEP?)

Nerve V *filled with Schwann cells*

Nerve VIII *boundary cap* (Schwann cell GEP?)

Nerve VIII *filled with Schwann cells*

Epithelium

Vestibulo-cochlear ganglion (VIII)

Trigeminal ganglion (V)

VIII ganglion neurons migrating from germinal source in otic epithelium?

54

TRIGEMINAL AND VESTIBULO-
COCHLEAR NERVE ENTRY ZONES

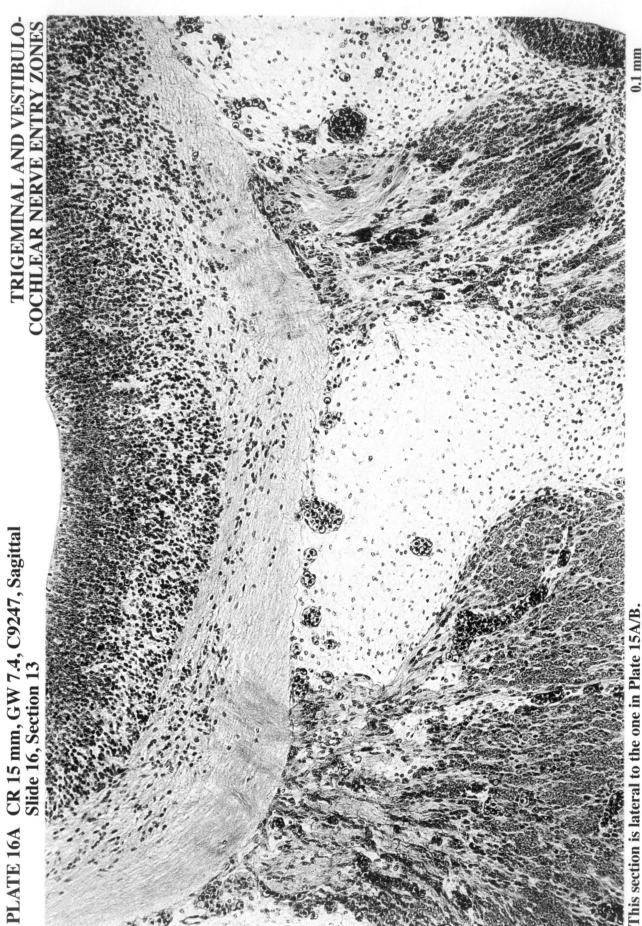

0.1 mm

This section is lateral to the one in Plate 15A/B.

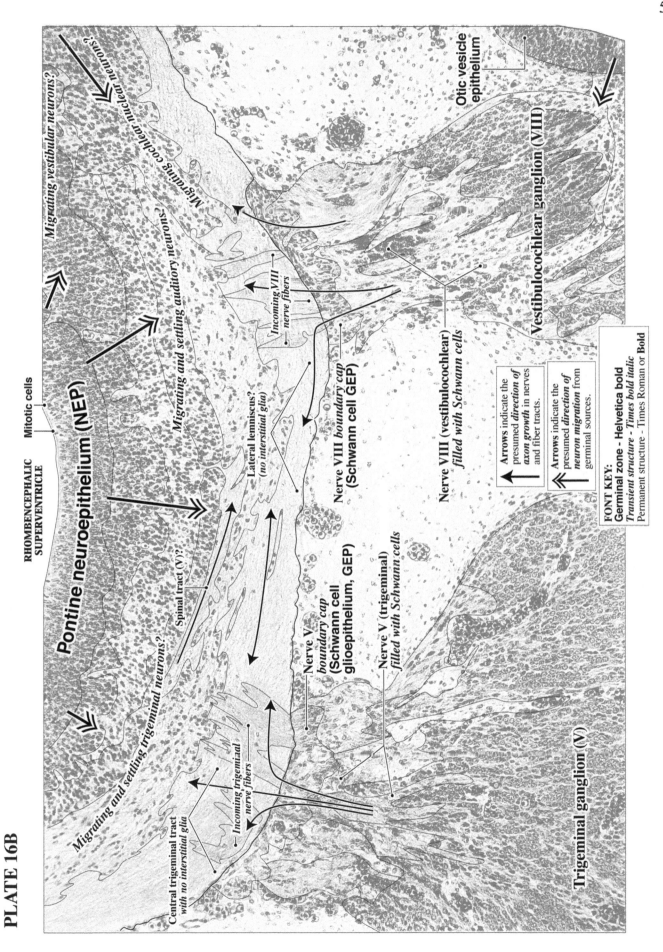

PLATE 16B

55

RHOMBENCEPHALIC
SUPERVENTRICLE

Mitotic cells

Pontine neuroepithelium (NEP)

Migrating vestibular neurons?

Migrating cochlear nuclear neurons?

Migrating and settling auditory neurons?

Migrating and settling trigeminal neurons?

Spinal tract (V)?

Lateral lemniscus?
(no interstitial glia)

Incoming VIII
nerve fibers

Otic vesicle
epithelium

Vestibulocochlear ganglion (VIII)

Nerve VIII boundary cap (Schwann cell GEP)

Nerve VIII (vestibulocochlear)
filled with Schwann cells

Arrows indicate the
presumed direction of
axon growth in nerves
and fiber tracts.

Arrows indicate the
presumed direction of
neuron migration from
germinal sources.

FONT KEY:
Germinal zone - Helvetica bold
Transient structure - Times bold italic
Permanent structure - Times Roman or Bold

Nerve V
boundary cap
(Schwann cell
glioepithelium, GEP)

Nerve V (trigeminal)
filled with Schwann cells

Central trigeminal tract
with no interstitial glia

Incoming trigeminal
nerve fibers

Trigeminal ganglion (V)

PLATE 17A   **CR 15 mm, GW 7.4, C9247, Sagittal**   **ENTRY ZONES OF**
Slide 18, Section 13   **NERVES IX AND X**

A.
Nerve IX

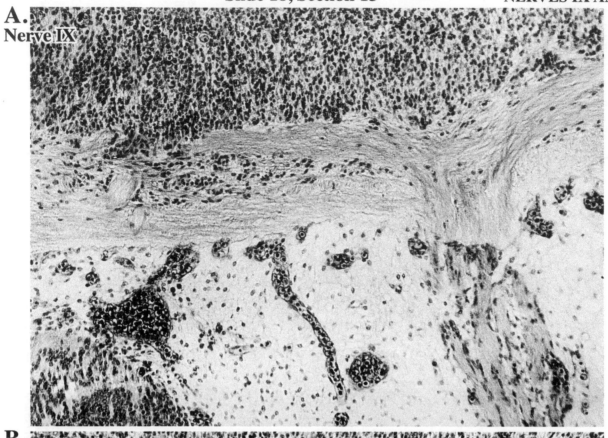

B.
Nerve X

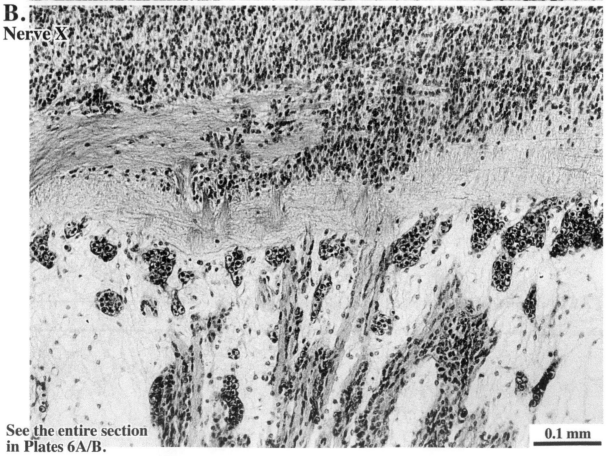

See the entire section
in Plates 6A/B.

0.1 mm

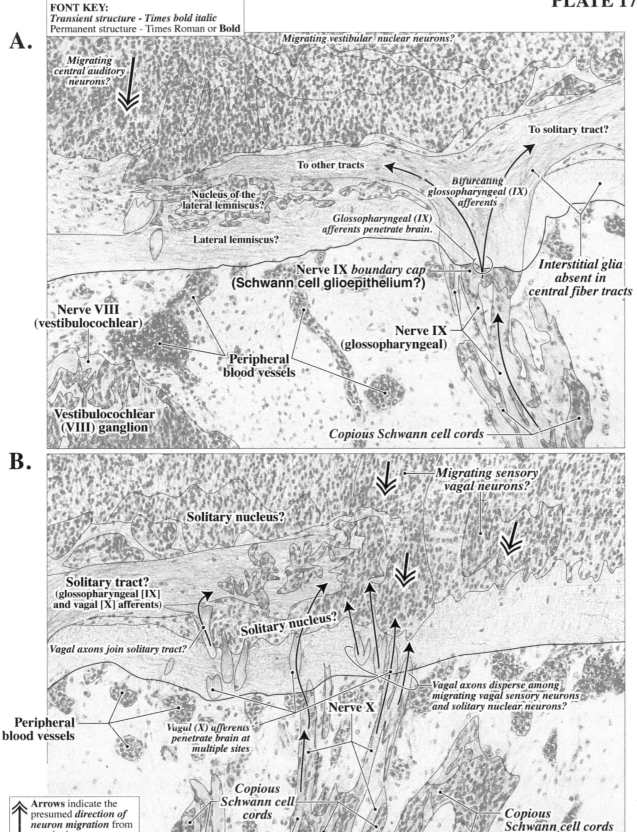

A.

FONT KEY:
Transient structure - Times bold italic
Permanent structure - Times Roman or **Bold**

Migrating vestibular nuclear neurons?

Migrating central auditory neurons?

To solitary tract?

To other tracts

Bifurcating glossopharyngeal (IX) afferents

Glossopharyngeal (IX) afferents penetrate brain.

Nucleus of the lateral lemniscus?

Interstitial glia absent in central fiber tracts

Lateral lemniscus?

Nerve IX *boundary cap*
(Schwann cell glioepithelium?)

**Nerve VIII
(vestibulocochlear)**

**Peripheral
blood vessels**

**Nerve IX
(glossopharyngeal)**

**Vestibulocochlear
(VIII) ganglion**

Copious Schwann cell cords

B.

*Migrating sensory
vagal neurons?*

Solitary nucleus?

Solitary tract?
(glossopharyngeal [IX]
and vagal [X] afferents)

Solitary nucleus?

Vagal axons join solitary tract?

*Vagal axons disperse among
migrating vagal sensory neurons
and solitary nuclear neurons?*

**Peripheral
blood vessels**

*Vagal (X) afferents
penetrate brain at
multiple sites*

Nerve X

*Copious
Schwann cell
cords*

*Copious
Schwann cell cords*

**Nerve X
(vagus)**

Arrows indicate the
presumed *direction of
neuron migration* from
germinal sources.

Arrows indicate the
presumed *direction of
axon growth* in nerves
and fiber tracts.

MEDIAL PONS AND MEDULLA

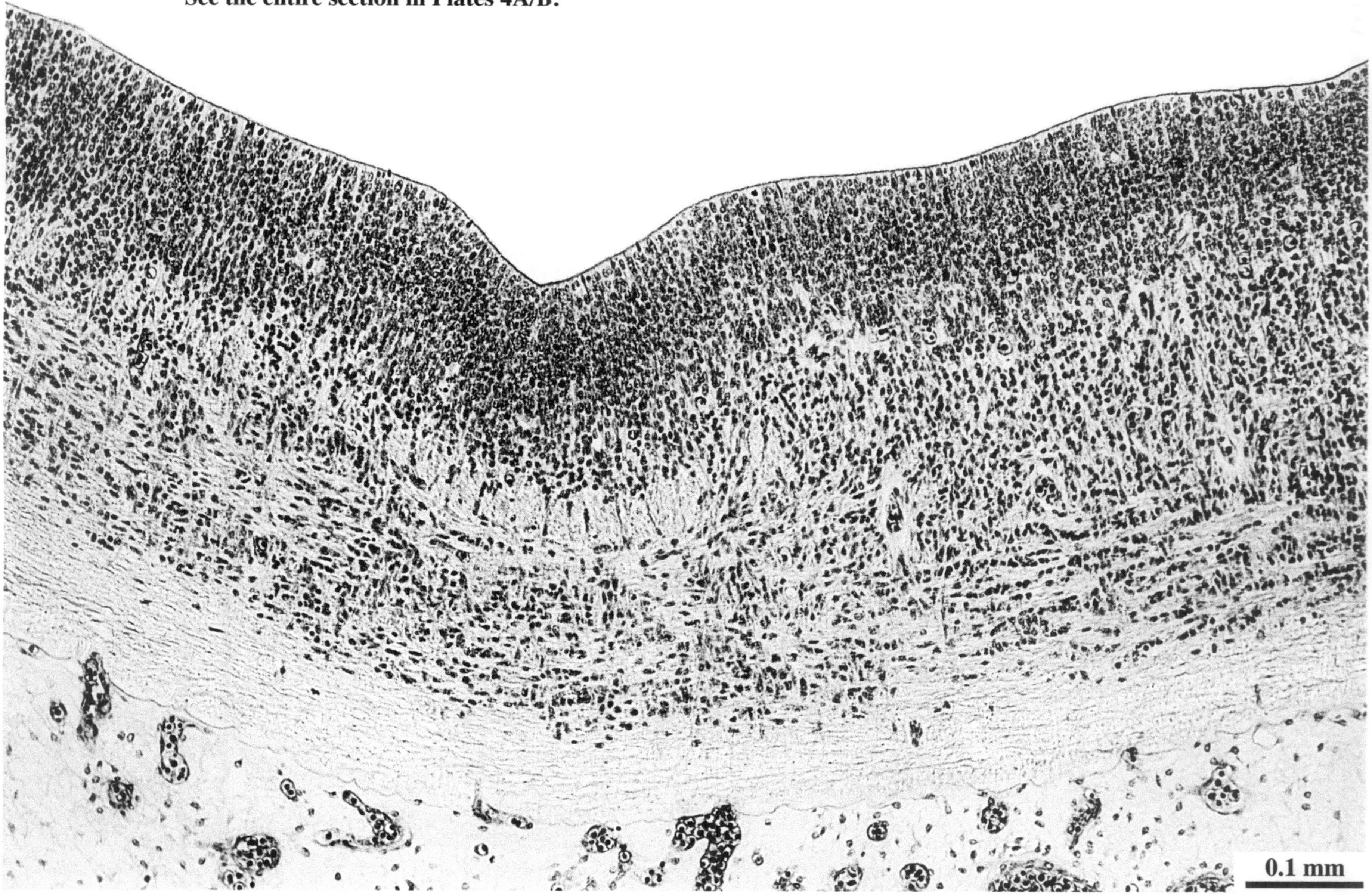

PLATE 18A   CR 15 mm, GW 7.4, C9247, Sagittal
            Slide 21, Section 8
            See the entire section in Plates 4A/B.

0.1 mm

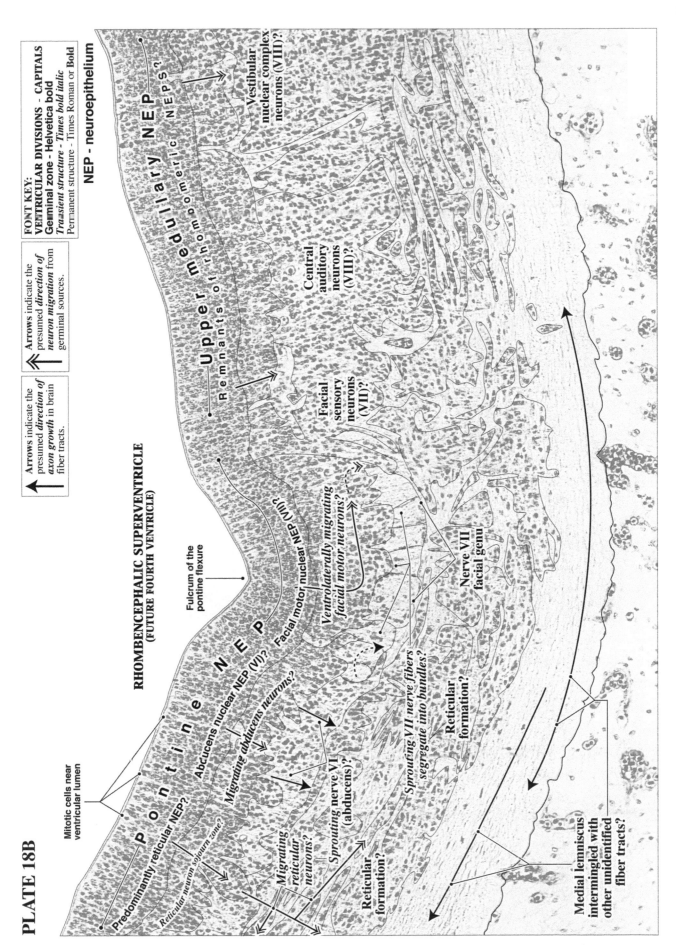

PLATE 18B

RHOMBENCEPHALIC SUPERVENTRICLE
(FUTURE FOURTH VENTRICLE)

Arrows indicate the presumed *direction of axon growth* in brain fiber tracts.

Arrows indicate the presumed *direction of neuron migration* from germinal sources.

FONT KEY:
VENTRICULAR DIVISIONS - CAPITALS
Geminal zone - Helvetica bold
Transient structure - Times bold italic
Permanent structure - Times Roman or Bold

NEP - neuroepithelium

Upper Medullary NEP

Remnants of rhombomeric NEPs?

Vestibular nuclear complex neurons (VIII)?

Central auditory neurons (VIII)?

Facial sensory neurons (VII)?

P o n t i n e   N E P

Fulcrum of the pontine flexure

Facial motor nuclear NEP (VII)?

Abducens nuclear NEP (VI)?

Ventrolaterally migrating facial motor neurons?

Migrating abducens neurons?

Sprouting nerve VI (abducens)?

Nerve VII facial genu

Sprouting VII nerve fibers segregate into bundles?

Reticular formation?

Mitotic cells near ventricular lumen

Predominantly reticular NEP?

Reticular neuron sojourn zone?

Migrating reticular neurons?

Reticular formation?

Medial lemniscus intermingled with other unidentified fiber tracts?

ISTHMUS AND MIDLINE RAPHE
GLIAL STRUCTURE

PLATE 19A

CR 15 mm, GW 7.4, C9247, Sagittal
Slide 27, Section 14

A. ISTHMUS

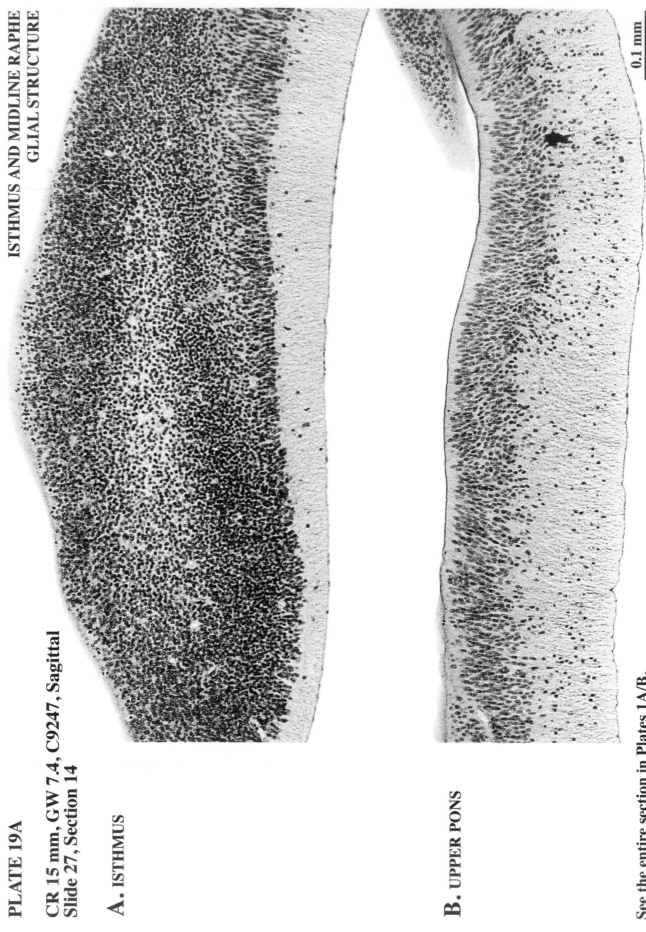

0.1 mm

B. UPPER PONS

See the entire section in Plates 1A/B.

PLATE 19B

A.

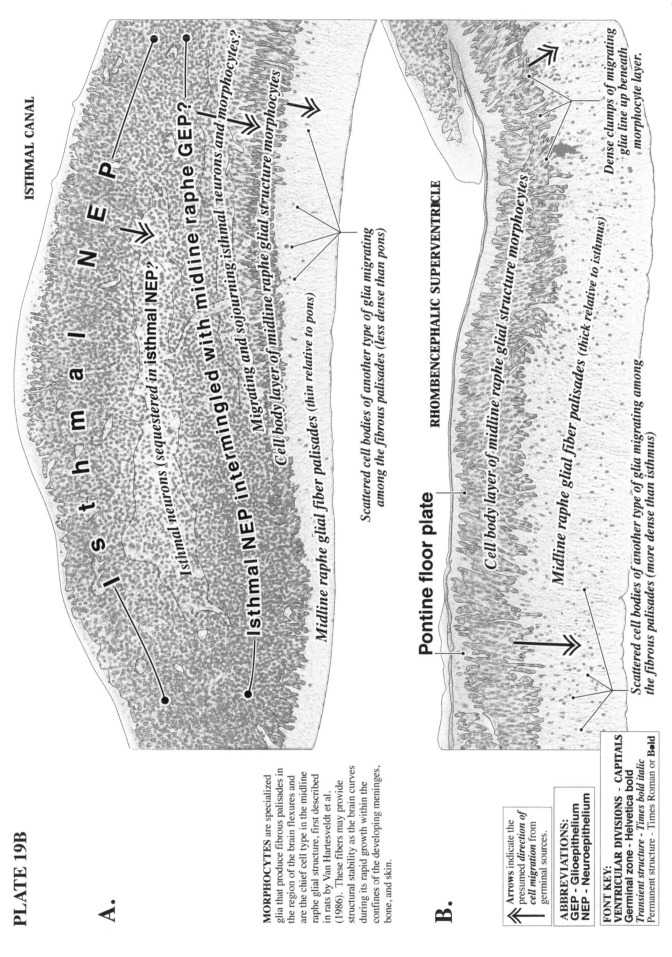

ISTHMAL CANAL

Isthmal NEP

Isthmal NEP intermingled with midline raphe GEP?

Isthmal neurons (sequestered in isthmal NEP?)

Migrating and sojourning isthmal neurons and morphocytes?

Cell body layer of midline raphe glial structure morphocytes

Scattered cell bodies of another type of glia migrating among the fibrous palisades (less dense than pons)

Midline raphe glial fiber palisades (thin relative to pons)

MORPHOCYTES are specialized glia that produce fibrous palisades in the region of the brain flexures and are the chief cell type in the midline raphe glial structure, first described in rats by Van Hartesveldt et al. (1986). These fibers may provide structural stability as the brain curves during its rapid growth within the confines of the developing meninges, bone, and skin.

B.

RHOMBENCEPHALIC SUPERVENTRICLE

Dense clumps of migrating glia line up beneath morphocyte layer.

Cell body layer of midline raphe glial structure morphocytes

Midline raphe glial fiber palisades (thick relative to isthmus)

Scattered cell bodies of another type of glia migrating among the fibrous palisades (more dense than isthmus)

Pontine floor plate

Arrows indicate the presumed *direction of cell migration* from germinal sources.

ABBREVIATIONS:
GEP - Glioepithelium
NEP - Neuroepithelium

FONT KEY:
VENTRICULAR DIVISIONS - **CAPITALS**
Germinal zone - **Helvetica bold**
Transient structure - Times bold italic
Permanent structure - Times Roman or **B**old

PLATE 20A

LOWER PONTINE NEUROEPITHELIUM AND MIDLINE RAPHE GLIAL STRUCTURE

CR 15 mm, GW 7.4, C9247, Sagittal
Slide 27, Section 14

PONTINE FLEXURE

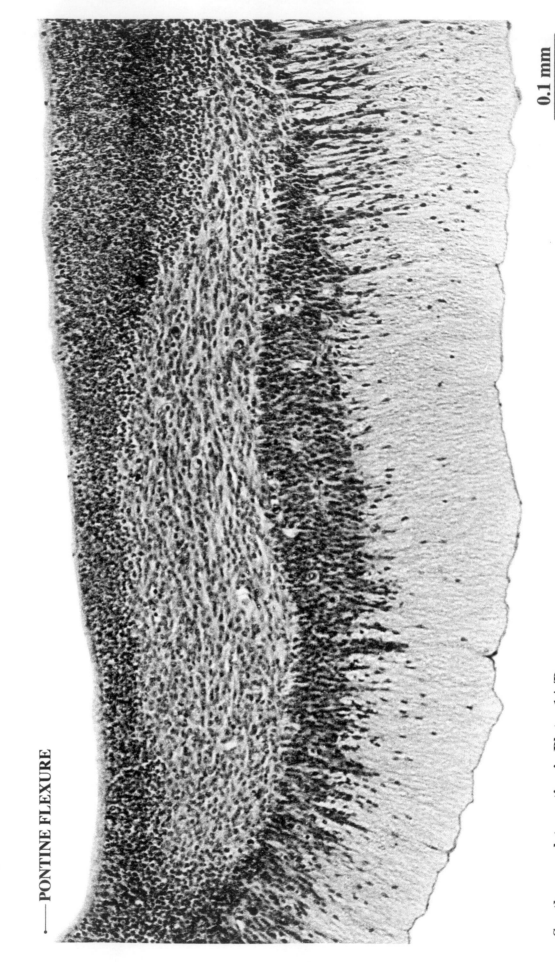

0.1 mm

See the complete section in Plates 1A/B.

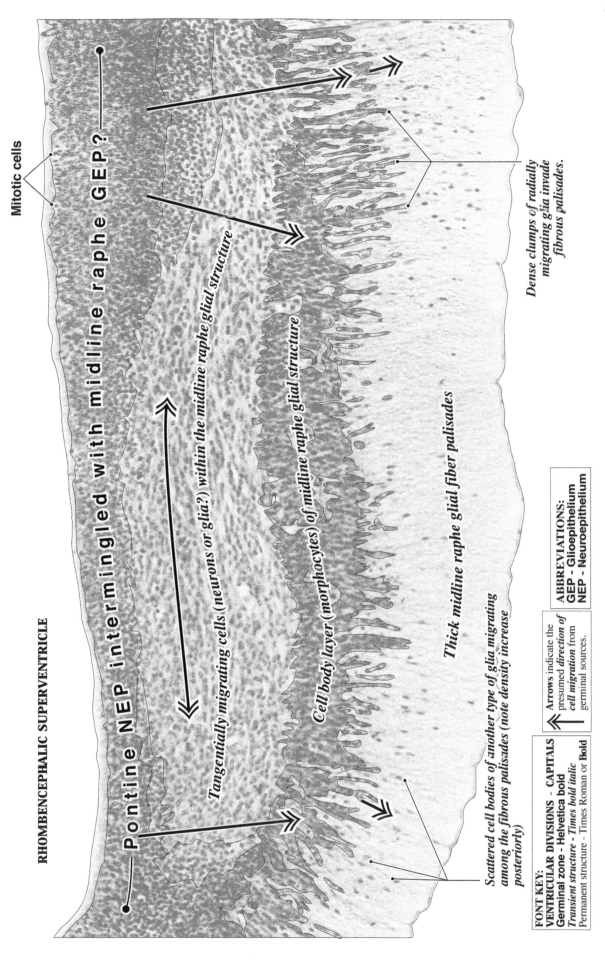

PLATE 20B

63

RHOMBENCEPHALIC SUPERVENTRICLE

Mitotic cells

Pontine NEP intermingled with midline raphe GEP?

Tangentially migrating cells (neurons or glia?) within the midline raphe glial structure

Cell body layer (morphocytes) of midline raphe glial structure

Thick midline raphe glial fiber palisades

Scattered cell bodies of another type of glia migrating among the fibrous palisades (note density increase posteriorly)

Dense clumps of radially migrating glia invade fibrous palisades.

Arrows indicate the presumed *direction of cell migration* from germinal sources.

ABBREVIATIONS:
GEP - Glioepithelium
NEP - Neuroepithelium

FONT KEY:
VENTRICULAR DIVISIONS - CAPITALS
Germinal zone - Helvetica bold
Transient structure - Times bold italic
Permanent structure - Times Roman or **Bold**

PLATE 21A

CR 15 mm, GW 7.4, C9247, Sagittal
Slide 27, Section 14

LOWER MEDULLARY NEUROEPITHELIUM
AND MIDLINE RAPHE GLIAL STRUCTURE

MEDULLA
(slightly anterior
to medullary
flexure)

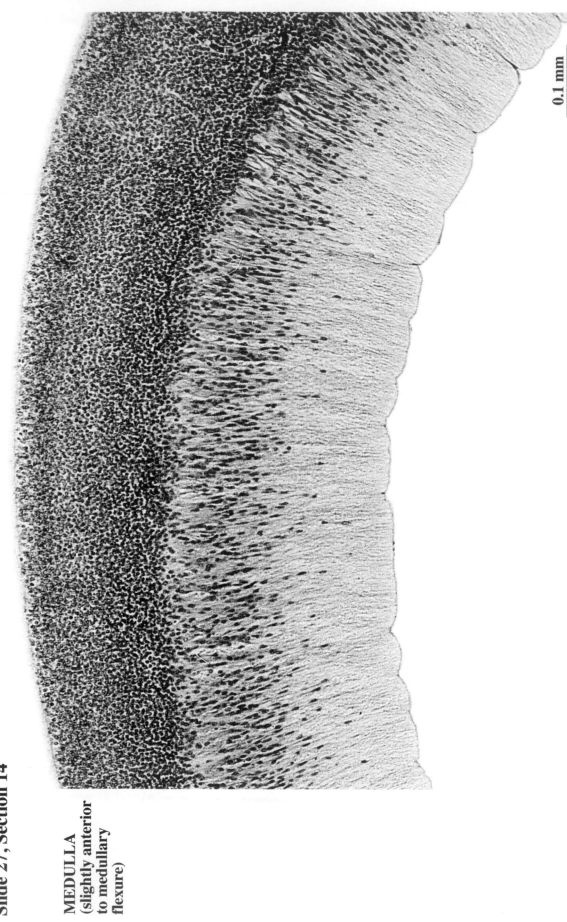

0.1 mm

See the complete section in Plates 1A/B.

65

PLATE 21B

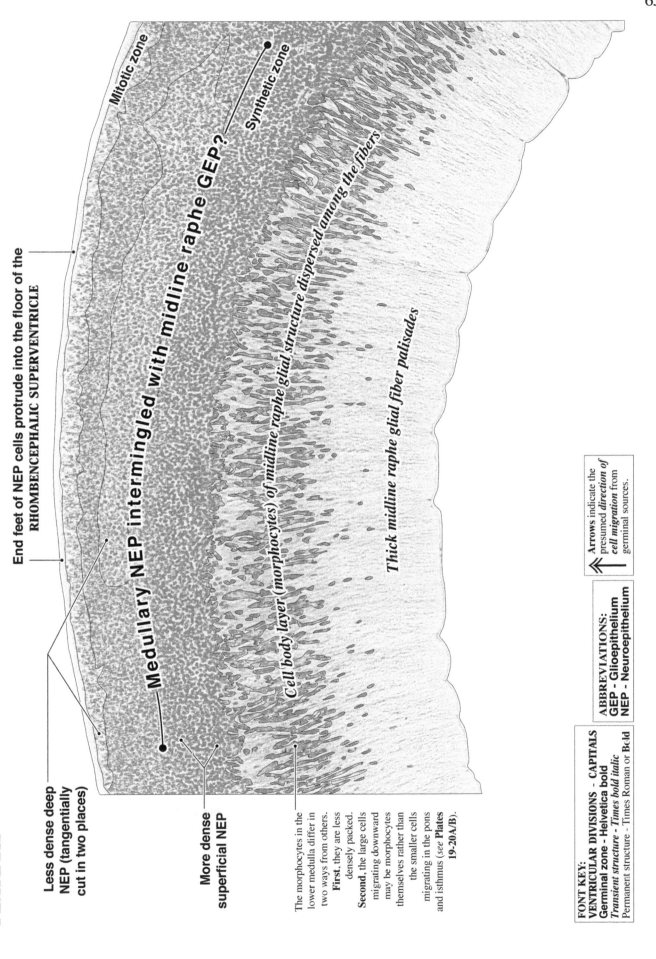

End feet of NEP cells protrude into the floor of the
RHOMBENCEPHALIC SUPERVENTRICLE

Mitotic zone

Synthetic zone

Medullary NEP intermingled with midline raphe GEP?

Cell body layer (morphocytes) of midline raphe glial structure dispersed among the fibers

Thick midline raphe glial fiber palisades

Less dense deep
NEP (tangentially
cut in two places)

More dense
superficial NEP

The morphocytes in the
lower medulla differ in
two ways from others.
First, they are less
densely packed.
Second, the large cells
migrating downward
may be morphocytes
themselves rather than
the smaller cells
migrating in the pons
and isthmus (*see* **Plates
19-20A/B**).

Arrows indicate the
presumed *direction of
cell migration* from
germinal sources.

ABBREVIATIONS:
GEP - Glioepithelium
NEP - Neuroepithelium

FONT KEY:
VENTRICULAR DIVISIONS - CAPITALS
Germinal zone - Helvetica bold
Transient structure - Times bold italic
Permanent structure - Times Roman or Bold

PART III: M2051
CR 15.0 mm (GW 7.4)
Frontal/Horizontal

This specimen is embryo #2051 in the Minot Collection, designated here as M2051. The crown-rump length (CR) is 15-mm estimated to be at gestational week (GW) 7.4 (similar to rat embryos on day E14 as discussed in the **Introduction**. Most of M2051's brain sections are cut (10 μm) in the frontal plane, but that shifts to predominantly horizontal in the pons and medulla. We photographed 87 sections at low magnification from the frontal prominence to the posterior tips of the mesencephalon and medulla. Seventeen of these sections are illustrated in **Plates 22-38A/B**. All photographs were used to produce computer-aided 3-D reconstructions of the external features of M2051's brain (**Figure 13**), and to show each illustrated section *in situ* (*insets*, **Plates 22-38A**). Each illustrated section shows the brain with all surrounding tissues. Labels in **A Plates** (normal-contrast images) identify non-neural and peripheral neural structures; labels in **B Plates** (low-contrast copies) identify central neural structures. **Plates 39A/B** show a high-magnification view of the neocortical neuroepithelium.

All parts of the cerebral cortical telencephalic neuroepithelium (NEP) are rapidly increasing their pool of neuronal and glial stem cells as they expand the shorelines of the enlarging telencephalic superventricle. Some pioneer Cajal-Retzius neurons have migrated into the cell-sparse primordial plexiform layer adjacent to the cerebral cortical NEP. Many neurons in layer VII (subplate) are already postmitotic. Some layer V-VI neurons are being generated but nearly all stem cells of cortical neuronal populations in layers IV-II are stockbuilding. The basal ganglionic and basal telencephalic NEPs are in the neurogenetic phase and are flanked by waves of migrating neurons that settle in the parenchyma.

The diencephalic NEP surrounds a narrowing superventricle. It is thinnest in the hypothalamic and subthalamic areas, where it is surrounded by densely packed waves of migrating neurons. It is postulated that these areas of the superventricle have shrinking shorelines as the NEPs "unload" their stock of neuronal precursors. In contrast, the thalamic superventricle shoreline is still thick and active, but most neuronal populations are being generated and are probably being sequestered in the NEP itself. That is why few cells are in the primordial plexiform layer outside the thalamic NEPs.

The mesencephalon contains neurogenetic-stage NEPs throughout its entire extent. Neurons are migrating out in large numbers and are accumulating in dense clumps that blur the basal border of the NEP in the tegmental areas. Some cells lie farther out in the tegmental and isthmal parenchyma and are more sparsely scattered adjacent to subpial fiber bands. Many of these neuronal groups are tentatively identified. There are also some prominent fiber tracts invading the tegmentum and a few are identified. The sprouting oculomotor nerve fibers are identified as well.

Both the pons and medulla have NEPs that are thick but many postmitotic neurons are sequestered in their basal parts. The expanding parenchyma has cells settling in longitudinal arrays at the pontine flexure. Cells are settling in the reticular formation throughout the pons and medulla. Some facial motor neurons are migrating from medial to lateral, leaving behind their axons in a small, but definite genu of the facial nerve. Trigeminal neurons are settling in the pons. Migrating inferior olive neurons are in the posterior intramural migratory stream outside the precerebellar NEP in the posterior lower rhombic lip, but very few neurons have settled in the inferior olive. Many neurons have settled in the solitary nucleus, surrounding a definite solitary tract. In spite of its thickness, the cerebellar NEP is in the neurogenetic phase, only adding late-generated Purkinje cells to the deep nuclear neurons that have already settled in the cerebellar transitional field. Purkinje cells are either sequestered in the basal cerebellar NEP or are sojourning in the deepest layer of the cerebellar transitional field.

M2051 Computer-aided 3-D Brain Reconstructions

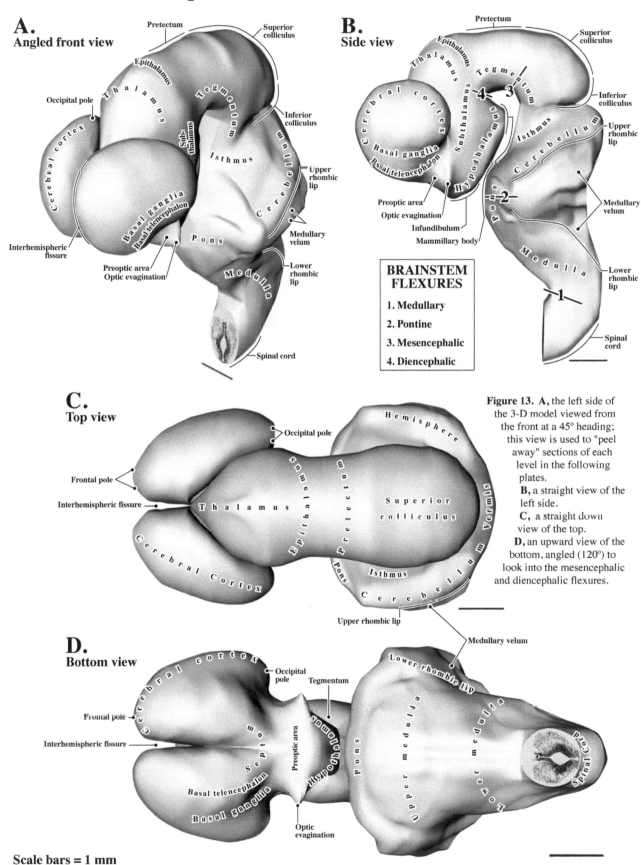

A. Angled front view

Pretectum
Superior colliculus
Epithalamus
Thalamus
Tegmentum
Occipital pole
Cerebral cortex
Sub-thalamus
Isthmus
Cerebellum
Inferior colliculus
Upper rhombic lip
Basal ganglia
Basal telencephalon
Interhemispheric fissure
Preoptic area
Optic evagination
Pons
Medulla
Medullary velum
Lower rhombic lip
Spinal cord

B. Side view

Pretectum
Superior colliculus
Epithalamus
Thalamus
Tegmentum
4 3
Cerebral cortex
Subthalamus
Inferior colliculus
Basal ganglia
Isthmus
Cerebellum
Upper rhombic lip
Basal telencephalon
Hypothalamus
Preoptic area
Optic evagination
Pons
2
Medullary velum
Infundibulum
Mammillary body
Medulla
Lower rhombic lip
1
Spinal cord

BRAINSTEM FLEXURES

1. Medullary
2. Pontine
3. Mesencephalic
4. Diencephalic

C. Top view

Hemisphere
Occipital pole
Frontal pole
Interhemispheric fissure
Thalamus
Epithalamus
Pretectum
Superior colliculus
Vermis
Cerebral Cortex
Pons
Isthmus
Cerebellum
Upper rhombic lip
Medullary velum

Figure 13. A, the left side of the 3-D model viewed from the front at a 45° heading; this view is used to "peel away" sections of each level in the following plates.

B, a straight view of the left side.

C, a straight down view of the top.

D, an upward view of the bottom, angled (120°) to look into the mesencephalic and diencephalic flexures.

D. Bottom view

Cerebral cortex
Occipital pole
Tegmentum
Frontal pole
Interhemispheric fissure
Septum
Preoptic area
Hypothalamus
Basal telencephalon
Basal ganglia
Optic evagination
Lower rhombic lip
Pons
Upper medulla
Lower medulla
Spinal Cord

Scale bars = 1 mm

PLATE 22A

CR 15.0 mm, GW7.4
M2051, Frontal/Horizontal
Section 66

Non-neural structures labeled

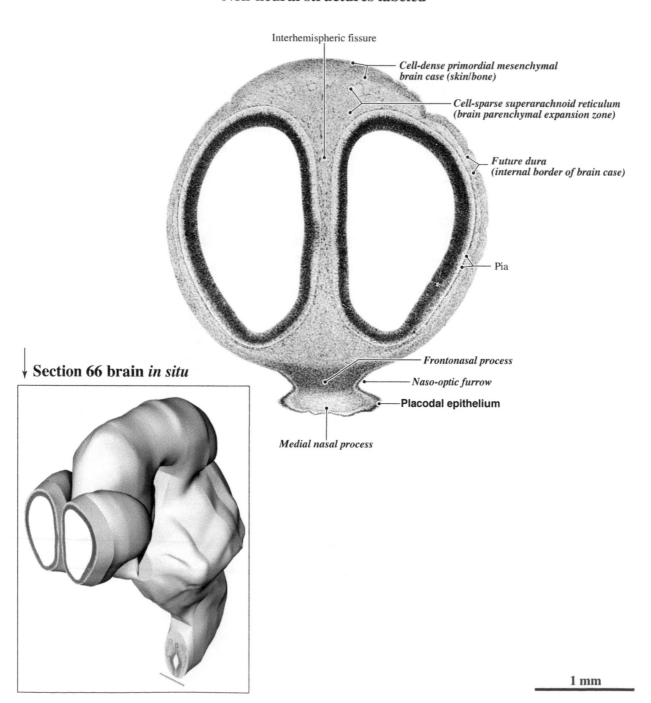

Interhemispheric fissure

Cell-dense primordial mesenchymal brain case (skin/bone)

Cell-sparse superarachnoid reticulum (brain parenchymal expansion zone)

Future dura (internal border of brain case)

Pia

Frontonasal process

Naso-optic furrow

Placodal epithelium

Medial nasal process

↓ **Section 66 brain *in situ***

1 mm

Neural structures labeled

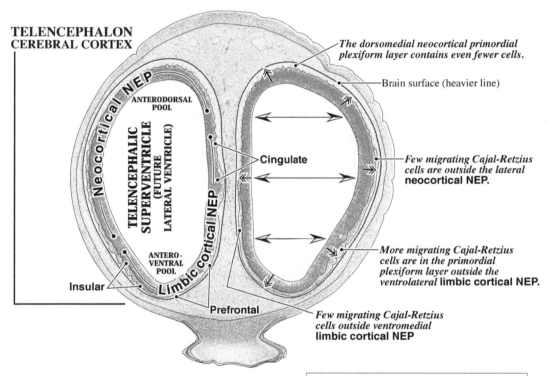

TELENCEPHALON
CEREBRAL CORTEX

Neocortical NEP

ANTERODORSAL
POOL

TELENCEPHALIC
SUPERVENTRICLE
(FUTURE
LATERAL VENTRICLE)

Cingulate

Limbic cortical NEP

ANTERO-
VENTRAL
POOL

Insular

Prefrontal

The dorsomedial neocortical primordial plexiform layer contains even fewer cells.

Brain surface (heavier line)

Few migrating Cajal-Retzius cells are outside the lateral neocortical NEP.

More migrating Cajal-Retzius cells are in the primordial plexiform layer outside the ventrolateral **limbic cortical NEP.**

Few migrating Cajal-Retzius cells outside ventromedial **limbic cortical NEP**

The density of migrating cells in the primordial plexiform layer indicates ventrolateral-to-dorsomedial and ventrolateral-to-ventromedial maturation gradients in the cerebral cortex.

NEP - Neuroepithelium

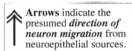

Arrows indicate the presumed *direction of neuron migration* from neuroepithelial sources.

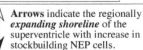

Arrows indicate the regionally *expanding shoreline* of the superventricle with increase in stockbuilding NEP cells.

FONT KEY:
VENTRICULAR DIVISIONS - CAPITALS
Germinal zone - Helvetica bold
Transient structure - Times bold italic
Permanent structure - Times Roman or **Bold**

PLATE 23A

CR 15.0 mm, GW7.4
M2051, Frontal/Horizontal
Section 107

Non-neural and peripheral neural structures labeled

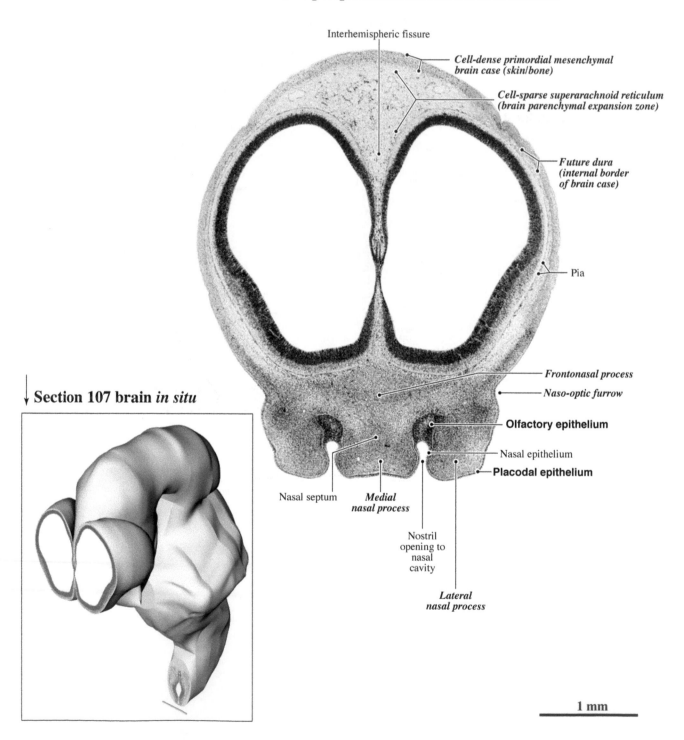

Interhemispheric fissure

Cell-dense primordial mesenchymal brain case (skin/bone)

Cell-sparse superarachnoid reticulum (brain parenchymal expansion zone)

Future dura (internal border of brain case)

Pia

Frontonasal process

Naso-optic furrow

Olfactory epithelium

Nasal epithelium

Placodal epithelium

Nasal septum

Medial nasal process

Nostril opening to nasal cavity

Lateral nasal process

↓ **Section 107 brain** *in situ*

1 mm

Central neural structures labeled

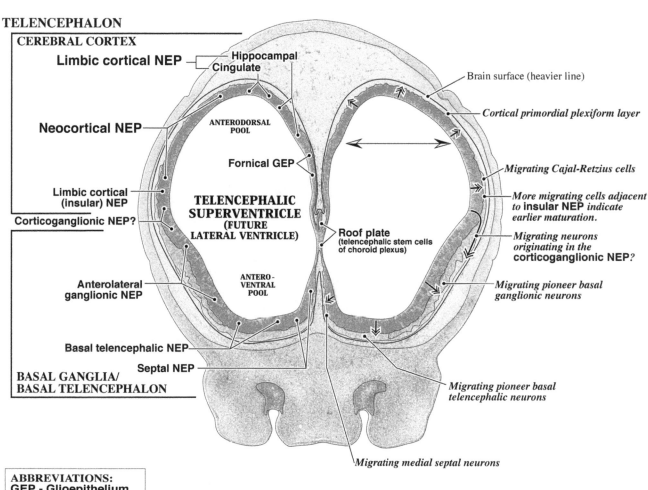

TELENCEPHALON

CEREBRAL CORTEX

Limbic cortical NEP

Hippocampal
Cingulate

Neocortical NEP

ANTERODORSAL
POOL

Fornical GEP

**Limbic cortical
(insular) NEP**

**TELENCEPHALIC
SUPERVENTRICLE
(FUTURE
LATERAL VENTRICLE)**

Corticoganglionic NEP?

**Anterolateral
ganglionic NEP**

ANTERO-
VENTRAL
POOL

Roof plate
(telencephalic stem cells
of choroid plexus)

Basal telencephalic NEP

Septal NEP

BASAL GANGLIA/
BASAL TELENCEPHALON

Brain surface (heavier line)

Cortical primordial plexiform layer

Migrating Cajal-Retzius cells

*More migrating cells adjacent
to* **insular NEP** *indicate
earlier maturation.*

*Migrating neurons
originating in the
corticoganglionic NEP?*

*Migrating pioneer basal
ganglionic neurons*

*Migrating pioneer basal
telencephalic neurons*

Migrating medial septal neurons

ABBREVIATIONS:
GEP - Glioepithelium
NEP - Neuroepithelium

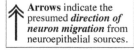

Arrows indicate the
presumed *direction of
neuron migration* from
neuroepithelial sources.

Arrows indicate the regionally
expanding shoreline of the
superventricle with increase in
stockbuilding NEP cells.

FONT KEY:
VENTRICULAR DIVISIONS - CAPITALS
Germinal zone - Helvetica bold
Transient structure - Times bold italic
Permanent structure - Times Roman or **Bold**

72

PLATE 24A

**CR 15.0 mm, GW7.4
M2051, Frontal/Horizontal
Section 130**

**See a high-magnification
view of the cerebral
cortex in Section 122
in Plates 39A/B.**

**Non-neural and peripheral
neural structures labeled**

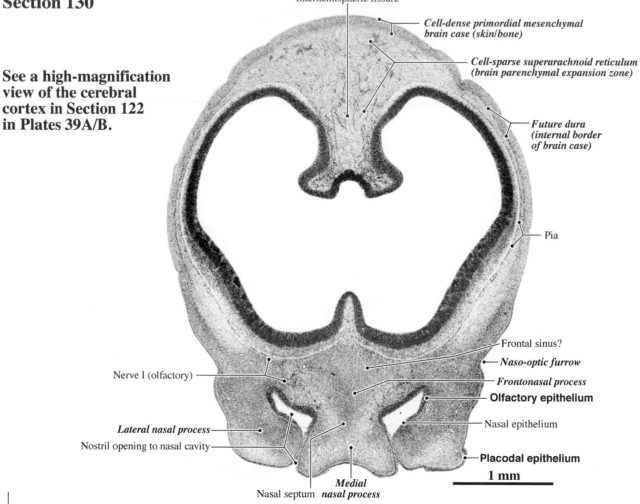

Interhemispheric fissure

*Cell-dense primordial mesenchymal
brain case (skin/bone)*

*Cell-sparse superarachnoid reticulum
(brain parenchymal expansion zone)*

*Future dura
(internal border
of brain case)*

Pia

Frontal sinus?

Naso-optic furrow

Frontonasal process

Olfactory epithelium

Nasal epithelium

Placodal epithelium

1 mm

Nerve I (olfactory)

Lateral nasal process

Nostril opening to nasal cavity

Nasal septum

*Medial
nasal process*

↓ **Section 130 brain *in situ***

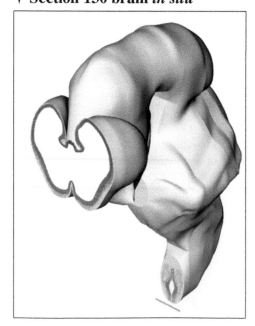

Central neural structures labeled

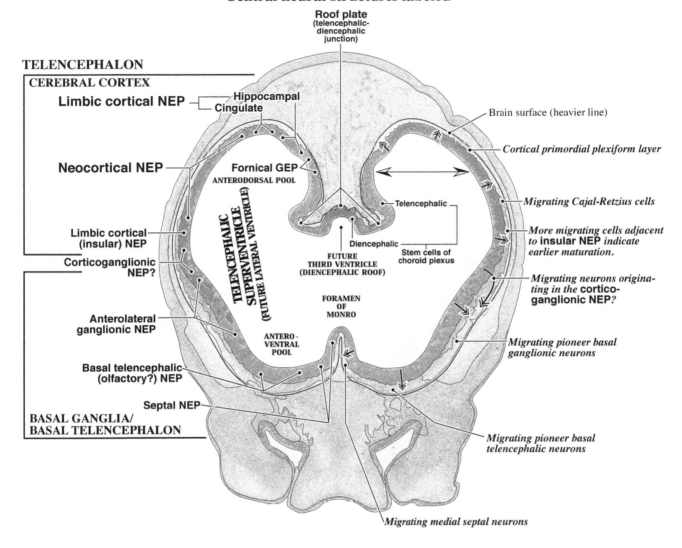

Roof plate
(telencephalic-
diencephalic
junction)

TELENCEPHALON

CEREBRAL CORTEX

Limbic cortical NEP — Hippocampal
Cingulate

Brain surface (heavier line)

Cortical primordial plexiform layer

Neocortical NEP — **Fornical GEP**
ANTERODORSAL POOL

Telencephalic

Migrating Cajal-Retzius cells

TELENCEPHALIC SUPERVENTRICLE
(FUTURE LATERAL VENTRICLE)

Diencephalic

More migrating cells adjacent to **insular NEP** *indicate earlier maturation.*

Limbic cortical (insular) NEP

FUTURE THIRD VENTRICLE (DIENCEPHALIC ROOF)

Stem cells of choroid plexus

Corticoganglionic NEP?

FORAMEN OF MONRO

Migrating neurons originating in the **corticoganglionic NEP?**

Anterolateral ganglionic NEP

ANTERO-VENTRAL POOL

Migrating pioneer basal ganglionic neurons

Basal telencephalic (olfactory?) NEP

Septal NEP

BASAL GANGLIA/
BASAL TELENCEPHALON

Migrating pioneer basal telencephalic neurons

Migrating medial septal neurons

ABBREVIATIONS:
GEP - Glioepithelium
NEP - Neuroepithelium

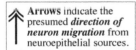

Arrows indicate the presumed *direction of neuron migration* from neuroepithelial sources.

Arrows indicate the regionally *expanding shoreline* of the superventricle with increase in stockbuilding NEP cells.

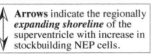

FONT KEY:
VENTRICULAR DIVISIONS - CAPITALS
Germinal zone - Helvetica bold
Transient structure - Times bold italic
Permanent structure - Times Roman or **Bold**

PLATE 25A

**CR 15.0 mm, GW7.4
M2051, Frontal/Horizontal
Section 159**

**Non-neural and peripheral
neural structures labeled**

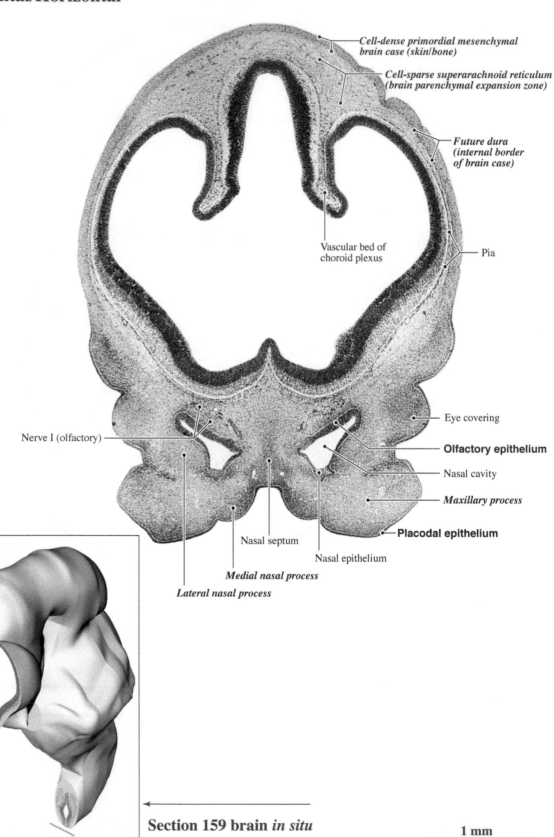

*Cell-dense primordial mesenchymal
brain case (skin/bone)*

*Cell-sparse superarachnoid reticulum
(brain parenchymal expansion zone)*

*Future dura
(internal border
of brain case)*

Vascular bed of
choroid plexus

Pia

Eye covering

Olfactory epithelium

Nerve I (olfactory)

Nasal cavity

Maxillary process

Placodal epithelium

Nasal septum

Nasal epithelium

Medial nasal process

Lateral nasal process

Section 159 brain *in situ*

1 mm

Central neural structures labeled

DIENCEPHALON
— THALAMUS

Thalamic NEP

Dorsal complex

Reticular nucleus

Anterior complex

Roof plate
(diencephalic stem cells of choroid plexus?)

Migrating pioneer thalamic neurons
Thalamic primordial plexiform layer

TELENCEPHALON

CEREBRAL CORTEX

Limbic cortical NEP — Hippocampal
— Cingulate

Neocortical NEP

Fornical GEP

DIENCEPHALIC SUPERVENTRICLE (THALAMIC POOL)

DORSAL POOL

TELENCEPHALIC SUPERVENTRICLE (FUTURE LATERAL VENTRICLE)

Brain surface (heavier line)

Cortical primordial plexiform layer

Migrating Cajal-Retzius cells

Roof plate
(telencephalic stem cells of choroid plexus)

More migrating cells adjacent to **insular NEP** *indicate earlier maturation.*

Limbic cortical (insular) NEP

Corticoganglionic NEP?

FORAMEN OF MONRO

Migrating neurons originating in the **corticoganglionic NEP**?

Anterolateral ganglionic NEP

Basal telencephalic (olfactory?) NEP

Septal NEP

VENTRAL POOL

Successive waves of migrating basal ganglionic neurons

Anteromedial ganglionic NEP

BASAL GANGLIA/
BASAL TELENCEPHALON

Migrating pioneer basal telencephalic neurons

Migrating medial septal neurons

ABBREVIATIONS:
GEP - Glioepithelium
NEP - Neuroepithelium

Arrows indicate the presumed *direction of neuron migration* from neuroepithelial sources.

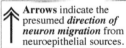

Arrows indicate the regionally *expanding shoreline* of the superventricle with increase in stockbuilding NEP cells.

FONT KEY:
VENTRICULAR DIVISIONS - CAPITALS
Germinal zone - Helvetica bold
Transient structure - Times bold italic
Permanent structure - Times Roman or **Bold**

PLATE 26A

CR 15.0 mm, GW7.4
M2051, Frontal/Horizontal
Section 190

Non-neural and peripheral
neural structures labeled

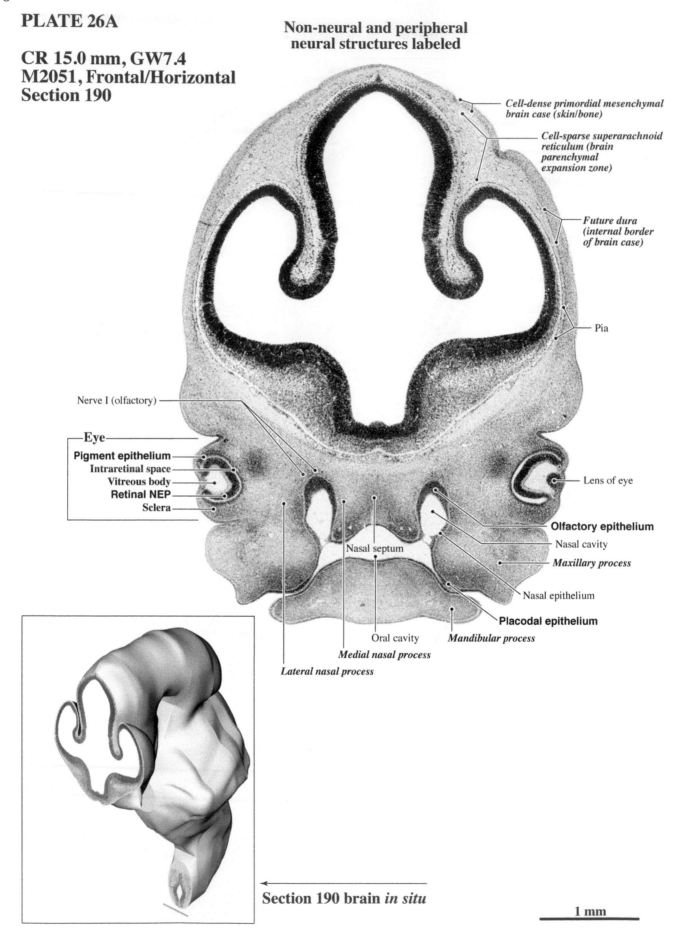

Cell-dense primordial mesenchymal brain case (skin/bone)

Cell-sparse superarachnoid reticulum (brain parenchymal expansion zone)

Future dura (internal border of brain case)

Pia

Nerve I (olfactory)

Eye
Pigment epithelium
Intraretinal space
Vitreous body
Retinal NEP
Sclera

Lens of eye

Olfactory epithelium

Nasal cavity

Maxillary process

Nasal septum

Nasal epithelium

Placodal epithelium

Oral cavity

Mandibular process

Medial nasal process

Lateral nasal process

Section 190 brain *in situ*

1 mm

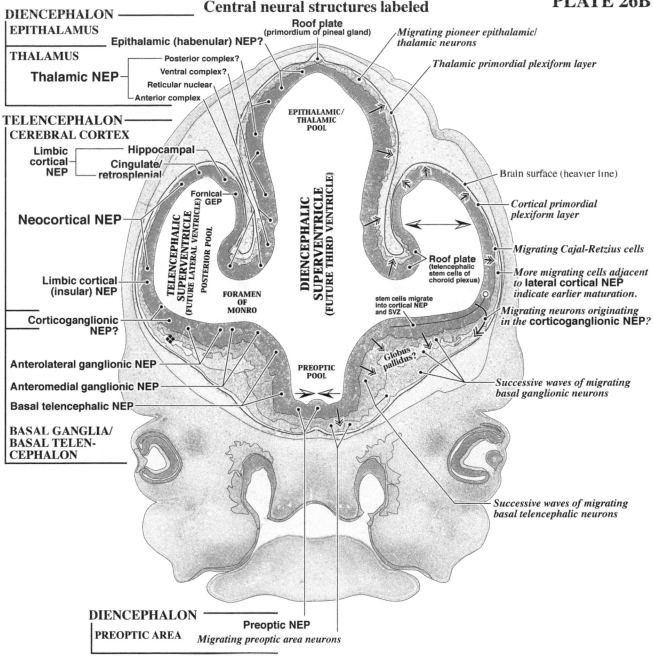

Central neural structures labeled

DIENCEPHALON
EPITHALAMUS
Epithalamic (habenular) NEP?
THALAMUS
Thalamic NEP
Posterior complex?
Ventral complex?
Reticular nuclear
Anterior complex

Roof plate
(primordium of pineal gland)

Migrating pioneer epithalamic/ thalamic neurons

Thalamic primordial plexiform layer

EPITHALAMIC/ THALAMIC POOL

TELENCEPHALON
CEREBRAL CORTEX
Limbic cortical NEP
Hippocampal
Cingulate/ retrosplenial

Neocortical NEP

Fornical GEP

TELENCEPHALIC SUPERVENTRICLE (FUTURE LATERAL VENTRICLE)
POSTERIOR POOL

DIENCEPHALIC SUPERVENTRICLE (FUTURE THIRD VENTRICLE)

Brain surface (heavier line)

Cortical primordial plexiform layer

Migrating Cajal-Retzius cells

More migrating cells adjacent to lateral cortical NEP *indicate earlier maturation.*

Limbic cortical (insular) NEP

Roof plate
(telencephalic stem cells of choroid plexus)

FORAMEN OF MONRO

stem cells migrate into cortical NEP and SVZ

Migrating neurons originating in the corticoganglionic NEP?

Corticoganglionic NEP?

Globus pallidus?

Anterolateral ganglionic NEP
Anteromedial ganglionic NEP
Basal telencephalic NEP

PREOPTIC POOL

Successive waves of migrating basal ganglionic neurons

BASAL GANGLIA/ BASAL TELEN-CEPHALON

Successive waves of migrating basal telencephalic neurons

DIENCEPHALON
PREOPTIC AREA

Preoptic NEP
Migrating preoptic area neurons

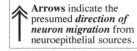

ABBREVIATIONS:
GEP - Glioepithelium
NEP - Neuroepithelium
SVZ - Subventricular zone

FONT KEY:
VENTRICULAR DIVISIONS - CAPITALS
Germinal zone - Helvetica bold
Transient structure - Times bold italic
Permanent structure - Times Roman or **Bold**

Arrows indicate the presumed *direction of neuron migration* from neuroepithelial sources.

Arrows indicate the regionally *expanding shoreline* of the superventricle with increase in stockbuilding NEP cells.

Arrows indicate the regionally *shrinking shoreline* of the superventricle as NEP cells are depleted while generating neurons.

PLATE 27A

**CR 15.0 mm, GW7.4
M2051, Frontal/Horizontal
Section 241**

**Non-neural and peripheral
neural structures labeled**

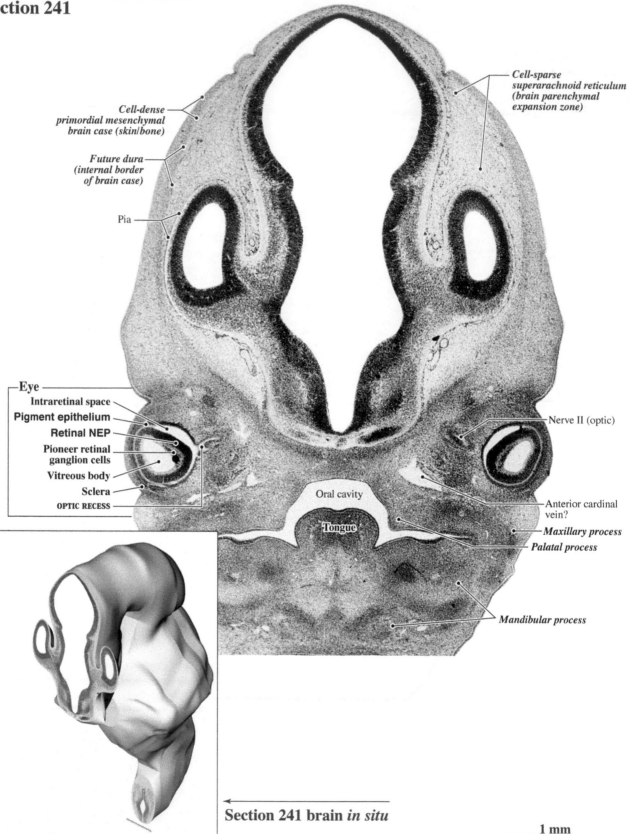

*Cell-sparse
superarachnoid reticulum
(brain parenchymal
expansion zone)*

*Cell-dense
primordial mesenchymal
brain case (skin/bone)*

*Future dura
(internal border
of brain case)*

Pia

Eye
Intraretinal space
Pigment epithelium
Retinal NEP
**Pioneer retinal
ganglion cells**
Vitreous body
Sclera
OPTIC RECESS

Nerve II (optic)

Oral cavity

Tongue

Anterior cardinal
vein?

Maxillary process
Palatal process

Mandibular process

Section 241 brain *in situ*

1 mm

Central neural structures labeled

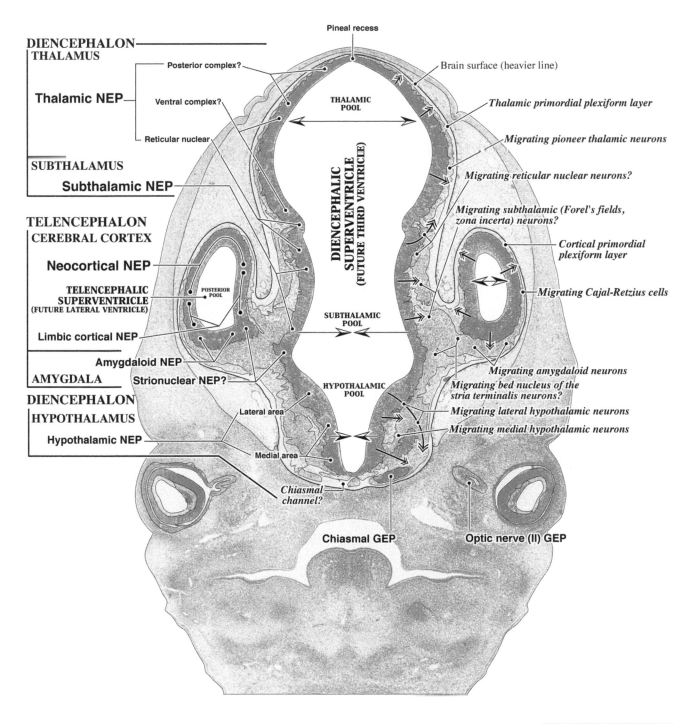

DIENCEPHALON
THALAMUS

Thalamic NEP

Posterior complex?

Ventral complex?

Reticular nuclear

Pineal recess

THALAMIC
POOL

Brain surface (heavier line)

Thalamic primordial plexiform layer

Migrating pioneer thalamic neurons

SUBTHALAMUS

Subthalamic NEP

DIENCEPHALIC
SUPERVENTRICLE
(FUTURE THIRD VENTRICLE)

Migrating reticular nuclear neurons?

Migrating subthalamic (Forel's fields, zona incerta) neurons?

TELENCEPHALON
CEREBRAL CORTEX

Neocortical NEP

TELENCEPHALIC
SUPERVENTRICLE
(FUTURE LATERAL VENTRICLE)

Limbic cortical NEP

POSTERIOR
POOL

Cortical primordial plexiform layer

Migrating Cajal-Retzius cells

SUBTHALAMIC
POOL

Amygdaloid NEP

Strionuclear NEP?

AMYGDALA

DIENCEPHALON
HYPOTHALAMUS

Hypothalamic NEP

Lateral area

Medial area

HYPOTHALAMIC
POOL

Migrating amygdaloid neurons

Migrating bed nucleus of the stria terminalis neurons?

Migrating lateral hypothalamic neurons

Migrating medial hypothalamic neurons

Chiasmal channel?

Chiasmal GEP

Optic nerve (II) GEP

ABBREVIATIONS:
GEP - Glioepithelium
NEP - Neuroepithelium

FONT KEY:
VENTRICULAR DIVISIONS - CAPITALS
Germinal zone - Helvetica bold
Transient structure - Times bold italic
Permanent structure - Times Roman or **Bold**

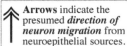

Arrows indicate the presumed *direction of neuron migration* from neuroepithelial sources.

Arrows indicate the regionally *expanding shoreline* of the superventricle with increase in stockbuilding NEP cells.

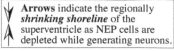

Arrows indicate the regionally *shrinking shoreline* of the superventricle as NEP cells are depleted while generating neurons.

PLATE 28A

CR 15.0 mm, GW7.4
M2051, Frontal/Horizontal
Section 258

Non-neural and peripheral
neural structures labeled

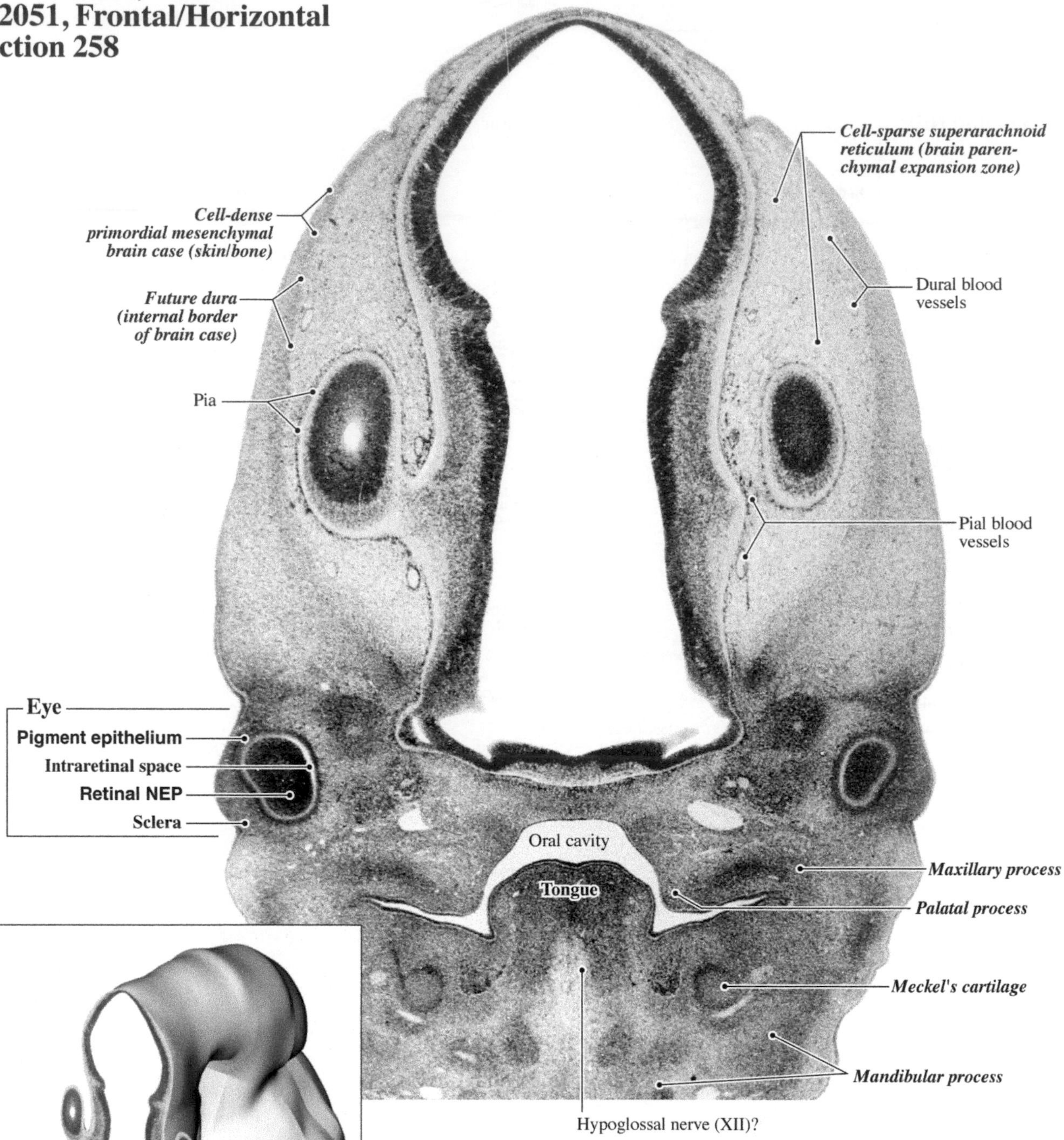

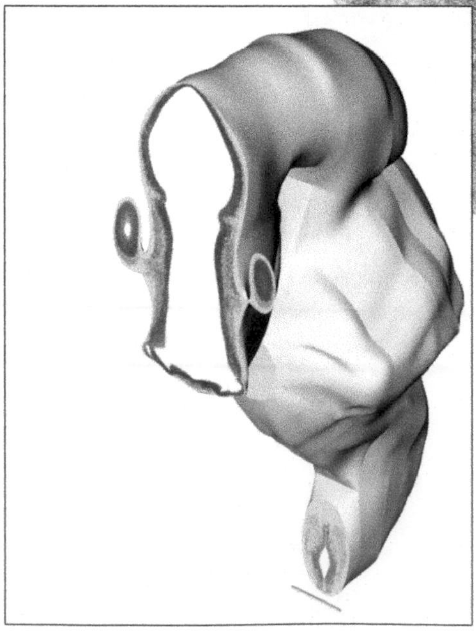

Cell-dense
primordial mesenchymal
brain case (skin/bone)

Future dura
(internal border
of brain case)

Pia

Cell-sparse superarachnoid
reticulum (brain paren-
chymal expansion zone)

Dural blood
vessels

Pial blood
vessels

Eye
Pigment epithelium
Intraretinal space
Retinal NEP
Sclera

Oral cavity

Tongue

Maxillary process

Palatal process

Meckel's cartilage

Mandibular process

Hypoglossal nerve (XII)?

Section 258 brain *in situ*

1 mm

81

Central neural structures labeled

PLATE 28B

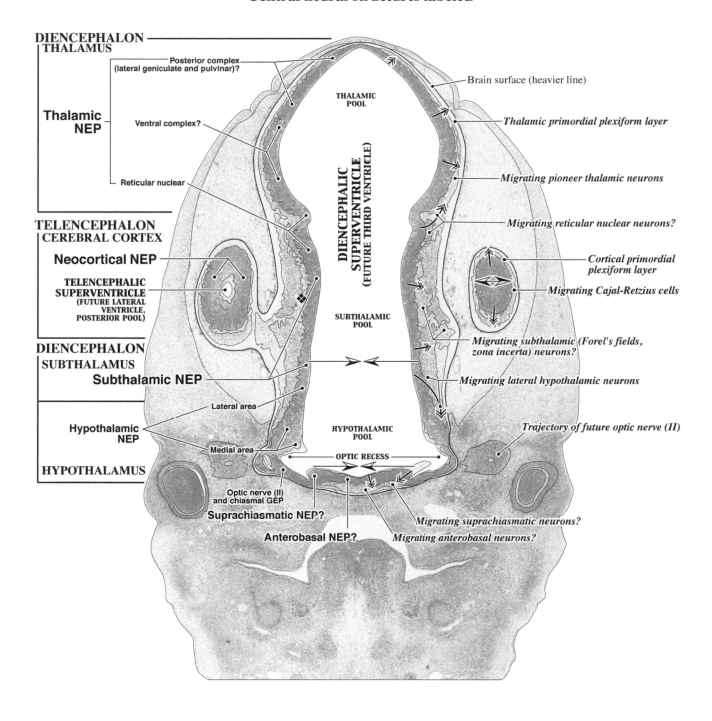

PLATE 29A

CR 15.0 mm, GW7.4
M2051, Frontal/Horizontal
Section 285

Non-neural and peripheral
neural structures labeled

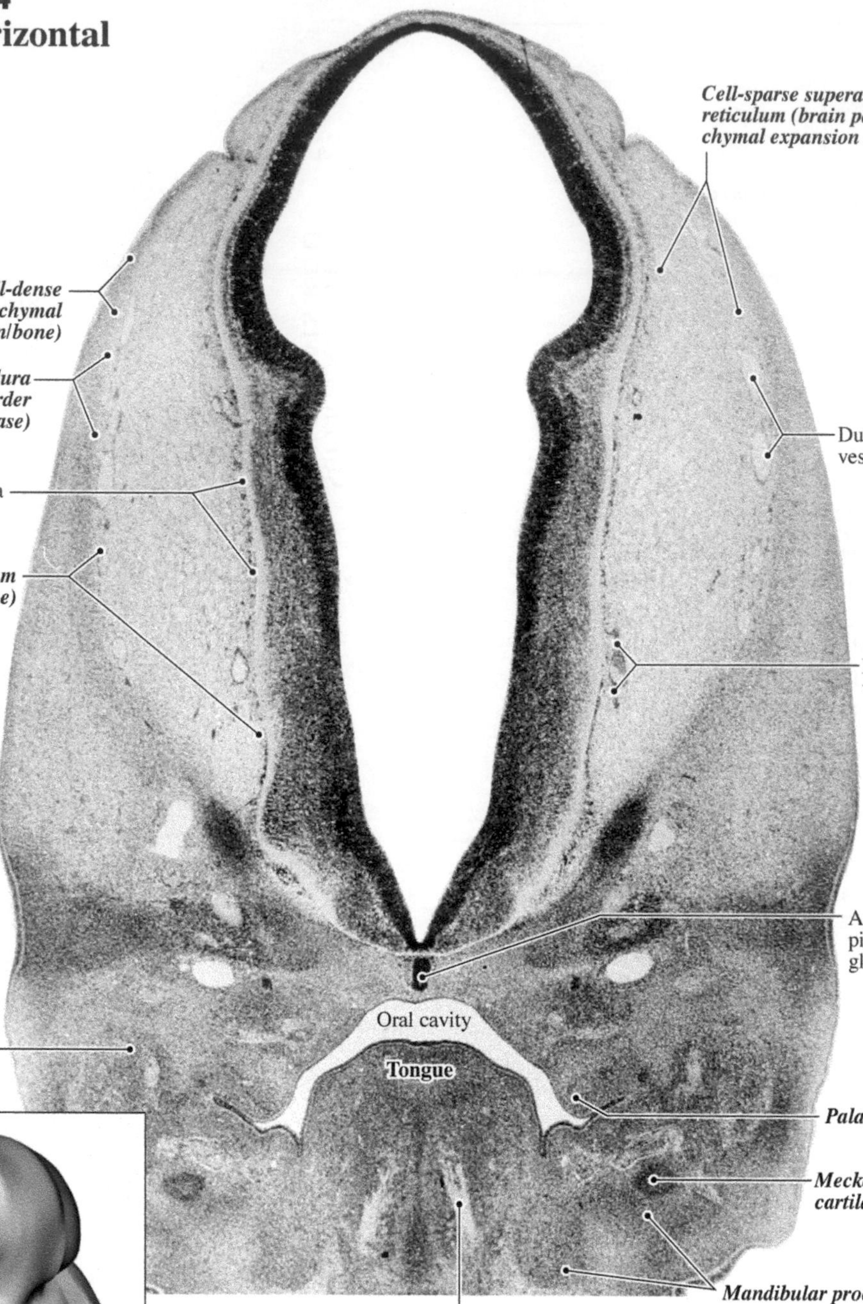

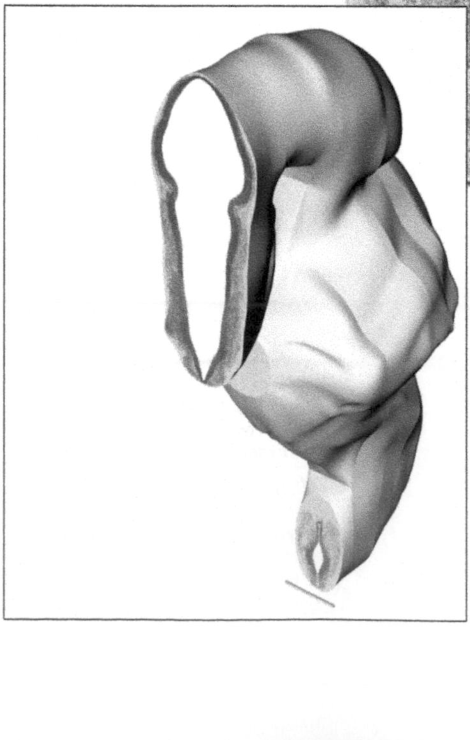

Cell-sparse superarachnoid reticulum (brain paren-chymal expansion zone)

Cell-dense primordial mesenchymal brain case (skin/bone)

Future dura (internal border of brain case)

Dural blood vessels

Pia

Cell-sparse superarachnoid reticulum (brain parenchymal expansion zone)

Pial blood vessels

Anterior pituitary gland

Oral cavity

Tongue

Maxillary process

Palatal process

Meckel's cartilage

Mandibular process

Hypoglossal nerve (XII)?

Section 285 brain *in situ*

1 mm

Central neural structures labeled

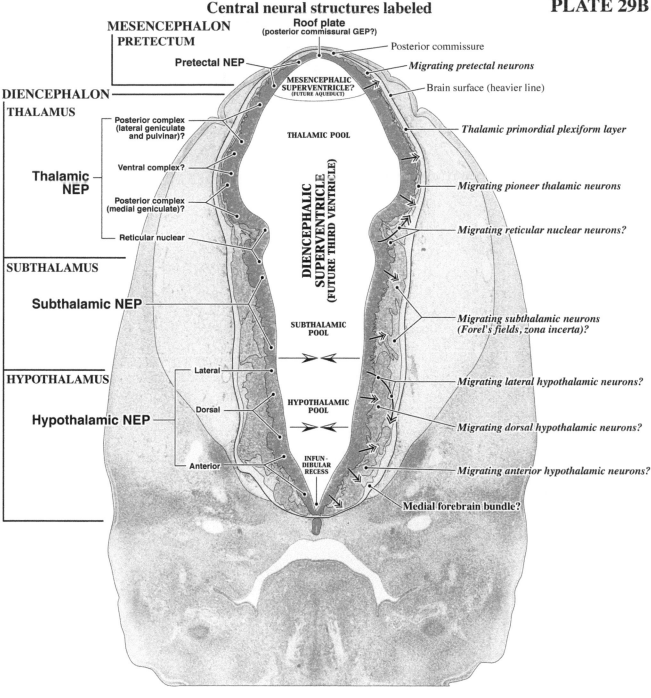

MESENCEPHALON
PRETECTUM

Roof plate
(posterior commissural GEP?)

Pretectal NEP

Posterior commissure

Migrating pretectal neurons

MESENCEPHALIC SUPERVENTRICLE?
(FUTURE AQUEDUCT)

Brain surface (heavier line)

DIENCEPHALON

THALAMUS

Posterior complex
(lateral geniculate
and pulvinar)?

THALAMIC POOL

Thalamic primordial plexiform layer

Thalamic NEP

Ventral complex?

Posterior complex
(medial geniculate)?

Reticular nuclear

DIENCEPHALIC SUPERVENTRICLE
(FUTURE THIRD VENTRICLE)

Migrating pioneer thalamic neurons

Migrating reticular nuclear neurons?

SUBTHALAMUS

Subthalamic NEP

SUBTHALAMIC POOL

Migrating subthalamic neurons
(Forel's fields, zona incerta)?

HYPOTHALAMUS

Lateral

HYPOTHALAMIC POOL

Migrating lateral hypothalamic neurons?

Hypothalamic NEP

Dorsal

Migrating dorsal hypothalamic neurons?

Anterior

INFUN-DIBULAR RECESS

Migrating anterior hypothalamic neurons?

Medial forebrain bundle?

ABBREVIATIONS:
GEP - Glioepithelium
NEP - Neuroepithelium

FONT KEY:
VENTRICULAR DIVISIONS - CAPITALS
Germinal zone - Helvetica bold
Transient structure - Times bold italic
Permanent structure - Times Roman or **Bold**

Arrows indicate the
presumed *direction of*
neuron migration from
neuroepithelial sources.

Arrows indicate the regionally
shrinking shoreline of the
superventricle as NEP cells are
depleted while generating neurons.

PLATE 30A

CR 15.0 mm, GW7.4
M2051, Frontal/Horizontal
Section 330

Non-neural and peripheral
neural structures labeled

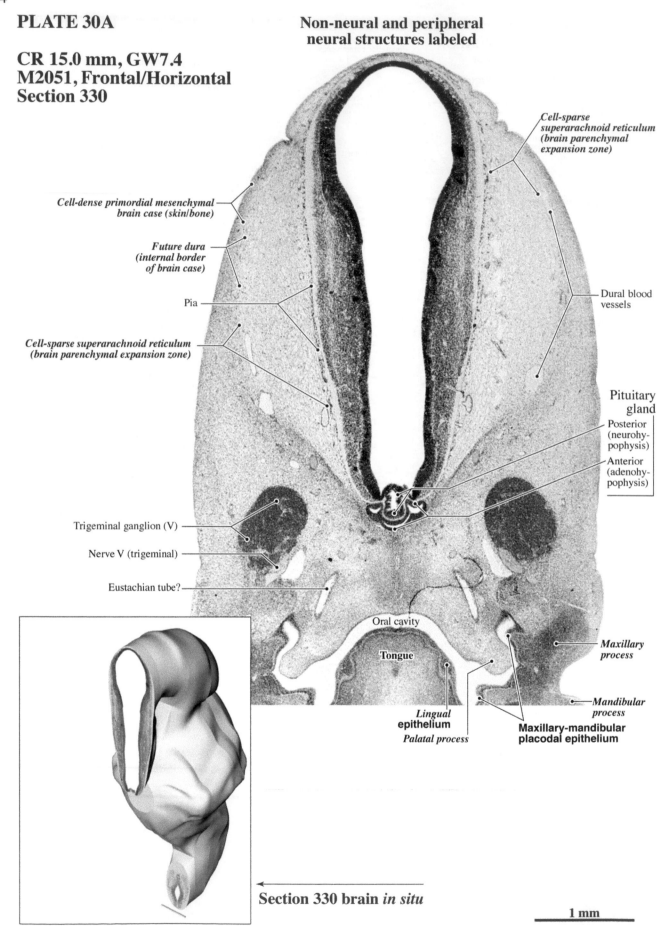

Cell-sparse
superarachnoid reticulum
(brain parenchymal
expansion zone)

Cell-dense primordial mesenchymal
brain case (skin/bone)

Future dura
(internal border
of brain case)

Pia

Dural blood
vessels

Cell-sparse superarachnoid reticulum
(brain parenchymal expansion zone)

Pituitary
gland

Posterior
(neurohy-
pophysis)

Anterior
(adenohy-
pophysis)

Trigeminal ganglion (V)

Nerve V (trigeminal)

Eustachian tube?

Oral cavity

Maxillary
process

Tongue

Mandibular
process

Lingual
epithelium

Maxillary-mandibular
placodal epithelium

Palatal process

Section 330 brain *in situ*

1 mm

Central neural structures labeled

MESENCEPHALON
PRETECTUM

Roof plate
(posterior commissural GEP?)

Pretectal NEP

Posterior commissure

Migrating pretectal neurons

MESENCEPHALIC
SUPERVENTRICLE?
(FUTURE AQUEDUCT)

Light areas are pockets of sprouting pretectal and thalamic axons

DIENCEPHALON
THALAMUS

Posterior complex
(lateral geniculate
and pulvinar)?

THALAMIC POOL

Thalamic primordial plexiform layer

Ventral complex?

Thalamic NEP

Successive waves of migrating thalamic neurons separated by sprouting axons

Posterior complex
(medial geniculate)?

DIENCEPHALIC
SUPERVENTRICLE
(FUTURE THIRD VENTRICLE)

Migrating reticular nuclear neurons?

Reticular nuclear

Brain surface (heavier line)

SUBTHALAMUS

Subthalamic NEP

SUB-
THALAMIC
POOL

Migrating subthalamic neurons (Forel's fields, zona incerta)?

Luysian migration (subthalamic nuclear neurons originating in **hypothalamic NEP***)?*

HYPOTHALAMUS

Lateral

HYPO-
THALAMIC
POOL

Migrating lateral hypothalamic neurons?

Medial forebrain bundle?

Hypothalamic NEP

Anterior/dorsal

INFUN-
DIBULAR
RECESS

Migrating dorsal and anterior hypothalamic neurons?

Migrating middle hypothalamic neurons?

Middle/infundibular

Migrating arcuate nuclear neurons?

Median eminence/
neurohypophysis
(pituicyte) GEP

ABBREVIATIONS:
GEP - Glioepithelium
NEP - Neuroepithelium

FONT KEY:
VENTRICULAR DIVISIONS - CAPITALS
Germinal zone - Helvetica bold
Transient structure - Times bold italic
Permanent structure - Times Roman or **Bold**

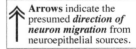

Arrows indicate the presumed *direction of neuron migration* from neuroepithelial sources.

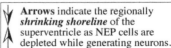

Arrows indicate the regionally *shrinking shoreline* of the superventricle as NEP cells are depleted while generating neurons.

PLATE 31A

CR 15.0 mm, GW7.4
M2051, Frontal/Horizontal
Section 357

Non-neural and peripheral
neural structures labeled

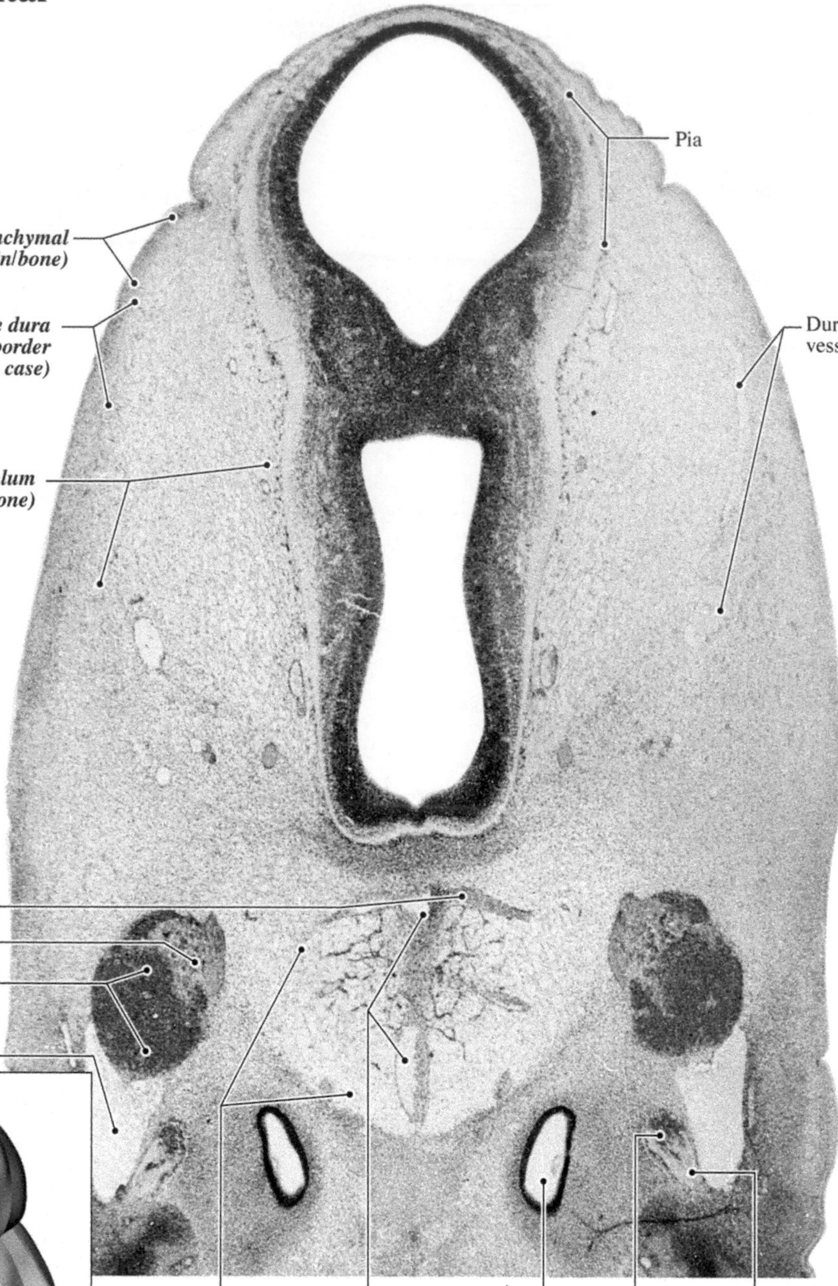

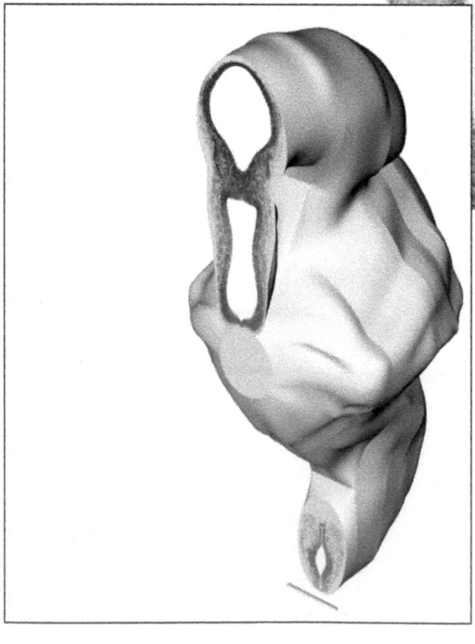

Pia

Cell-dense primordial mesenchymal
brain case (skin/bone)

Future dura
(internal border
of brain case)

Dural blood
vessels

Cell-sparse superarachnoid reticulum
(brain parenchymal expansion zone)

Posterior cerebral artery?

Nerve V (trigeminal)

Trigeminal ganglion (V)

Anterior cardinal vein?

Basilar
artery

Otic
vesicle

Facial
ganglion
(VII)?

Nerve
VII
(facial)?

Petrous
temporal
bone

Cell-sparse
superarachnoid reticulum
(brain parenchymal
expansion zone)

← **Section 357 brain *in situ***

1 mm

PLATE 31B

Central neural structures labeled

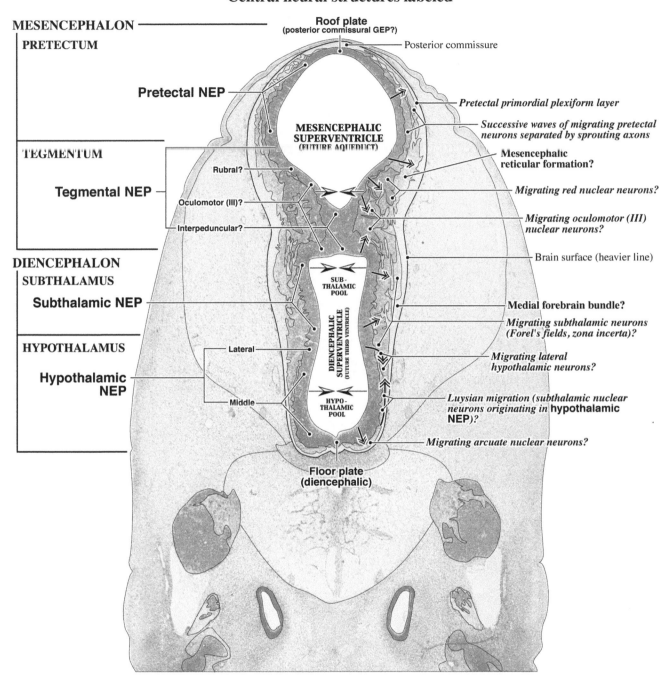

MESENCEPHALON

PRETECTUM

Roof plate
(posterior commissural GEP?)

Posterior commissure

Pretectal NEP

Pretectal primordial plexiform layer

Successive waves of migrating pretectal
neurons separated by sprouting axons

MESENCEPHALIC
SUPERVENTRICLE
(FUTURE AQUEDUCT)

Mesencephalic
reticular formation?

TEGMENTUM

Rubral?

Migrating red nuclear neurons?

Tegmental NEP

Oculomotor (III)?

Migrating oculomotor (III)
nuclear neurons?

Interpeduncular?

Brain surface (heavier line)

DIENCEPHALON

SUBTHALAMUS

SUB-
THALAMIC
POOL

Subthalamic NEP

DIENCEPHALIC
SUPERVENTRICLE
(FUTURE THIRD VENTRICLE)

Medial forebrain bundle?

Migrating subthalamic neurons
(Forel's fields, zona incerta)?

HYPOTHALAMUS

Lateral

Migrating lateral
hypothalamic neurons?

Hypothalamic
NEP

Middle

HYPO-
THALAMIC
POOL

Luysian migration (subthalamic nuclear
neurons originating in hypothalamic
NEP)?

Migrating arcuate nuclear neurons?

Floor plate
(diencephalic)

ABBREVIATIONS:
GEP - Glioepithelium
NEP - Neuroepithelium

FONT KEY:
VENTRICULAR DIVISIONS - CAPITALS
Germinal zone - Helvetica bold
Transient structure - Times bold italic
Permanent structure - Times Roman or **Bold**

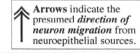

Arrows indicate the
presumed *direction of
neuron migration* from
neuroepithelial sources.

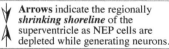

Arrows indicate the regionally
shrinking shoreline of the
superventricle as NEP cells are
depleted while generating neurons.

88

PLATE 32A

**CR 15.0 mm, GW7.4
M2051, Frontal/Horizontal
Section 384**

**Non-neural and peripheral
neural structures labeled**

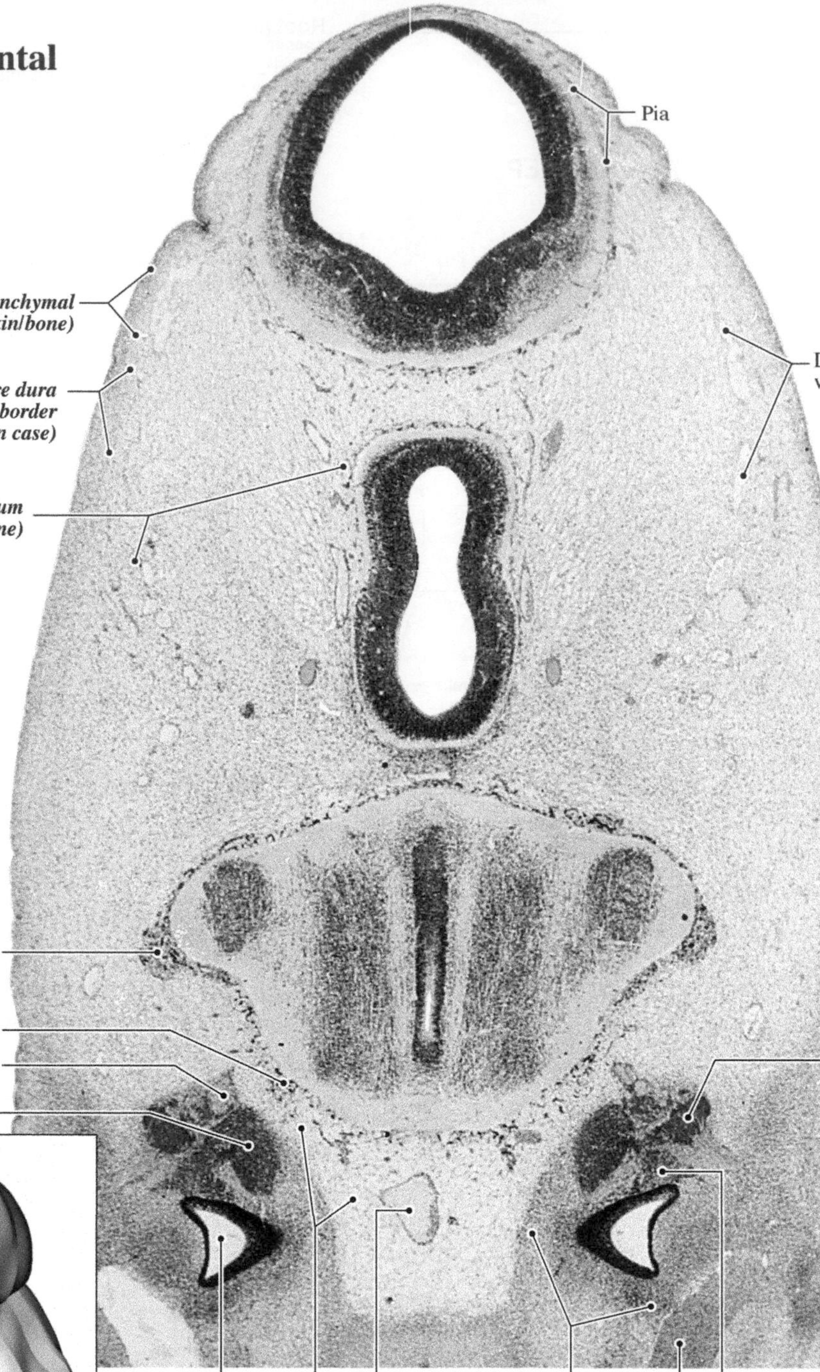

Pia

*Cell-dense primordial mesenchymal
brain case (skin/bone)*

*Future dura
(internal border
of brain case)*

Dural blood
vessels

*Cell-sparse superarachnoid reticulum
(brain parenchymal expansion zone)*

Trigeminal *boundary cap**

Facial *boundary cap**

Nerve VII (facial)?

Vestibular ganglion (VIII)

Facial
ganglion
(VII)?

Temporal bone
labyrinth
(otic vesicle)

Basilar
artery

Petrous temporal
bone

*Spiral ganglion (VIII)
budding from* **otic
vesicle epithelium***?*

*Cell-sparse
superarachnoid reticulum
(brain parenchymal
expansion zone)*

Anterior
cardinal
vein?

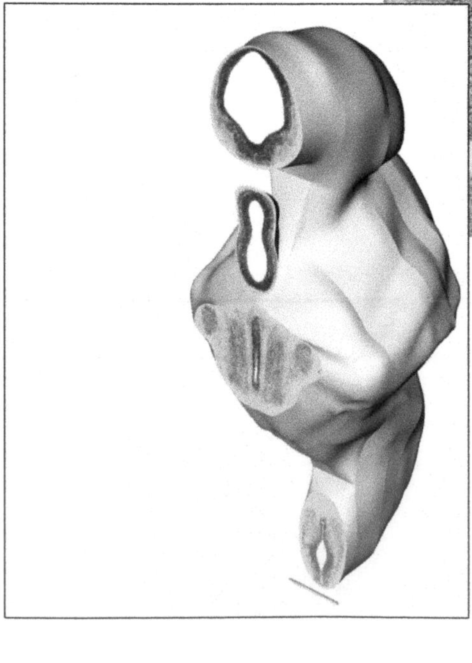

***Boundary caps are
Schwann cell GEPs?**

Section 384 brain *in situ*

1 mm

Central neural structures labeled **PLATE 32B**

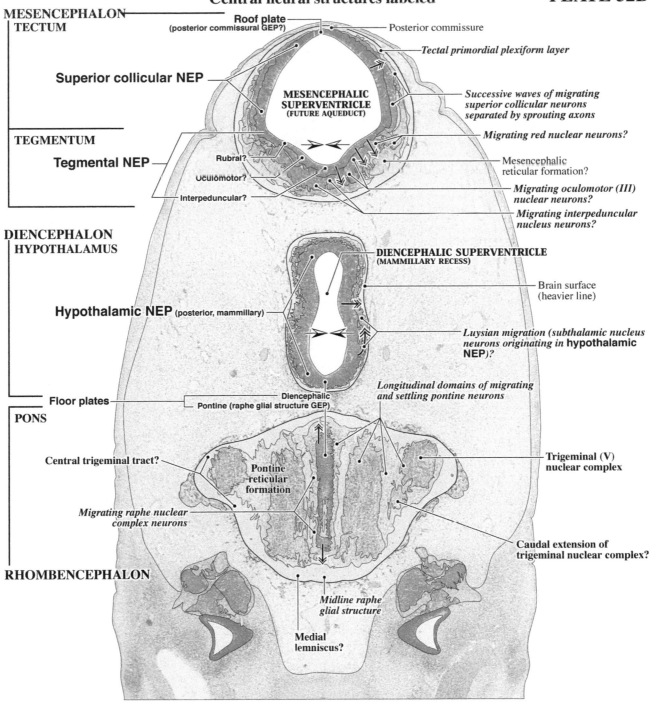

MESENCEPHALON
TECTUM

Roof plate
(posterior commissural GEP?) — Posterior commissure

Tectal primordial plexiform layer

Superior collicular NEP

MESENCEPHALIC SUPERVENTRICLE (FUTURE AQUEDUCT)

Successive waves of migrating superior collicular neurons separated by sprouting axons

Migrating red nuclear neurons?

TEGMENTUM

Tegmental NEP

Rubral?

Oculomotor?

Interpeduncular?

Mesencephalic reticular formation?

Migrating oculomotor (III) nuclear neurons?

Migrating interpeduncular nucleus neurons?

DIENCEPHALON
HYPOTHALAMUS

DIENCEPHALIC SUPERVENTRICLE (MAMMILLARY RECESS)

Brain surface (heavier line)

Hypothalamic NEP (posterior, mammillary)

*Luysian migration (subthalamic nucleus neurons originating in **hypothalamic NEP**)?*

Longitudinal domains of migrating and settling pontine neurons

Floor plates

Diencephalic

Pontine (raphe glial structure GEP)

PONS

Central trigeminal tract?

Pontine reticular formation

Trigeminal (V) nuclear complex

Migrating raphe nuclear complex neurons

Caudal extension of trigeminal nuclear complex?

RHOMBENCEPHALON

Midline raphe glial structure

Medial lemniscus?

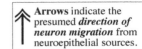

ABBREVIATIONS:
GEP - Glioepithelium
NEP - Neuroepithelium

FONT KEY:
VENTRICULAR DIVISIONS - CAPITALS
Germinal zone - Helvetica bold
Transient structure - Times bold italic
Permanent structure - Times Roman or **Bold**

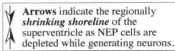

Arrows indicate the presumed *direction of neuron migration* from neuroepithelial sources.

Arrows indicate the regionally *shrinking shoreline* of the superventricle as NEP cells are depleted while generating neurons.

PLATE 33A

Non-neural and peripheral neural structures labeled

**CR 15.0 mm, GW7.4
M2051, Frontal/Horizontal
Section 420**

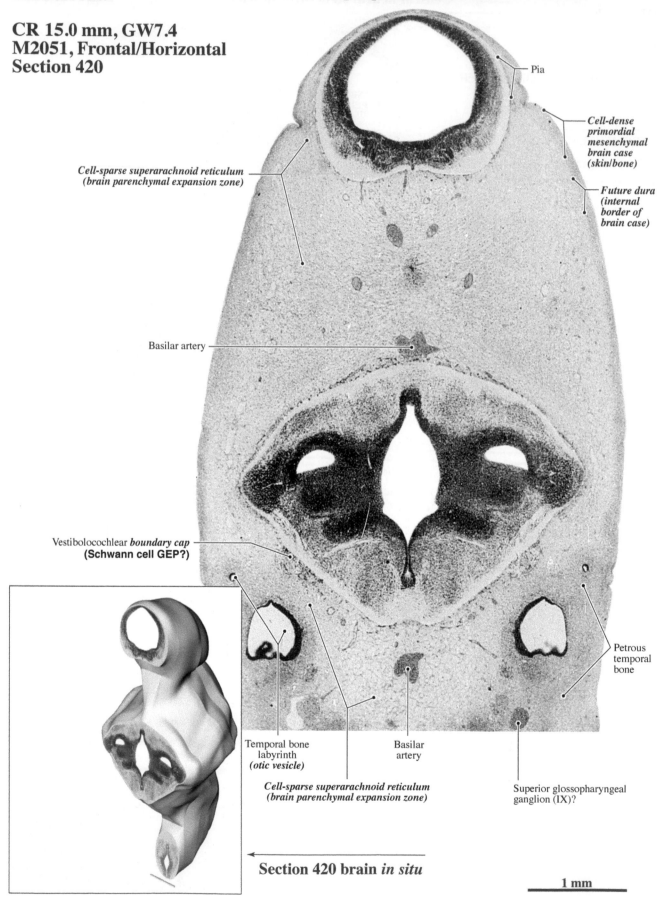

Pia

*Cell-dense
primordial
mesenchymal
brain case
(skin/bone)*

*Future dura
(internal
border of
brain case)*

*Cell-sparse superarachnoid reticulum
(brain parenchymal expansion zone)*

Basilar artery

Vestibolocochlear *boundary cap*
(Schwann cell GEP?)

Petrous
temporal
bone

Temporal bone
labyrinth
(otic vesicle)

Basilar
artery

*Cell-sparse superarachnoid reticulum
(brain parenchymal expansion zone)*

Superior glossopharyngeal
ganglion (IX)?

Section 420 brain *in situ*

1 mm

Central neural structures labeled   PLATE 33B

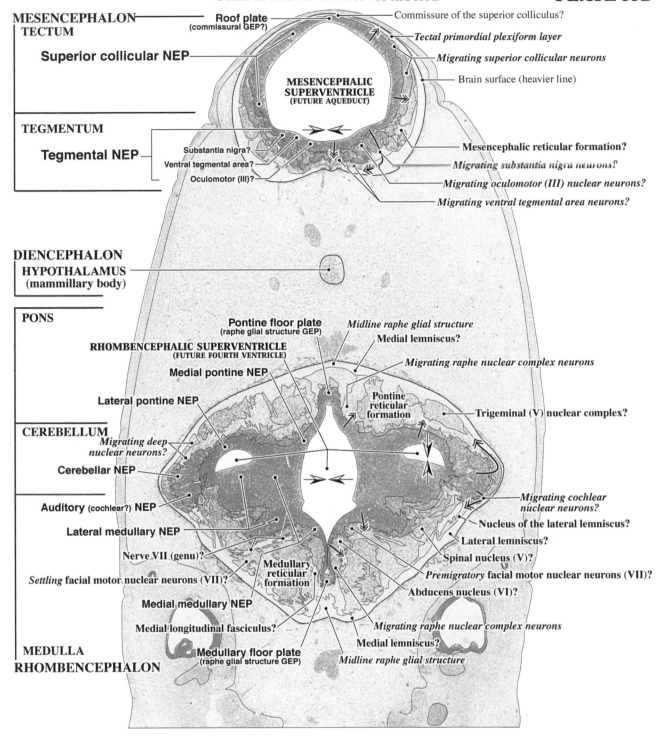

MESENCEPHALON
  TECTUM

Roof plate
(commissural GEP?)

Commissure of the superior colliculus?

Tectal primordial plexiform layer

Migrating superior collicular neurons

Superior collicular NEP

**MESENCEPHALIC SUPERVENTRICLE
(FUTURE AQUEDUCT)**

Brain surface (heavier line)

TEGMENTUM

  Tegmental NEP

Substantia nigra?
Ventral tegmental area?
Oculomotor (III)?

Mesencephalic reticular formation?

Migrating substantia nigra neurons?

Migrating oculomotor (III) nuclear neurons?

Migrating ventral tegmental area neurons?

DIENCEPHALON
  HYPOTHALAMUS
   (mammillary body)

PONS

Pontine floor plate
(raphe glial structure GEP)

Midline raphe glial structure
Medial lemniscus?

RHOMBENCEPHALIC SUPERVENTRICLE
(FUTURE FOURTH VENTRICLE)

Migrating raphe nuclear complex neurons

Medial pontine NEP

Pontine
reticular
formation

Lateral pontine NEP

Trigeminal (V) nuclear complex?

CEREBELLUM

*Migrating deep
nuclear neurons?*

Cerebellar NEP

*Migrating cochlear
nuclear neurons?*

Auditory (cochlear?) **NEP**

Nucleus of the lateral lemniscus?

Lateral medullary NEP

Lateral lemniscus?

Nerve VII (genu)?

Spinal nucleus (V)?

Settling facial motor nuclear neurons (VII)?

Medullary
reticular
formation

Premigratory facial motor nuclear neurons (VII)?

Abducens nucleus (VI)?

Medial medullary NEP

Medial longitudinal fasciculus?

Migrating raphe nuclear complex neurons

Medial lemniscus?

Medullary floor plate
(raphe glial structure GEP)

Midline raphe glial structure

MEDULLA
RHOMBENCEPHALON

ABBREVIATIONS:
GEP - Glioepithelium
NEP - Neuroepithelium

FONT KEY:
VENTRICULAR DIVISIONS - CAPITALS
Germinal zone - Helvetica bold
Transient structure - Times bold italic
Permanent structure - Times Roman or **Bold**

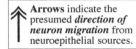

Arrows indicate the
presumed *direction of
neuron migration* from
neuroepithelial sources.

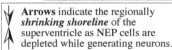

Arrows indicate the regionally
shrinking shoreline of the
superventricle as NEP cells are
depleted while generating neurons.

PLATE 34A

**CR 15.0 mm, GW7.4
M2051, Frontal/Horizontal
Section 438**

**Non-neural and
peripheral neural
structures labeled**

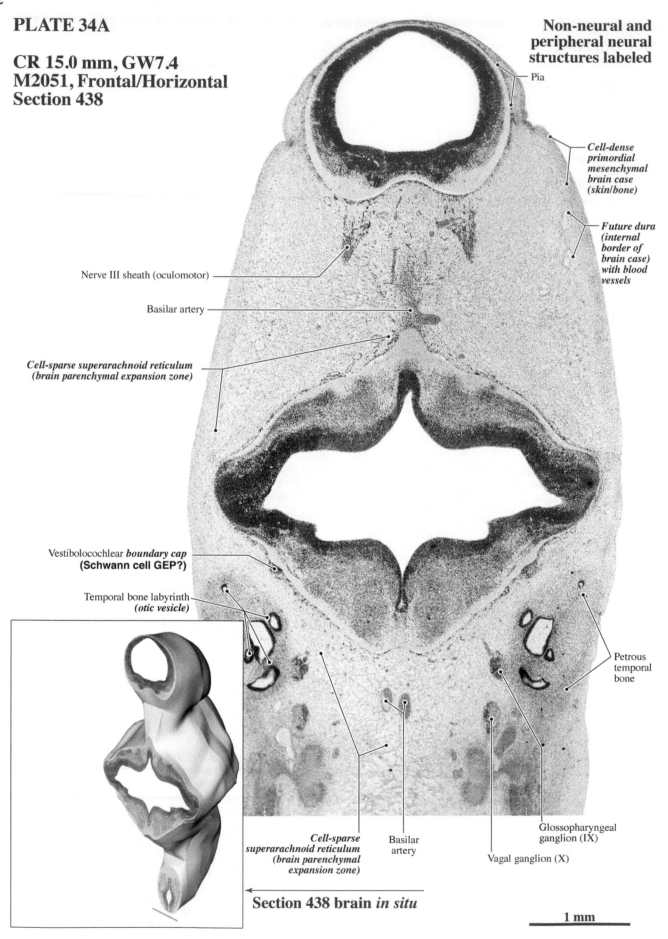

Pia

*Cell-dense
primordial
mesenchymal
brain case
(skin/bone)*

*Future dura
(internal
border of
brain case)
with blood
vessels*

Nerve III sheath (oculomotor)

Basilar artery

*Cell-sparse superarachnoid reticulum
(brain parenchymal expansion zone)*

Vestibolocochlear *boundary cap*
(Schwann cell GEP?)

Temporal bone labyrinth
(otic vesicle)

Petrous
temporal
bone

*Cell-sparse
superarachnoid reticulum
(brain parenchymal
expansion zone)*

Basilar
artery

Glossopharyngeal
ganglion (IX)

Vagal ganglion (X)

Section 438 brain *in situ*

1 mm

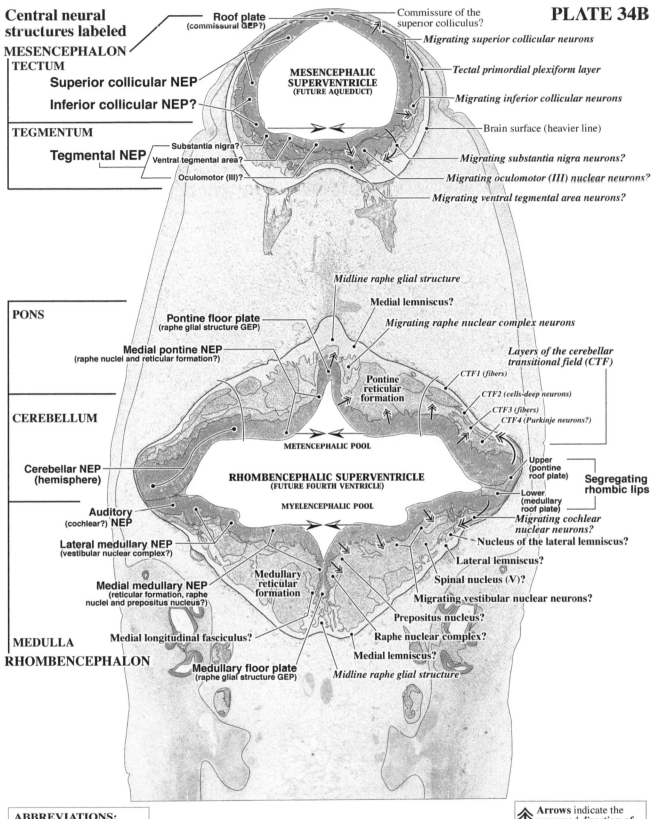

PLATE 34B

Central neural
structures labeled

MESENCEPHALON

TECTUM

Superior collicular NEP

Inferior collicular NEP?

TEGMENTUM

Tegmental NEP

Roof plate
(commissural GEP?)

Commissure of the
superior colliculus?

Migrating superior collicular neurons

MESENCEPHALIC
SUPERVENTRICLE
(FUTURE AQUEDUCT)

Tectal primordial plexiform layer

Migrating inferior collicular neurons

Brain surface (heavier line)

Substantia nigra?

Ventral tegmental area?

Oculomotor (III)?

Migrating substantia nigra neurons?

Migrating oculomotor (III) nuclear neurons?

Migrating ventral tegmental area neurons?

PONS

Midline raphe glial structure

Medial lemniscus?

Pontine floor plate
(raphe glial structure GEP)

Medial pontine NEP
(raphe nuclei and reticular formation?)

Migrating raphe nuclear complex neurons

*Layers of the cerebellar
transitional field (CTF)*

Pontine
reticular
formation

CTF1 (fibers)

CTF2 (cells-deep neurons)

CTF3 (fibers)

CTF4 (Purkinje neurons?)

CEREBELLUM

Cerebellar NEP
(hemisphere)

METENCEPHALIC POOL

RHOMBENCEPHALIC SUPERVENTRICLE
(FUTURE FOURTH VENTRICLE)

MYELENCEPHALIC POOL

Upper
(pontine
roof plate)

Lower
(medullary
roof plate)

**Segregating
rhombic lips**

Auditory
(cochlear?) NEP

Lateral medullary NEP
(vestibular nuclear complex?)

Medial medullary NEP
(reticular formation, raphe
nuclei and prepositus nucleus?)

Medullary
reticular
formation

*Migrating cochlear
nuclear neurons?*

Nucleus of the lateral lemniscus?

Lateral lemniscus?

Spinal nucleus (V)?

Migrating vestibular nuclear neurons?

Prepositus nucleus?

Raphe nuclear complex?

Medial lemniscus?

Medial longitudinal fasciculus?

MEDULLA

RHOMBENCEPHALON

Medullary floor plate
(raphe glial structure GEP)

Midline raphe glial structure

ABBREVIATIONS:
GEP - Glioepithelium
NEP - Neuroepithelium

FONT KEY:
VENTRICULAR DIVISIONS - CAPITALS
Germinal zone - Helvetica bold
Transient structure - Times bold italic
Permanent structure - Times Roman or **Bold**

Arrows indicate the
presumed *direction of
neuron migration* from
neuroepithelial sources.

Arrows indicate the regionally
shrinking shoreline of the
superventricle as NEP cells are
depleted while generating neurons.

PLATE 35A

CR 15.0 mm, GW7.4
M2051, Frontal/Horizontal
Section 488

Non-neural and
peripheral neural
structures labeled

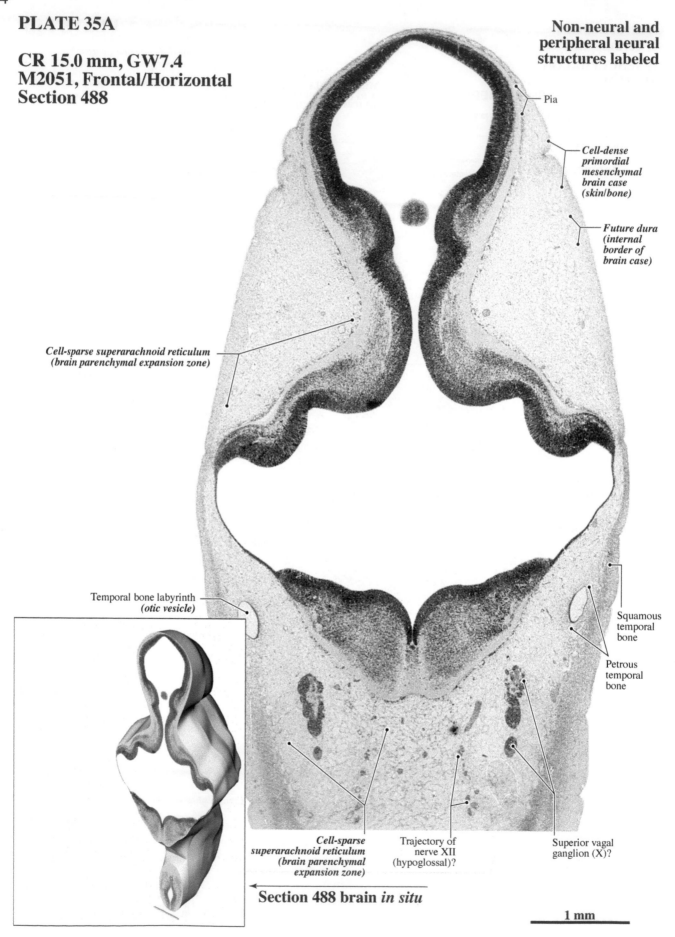

Pia

Cell-dense
primordial
mesenchymal
brain case
(skin/bone)

Future dura
(internal
border of
brain case)

Cell-sparse superarachnoid reticulum
(brain parenchymal expansion zone)

Temporal bone labyrinth
(otic vesicle)

Squamous
temporal
bone

Petrous
temporal
bone

Cell-sparse
superarachnoid reticulum
(brain parenchymal
expansion zone)

Trajectory of
nerve XII
(hypoglossal)?

Superior vagal
ganglion (X)?

Section 488 brain *in situ*

1 mm

Central neural structures labeled

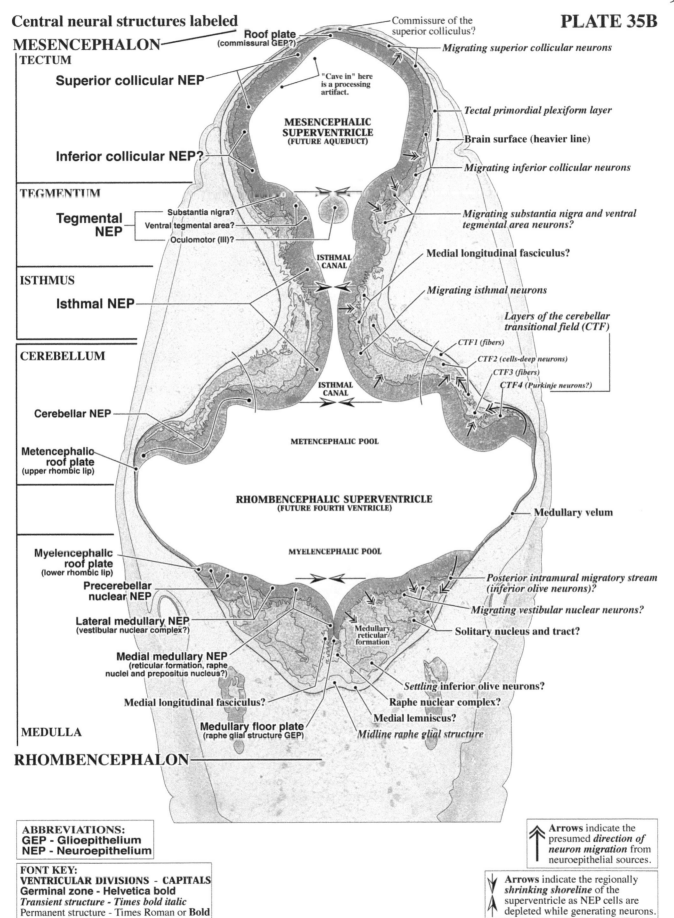

MESENCEPHALON

TECTUM

Superior collicular NEP

Inferior collicular NEP?

TEGMENTUM

Tegmental NEP

- Substantia nigra?
- Ventral tegmental area?
- Oculomotor (III)?

ISTHMUS

Isthmal NEP

CEREBELLUM

Cerebellar NEP

Metencephalic roof plate
(upper rhombic lip)

Myelencephalic roof plate
(lower rhombic lip)

Precerebellar nuclear NEP

Lateral medullary NEP
(vestibular nuclear complex?)

Medial medullary NEP
(reticular formation, raphe nuclei and prepositus nucleus?)

Medial longitudinal fasciculus?

MEDULLA

Medullary floor plate
(raphe glial structure GEP)

RHOMBENCEPHALON

Roof plate
(commissural GEP?)

Commissure of the superior colliculus?

"Cave in" here is a processing artifact.

MESENCEPHALIC SUPERVENTRICLE
(FUTURE AQUEDUCT)

ISTHMAL CANAL

ISTHMAL CANAL

METENCEPHALIC POOL

RHOMBENCEPHALIC SUPERVENTRICLE
(FUTURE FOURTH VENTRICLE)

MYELENCEPHALIC POOL

Medullary reticular formation

Migrating superior collicular neurons

Tectal primordial plexiform layer

Brain surface (heavier line)

Migrating inferior collicular neurons

Migrating substantia nigra and ventral tegmental area neurons?

Medial longitudinal fasciculus?

Migrating isthmal neurons

Layers of the cerebellar transitional field (CTF)

CTF1 (fibers)
CTF2 (cells-deep neurons)
CTF3 (fibers)
CTF4 (Purkinje neurons?)

Medullary velum

Posterior intramural migratory stream (inferior olive neurons)?

Migrating vestibular nuclear neurons?

Solitary nucleus and tract?

Settling inferior olive neurons?

Raphe nuclear complex?

Medial lemniscus?

Midline raphe glial structure

ABBREVIATIONS:
GEP - Glioepithelium
NEP - Neuroepithelium

FONT KEY:
VENTRICULAR DIVISIONS - CAPITALS
Germinal zone - Helvetica bold
Transient structure - Times bold italic
Permanent structure - Times Roman or **Bold**

Arrows indicate the presumed *direction of neuron migration* from neuroepithelial sources.

Arrows indicate the regionally *shrinking shoreline* of the superventricle as NEP cells are depleted while generating neurons.

PLATE 36A

CR 15.0 mm, GW7.4
M2051, Frontal/Horizontal
Section 553

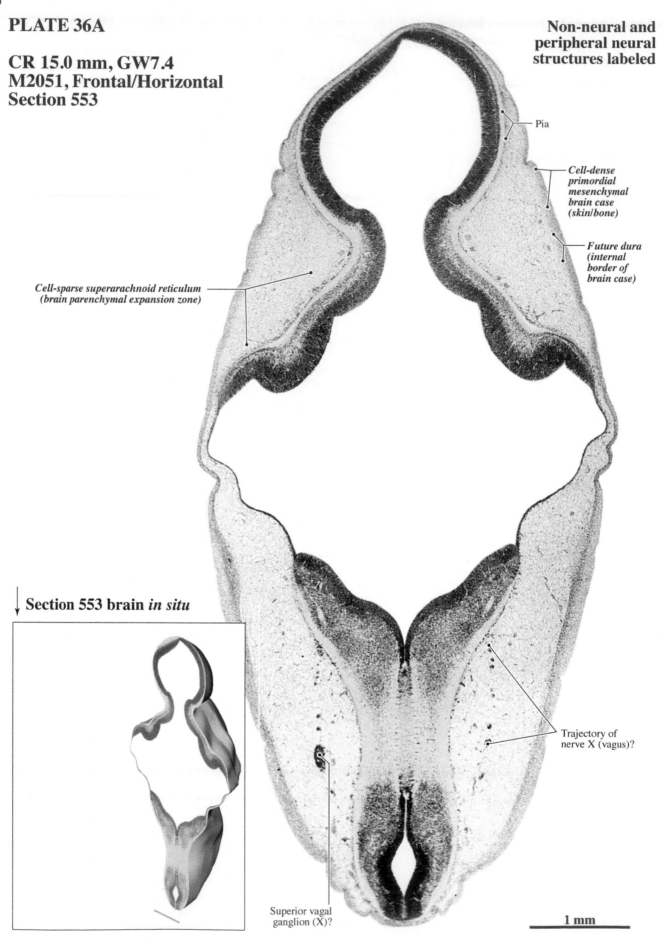

Pia

*Cell-dense
primordial
mesenchymal
brain case
(skin/bone)*

*Future dura
(internal
border of
brain case)*

*Cell-sparse superarachnoid reticulum
(brain parenchymal expansion zone)*

↓ **Section 553 brain *in situ***

Trajectory of
nerve X (vagus)?

Superior vagal
ganglion (X)?

1 mm

Central neural structures labeled

PLATE 36B

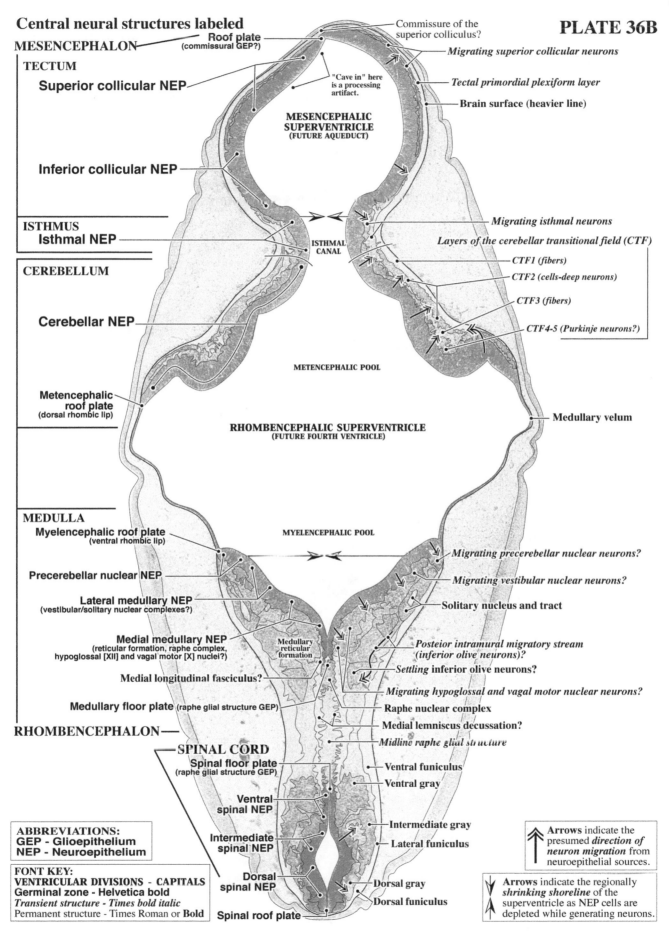

Roof plate (commissural GEP?)

Commissure of the superior colliculus?

Migrating superior collicular neurons

MESENCEPHALON

TECTUM

Superior collicular NEP

"Cave in" here is a processing artifact.

Tectal primordial plexiform layer

Brain surface (heavier line)

MESENCEPHALIC SUPERVENTRICLE (FUTURE AQUEDUCT)

Inferior collicular NEP

ISTHMUS

Isthmal NEP

ISTHMAL CANAL

Migrating isthmal neurons

Layers of the cerebellar transitional field (CTF)

CTF1 (fibers)

CTF2 (cells-deep neurons)

CTF3 (fibers)

CTF4-5 (Purkinje neurons?)

CEREBELLUM

Cerebellar NEP

METENCEPHALIC POOL

Metencephalic roof plate (dorsal rhombic lip)

RHOMBENCEPHALIC SUPERVENTRICLE (FUTURE FOURTH VENTRICLE)

Medullary velum

MEDULLA

MYELENCEPHALIC POOL

Myelencephalic roof plate (ventral rhombic lip)

Migrating precerebellar nuclear neurons?

Precerebellar nuclear NEP

Migrating vestibular nuclear neurons?

Lateral medullary NEP (vestibular/solitary nuclear complexes?)

Solitary nucleus and tract

Medial medullary NEP (reticular formation, raphe complex, hypoglossal [XII] and vagal motor [X] nuclei?)

Medullary reticular formation

Posteior intramural migratory stream (inferior olive neurons)?

Settling inferior olive neurons?

Medial longitudinal fasciculus?

Migrating hypoglossal and vagal motor nuclear neurons?

Medullary floor plate (raphe glial structure GEP)

Raphe nuclear complex

Medial lemniscus decussation?

RHOMBENCEPHALON

Midline raphe glial structure

SPINAL CORD

Spinal floor plate (raphe glial structure GEP)

Ventral funiculus

Ventral gray

Ventral spinal NEP

Intermediate gray

Intermediate spinal NEP

Lateral funiculus

Dorsal spinal NEP

Dorsal gray

Dorsal funiculus

Spinal roof plate

ABBREVIATIONS:
GEP - Glioepithelium
NEP - Neuroepithelium

FONT KEY:
VENTRICULAR DIVISIONS - CAPITALS
Germinal zone - Helvetica bold
Transient structure - Times bold italic
Permanent structure - Times Roman or **Bold**

Arrows indicate the presumed *direction of neuron migration* from neuroepithelial sources.

Arrows indicate the regionally *shrinking shoreline* of the superventricle as NEP cells are depleted while generating neurons.

PLATE 37A

CR 15.0 mm, GW7.4
M2051, Frontal/Horizontal
Section 583

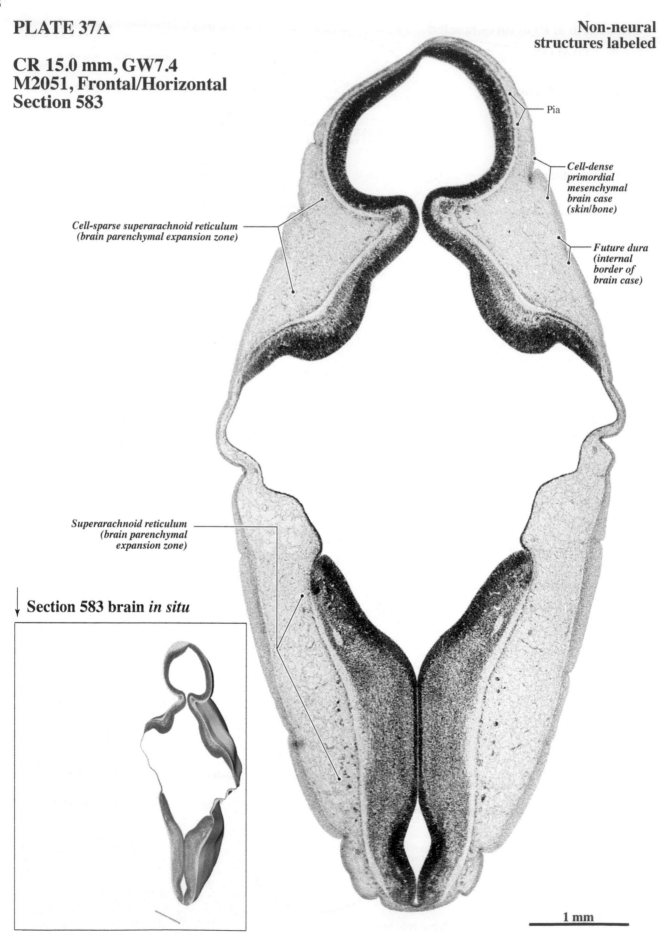

Pia

*Cell-dense
primordial
mesenchymal
brain case
(skin/bone)*

*Future dura
(internal
border of
brain case)*

*Cell-sparse superarachnoid reticulum
(brain parenchymal expansion zone)*

*Superarachnoid reticulum
(brain parenchymal
expansion zone)*

↓ **Section 583 brain *in situ***

1 mm

Neural structures labeled

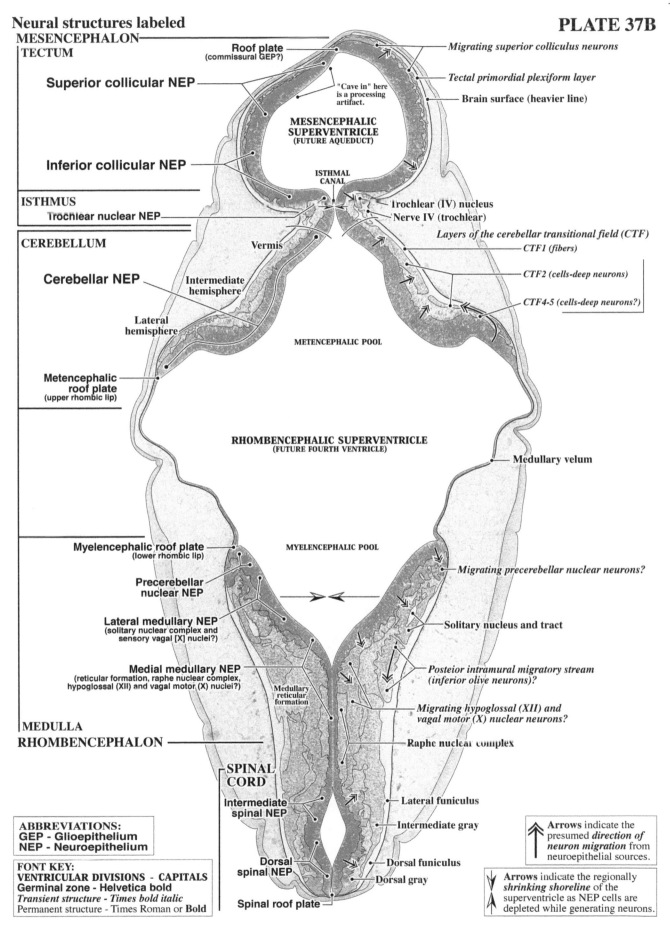

MESENCEPHALON

TECTUM

Roof plate
(commissural GEP?)

Migrating superior colliculus neurons

Superior collicular NEP

"Cave in" here
is a processing
artifact.

Tectal primordial plexiform layer

Brain surface (heavier line)

**MESENCEPHALIC
SUPERVENTRICLE
(FUTURE AQUEDUCT)**

Inferior collicular NEP

ISTHMAL
CANAL

Trochlear (IV) nucleus

Nerve IV (trochlear)

ISTHMUS

Trochlear nuclear NEP

Layers of the cerebellar transitional field (CTF)

CEREBELLUM

Vermis

CTF1 *(fibers)*

Cerebellar NEP

Intermediate
hemisphere

CTF2 *(cells-deep neurons)*

Lateral
hemisphere

CTF4-5 *(cells-deep neurons?)*

METENCEPHALIC POOL

**Metencephalic
roof plate**
(upper rhombic lip)

**RHOMBENCEPHALIC SUPERVENTRICLE
(FUTURE FOURTH VENTRICLE)**

Medullary velum

Myelencephalic roof plate
(lower rhombic lip)

MYELENCEPHALIC POOL

Migrating precerebellar nuclear neurons?

**Precerebellar
nuclear NEP**

Solitary nucleus and tract

Lateral medullary NEP
(solitary nuclear complex and
sensory vagal [X] nuclei?)

*Posteior intramural migratory stream
(inferior olive neurons)?*

Medial medullary NEP
(reticular formation, raphe nuclear complex,
hypoglossal (XII) and vagal motor (X) nuclei?)

Medullary
reticular
formation

*Migrating hypoglossal (XII) and
vagal motor (X) nuclear neurons?*

MEDULLA

Raphe nuclear complex

RHOMBENCEPHALON

**SPINAL
CORD**

Lateral funiculus

**Intermediate
spinal NEP**

Intermediate gray

**Dorsal
spinal NEP**

Dorsal funiculus

Dorsal gray

Spinal roof plate

**ABBREVIATIONS:
GEP - Glioepithelium
NEP - Neuroepithelium**

**FONT KEY:
VENTRICULAR DIVISIONS - CAPITALS
Germinal zone - Helvetica bold
Transient structure - Times bold italic
Permanent structure - Times Roman or Bold**

Arrows indicate the
presumed *direction of
neuron migration* from
neuroepithelial sources.

Arrows indicate the regionally
shrinking shoreline of the
superventricle as NEP cells are
depleted while generating neurons.

PLATE 38A

CR 15.0 mm, GW7.4
M2051, Frontal/Horizontal
Section 643

Non-neural structures labeled

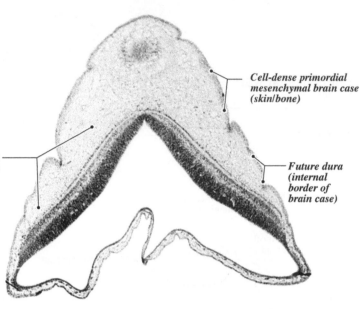

Cell-dense primordial
mesenchymal brain case
(skin/bone)

Cell-sparse superarachnoid reticulum
(brain parenchymal expansion zone)

Future dura
(internal
border of
brain case)

Cell-sparse superarachnoid reticulum
(brain parenchymal expansion zone)

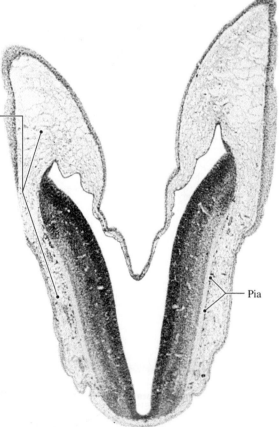

Pia

↓ **Section 643 brain *in situ***

1 mm

Neural structures labeled

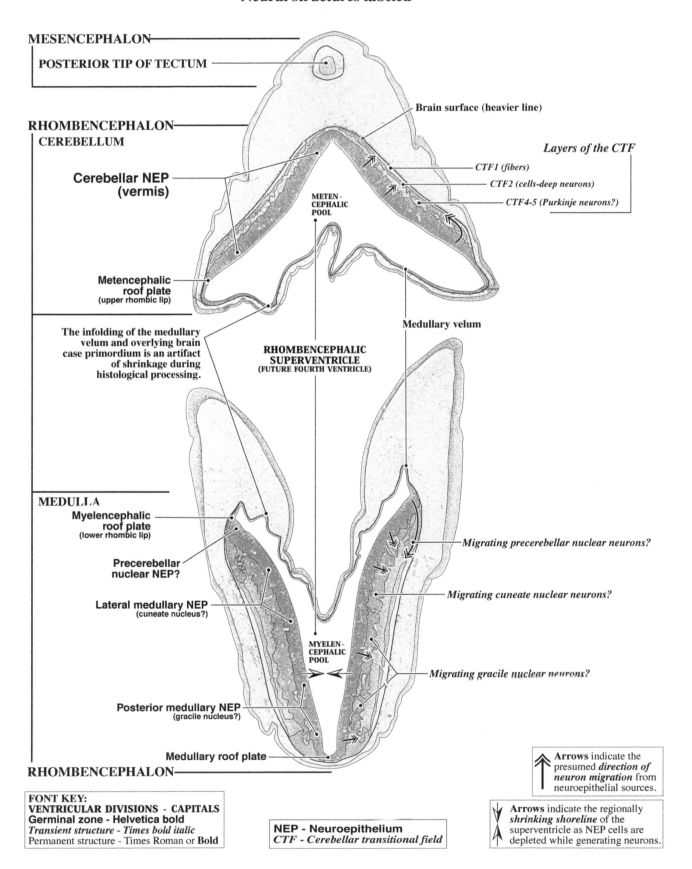

MESENCEPHALON

POSTERIOR TIP OF TECTUM

Brain surface (heavier line)

RHOMBENCEPHALON
CEREBELLUM

Layers of the CTF

Cerebellar NEP
(vermis)

CTF1 (fibers)
CTF2 (cells-deep neurons)
CTF4-5 (Purkinje neurons?)

METEN-
CEPHALIC
POOL

Metencephalic
roof plate
(upper rhombic lip)

The infolding of the medullary
velum and overlying brain
case primordium is an artifact
of shrinkage during
histological processing.

Medullary velum

**RHOMBENCEPHALIC
SUPERVENTRICLE
(FUTURE FOURTH VENTRICLE)**

MEDULLA

Myelencephalic
roof plate
(lower rhombic lip)

Precerebellar
nuclear NEP?

Lateral medullary NEP
(cuneate nucleus?)

Migrating precerebellar nuclear neurons?

Migrating cuneate nuclear neurons?

MYELEN-
CEPHALIC
POOL

Posterior medullary NEP
(gracile nucleus?)

Migrating gracile nuclear neurons?

Medullary roof plate

RHOMBENCEPHALON

FONT KEY:
VENTRICULAR DIVISIONS - CAPITALS
Germinal zone - Helvetica bold
Transient structure - Times bold italic
Permanent structure - Times Roman or **Bold**

NEP - Neuroepithelium
CTF - Cerebellar transitional field

Arrows indicate the
presumed *direction of
neuron migration* from
neuroepithelial sources.

Arrows indicate the regionally
shrinking shoreline of the
superventricle as NEP cells are
depleted while generating neurons.

102

CEREBRAL CORTEX
FUTURE PARACENTRAL LOBULE

PLATE 39A
CR 15.0 mm, GW7.4
M2051, Frontal/Horizontal
Section 122

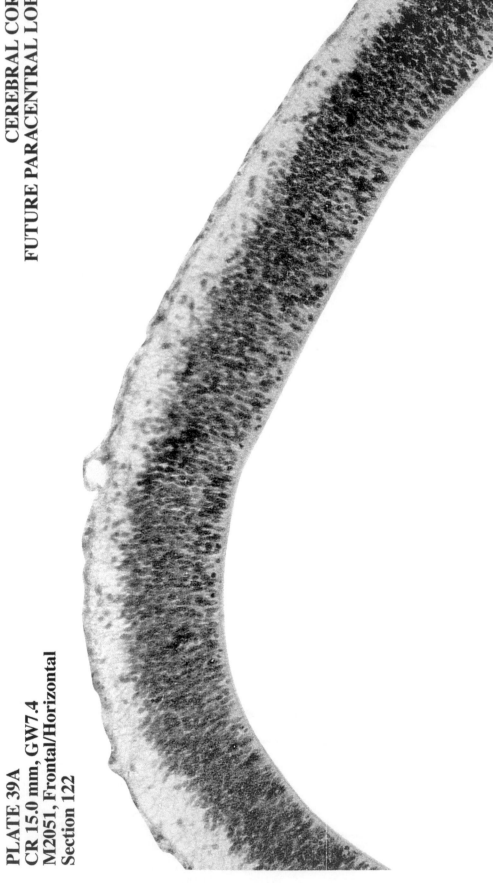

0.1 mm

See the entire section 130 in Plate 3A/B.

PLATE 39B

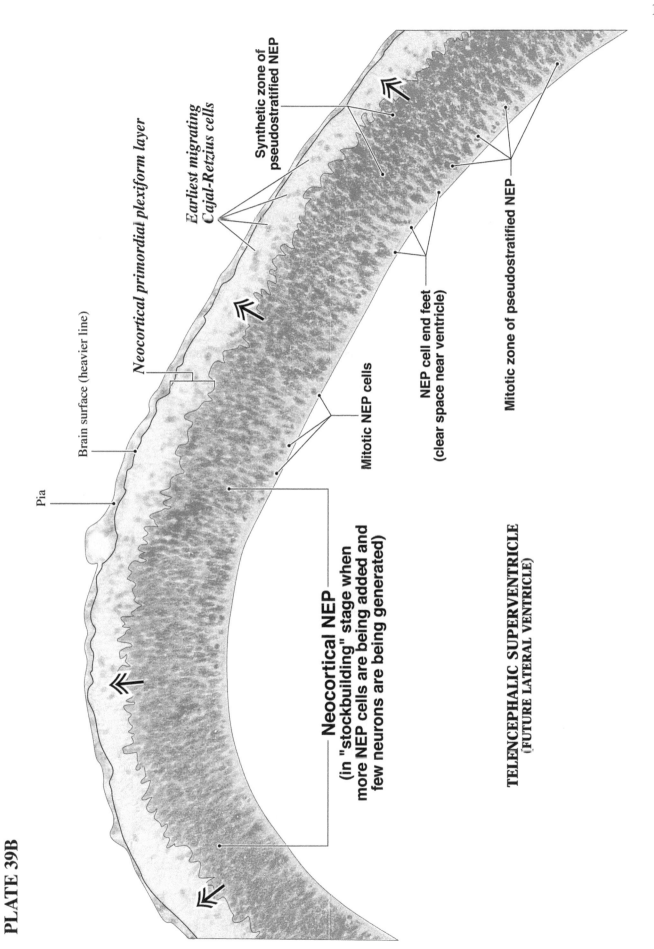

Brain surface (heavier line)

Pia

Neocortical primordial plexiform layer

Earliest migrating Cajal-Retzius cells

Synthetic zone of pseudostratified NEP

Mitotic NEP cells

NEP cell end feet
(clear space near ventricle)

Mitotic zone of pseudostratified NEP

Neocortical NEP
(in "stockbuilding" stage when
more NEP cells are being added and
few neurons are being generated)

TELENCEPHALIC SUPERVENTRICLE
(FUTURE LATERAL VENTRICLE)

PART IV: C492
CR 15.8 mm (GW 7.5)
Horizontal

Carnegie Collection specimen #492 (designated here as C492) was obtained in 1911 after a miscarriage. It has a crown-rump length (CR) of 15.8 mm and is estimated to be at gestational week (GW) 7.5. C492 was preserved in Zenker's fixative, embedded in paraffin, and was cut in 40-μm sections that were stained with aluminum cochineal. The dorsal diencephalon and mesencephalic tectum are cut in the horizontal section plane. The plane shifts to predominantly transverse in the pons and medulla. In general, C492's sections are cut perpendicular to M2155's sections (**Part 2**). We photographed 101 sections at low magnification from the uppermost tip of the pretectum through the spinal cord. Seventeen of these sections are illustrated in **Plates 40A/B to 56A/B**. All photographs containing the brain were used to produce computer-aided 3-D reconstructions of the external features of C492's brain (**Figure 14**), and to show each illustrated section *in situ* (*insets,* **Plates 40-56A**). Each illustrated section shows the brain with all surrounding tissues. Labels in **A Plates** (normal-contrast images) identify non-neural and peripheral neural structures; labels in **B Plates** (low-contrast images) identify central neural structures.

A "stockbuilding" cortical telencephalic neuroepithelium (NEP) surrounds the enlarging roof of the telencephalic superventricle. Few migrating neurons are adjacent to the cortical NEP while there are many adjacent to the basal ganglionic and basal telencephalic NEPs. The plane of C492's sections are ideal to show two features of the telencephalic/diencephalic junction that are not seen as clearly in the other GW7.4 specimens. First, the telencephalic and diencephalic superventricles are continuous at the very wide foramen of Monro. Second, the posterior basal ganglionic NEP forms a continuum with the ventral diencephalic NEP, making it difficult to distinguish telencephalic from diencephalic structures.

The thalamic NEP, in spite of its thickness, is well into its neurogenetic stage but it is clearly the least mature diencephalic component. There are few neurons outside the thalamic NEPs, except in ventral areas where pioneer reticular nuclear neurons are migrating. The subthalamic NEP is considerably thinner as neuron progenitors unload cells into a thick parenchyma, and the subthalamic pool in the third ventricle narrows. The hypothalamic NEPs are somewhat thicker, but many neurons have been offloaded into a thickening parenchyma.

The roof (tectum and pretectum) of the mesencephalon contains a neurogenetic NEP adjacent to a very thin layer of pioneer migrating neurons. Postmitotic neurons are sequestered in the basal parts. The bundles of fibers in the posterior commissure are very distinct in anterior sections of the pretectum. The tegmental and isthmal NEPs are thinning as their neuronal progeny migrate out and form neuronal clumps in the parenchyma. Many of these groups are tentatively identified. Similar to the other GW7.4 specimens, there is a very thick subpial fiber band; no doubt, these are fibers from sources outside the mesencephalon and isthmus, but a few central local fibers are identified, such as the oculomotor nerve fibers.

As in other GW7.4 specimens, the pons and medulla have NEPs that are shrinking as stem cells unload their neuronal and glial progeny into an expanding parenchyma. For the most part, nuclear subdivisions are indistinct but some tentative identifications are made. The superior olivary complex, facial motor nucleus, inferior olivary complex, and solitary nucleus can be tentatively identified. The subpial fiber band is thick and prominent, and fiber tracts like the medial lemniscus and spinocerebellar tracts are tentatively identified.

The cerebellar NEP sharply juts into the rhombencephalic superventricle. The bands in the cerebellar transitional fields are similar to other GW7.4 specimens. These bands are prominent in the future hemisphere and are absent in the future vermis.

C492 Computer-aided 3-D Brain Reconstructions

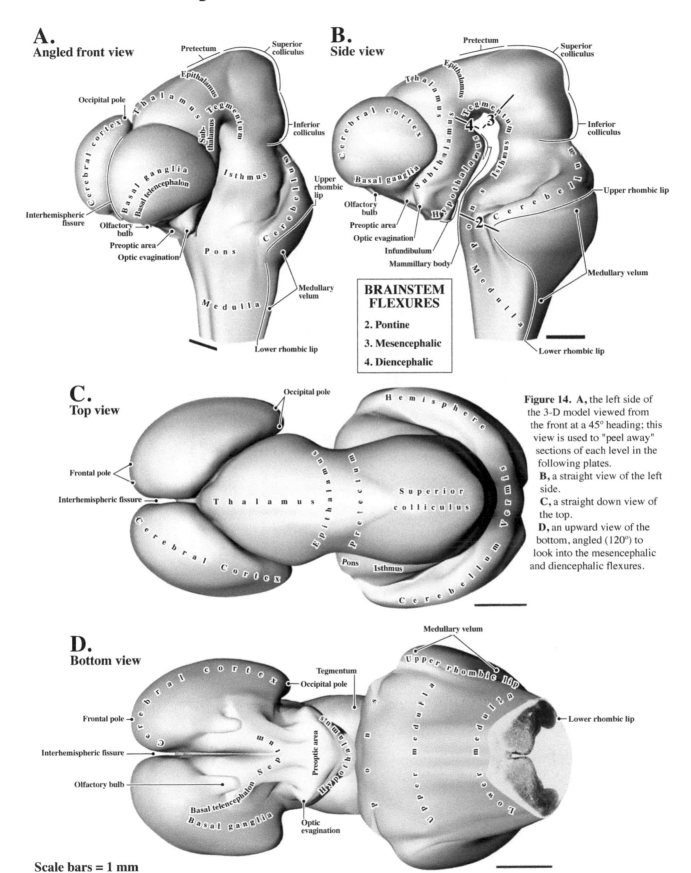

A.
Angled front view

Pretectum
Superior colliculus
Epithalamus
Occipital pole
Thalamus
Tegmentum
Sub-thalamus
Inferior colliculus
Cerebral cortex
Basal ganglia
Basal telencephalon
Isthmus
Cerebellum
Interhemispheric fissure
Olfactory bulb
Upper rhombic lip
Preoptic area
Optic evagination
Pons
Cerebellum
Medulla
Medullary velum
Lower rhombic lip

B.
Side view

Pretectum
Superior colliculus
Epithalamus
Thalamus
Tegmentum
4 3
Cerebral cortex
Subthalamus
Inferior colliculus
Isthmus
Basal ganglia
Hypothalamus
Cerebellum
Upper rhombic lip
Olfactory bulb
Preoptic area
Optic evagination
2
Cerebellum
Infundibulum
Medullary velum
Mammillary body
Medulla
Lower rhombic lip

BRAINSTEM FLEXURES

2. Pontine
3. Mesencephalic
4. Diencephalic

C.
Top view

Occipital pole
Hemisphere
Frontal pole
Epithalamus
Pretectum
Cerebellum Vermis
Interhemispheric fissure
Thalamus
Superior colliculus
Cerebral Cortex
Pons
Isthmus
Cerebellum Vermis

Figure 14. A, the left side of the 3-D model viewed from the front at a 45° heading; this view is used to "peel away" sections of each level in the following plates.
B, a straight view of the left side.
C, a straight down view of the top.
D, an upward view of the bottom, angled (120°) to look into the mesencephalic and diencephalic flexures.

D.
Bottom view

Medullary velum
Cerebral cortex
Tegmentum
Upper rhombic lip
Occipital pole
Frontal pole
Hypothalamus
Preoptic area
Septum
Medulla
Lower rhombic lip
Interhemispheric fissure
Olfactory bulb
Upper medulla
Basal telencephalon
Basal ganglia
Optic evagination
Lower medulla

Scale bars = 1 mm

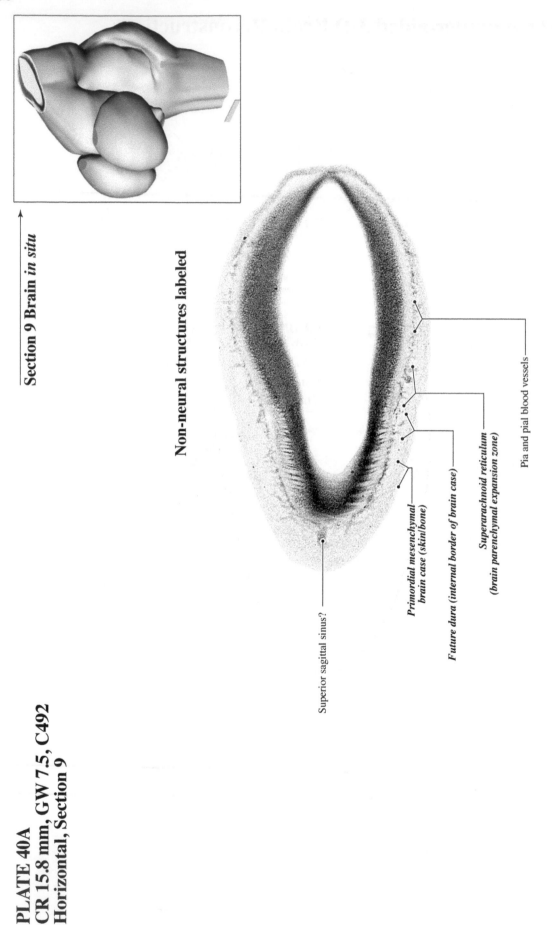

PLATE 40A
CR 15.8 mm, GW 7.5, C492
Horizontal, Section 9

Section 9 Brain *in situ*

Non-neural structures labeled

1 mm

Superior sagittal sinus?

Primordial mesenchymal brain case (skin/bone)

Future dura (internal border of brain case)

Superarachnoid reticulum (brain parenchymal expansion zone)

Pia and pial blood vessels

PLATE 40B

Neural structures labeled

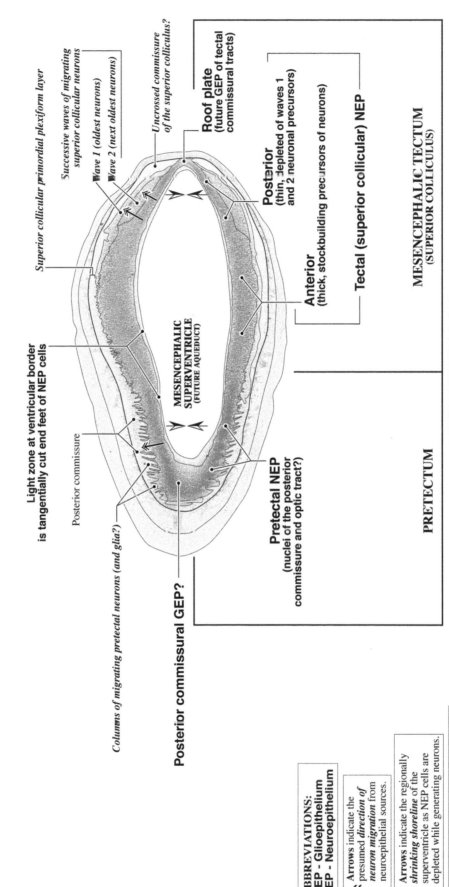

Superior collicular primordial plexiform layer

Successive waves of migrating superior collicular neurons

◄ *Wave 1 (oldest neurons)*
◄ *Wave 2 (next oldest neurons)*

Uncrossed commissure of the superior colliculus?

Roof plate
(future GEP of tectal commissural tracts)

Posterior
(thin, depleted of waves 1 and 2 neuronal precursors)

Anterior
(thick, stockbuilding precursors of neurons)

Tectal (superior collicular) NEP

MESENCEPHALIC TECTUM
(SUPERIOR COLLICULUS)

Light zone at ventricular border
is tangentially cut end feet of NEP cells

Posterior commissure

Columns of migrating pretectal neurons (and glia?)

Posterior commissural GEP?

MESENCEPHALIC
SUPERVENTRICLE
(FUTURE AQUEDUCT)

Pretectal NEP
(nuclei of the posterior commissure and optic tract?)

PRETECTUM

MESENCEPHALON

Section 29 Brain *in situ*

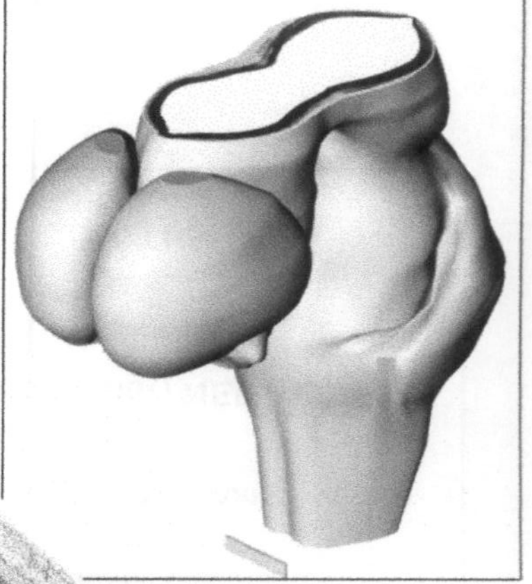

1 mm

PLATE 41A
CR 15.8 mm, GW 7.5, C492
Horizontal, Section 29

Non-neural structures labeled

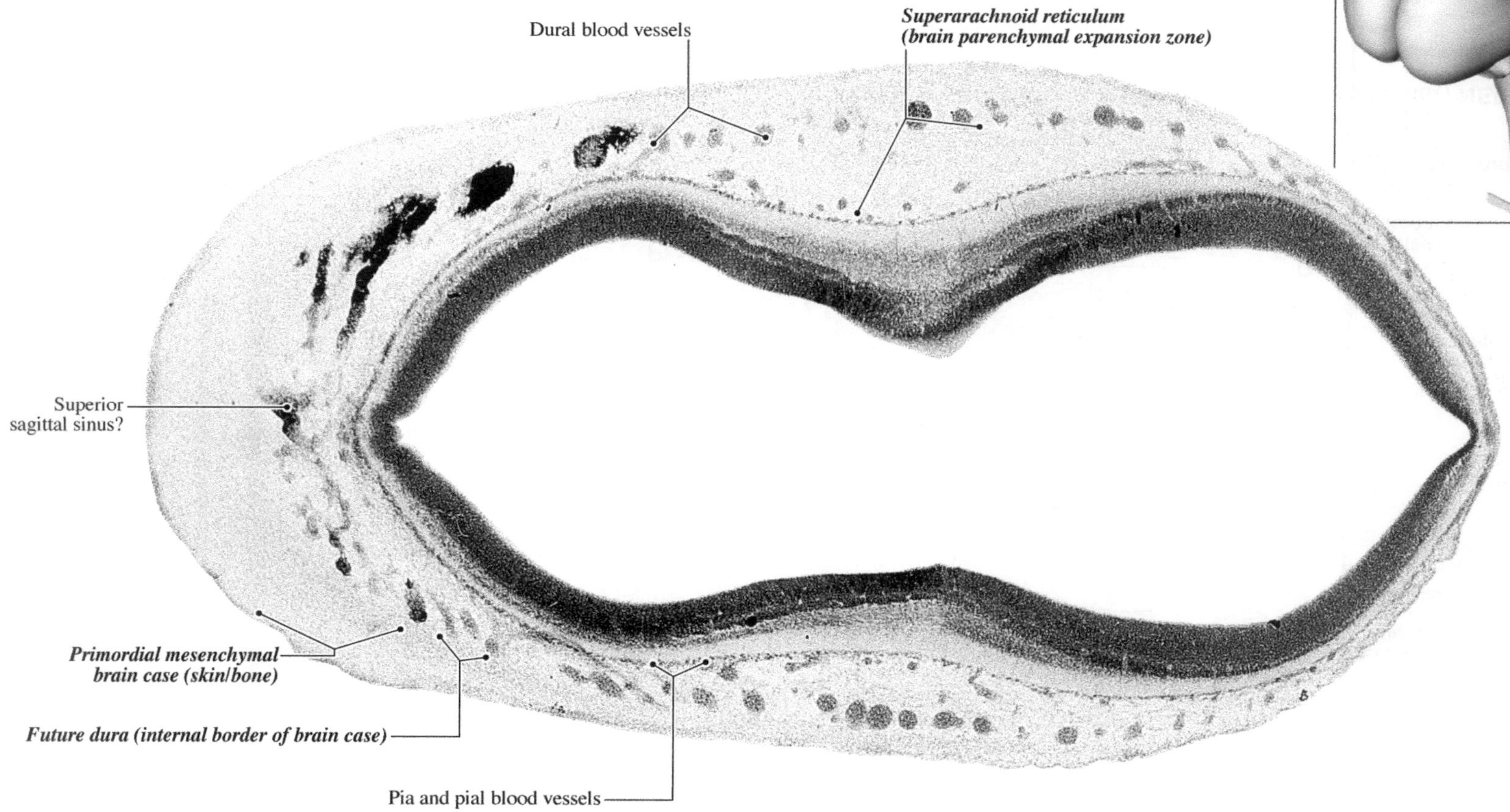

Superarachnoid reticulum
(brain parenchymal expansion zone)

Dural blood vessels

Superior
sagittal sinus?

Pia and pial blood vessels

Primordial mesenchymal
brain case (skin/bone)

Future dura (internal border of brain case)

Dark stain in some blood vessels is injected ink.

PLATE 41B

Neural structures labeled

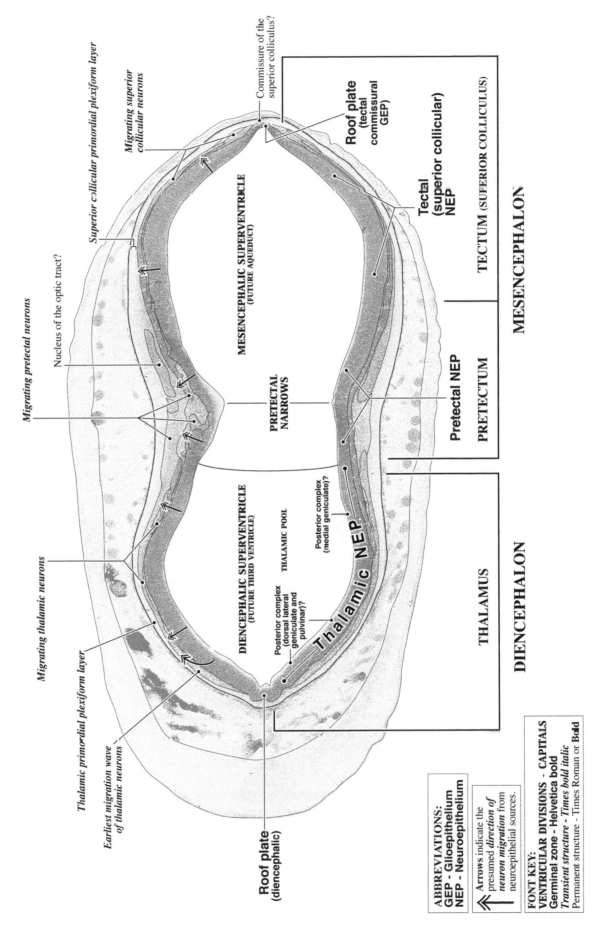

110

PLATE 42A
CR 15.8 mm, GW 7.5, C492
Horizontal, Section 36

Non-neural structures labeled

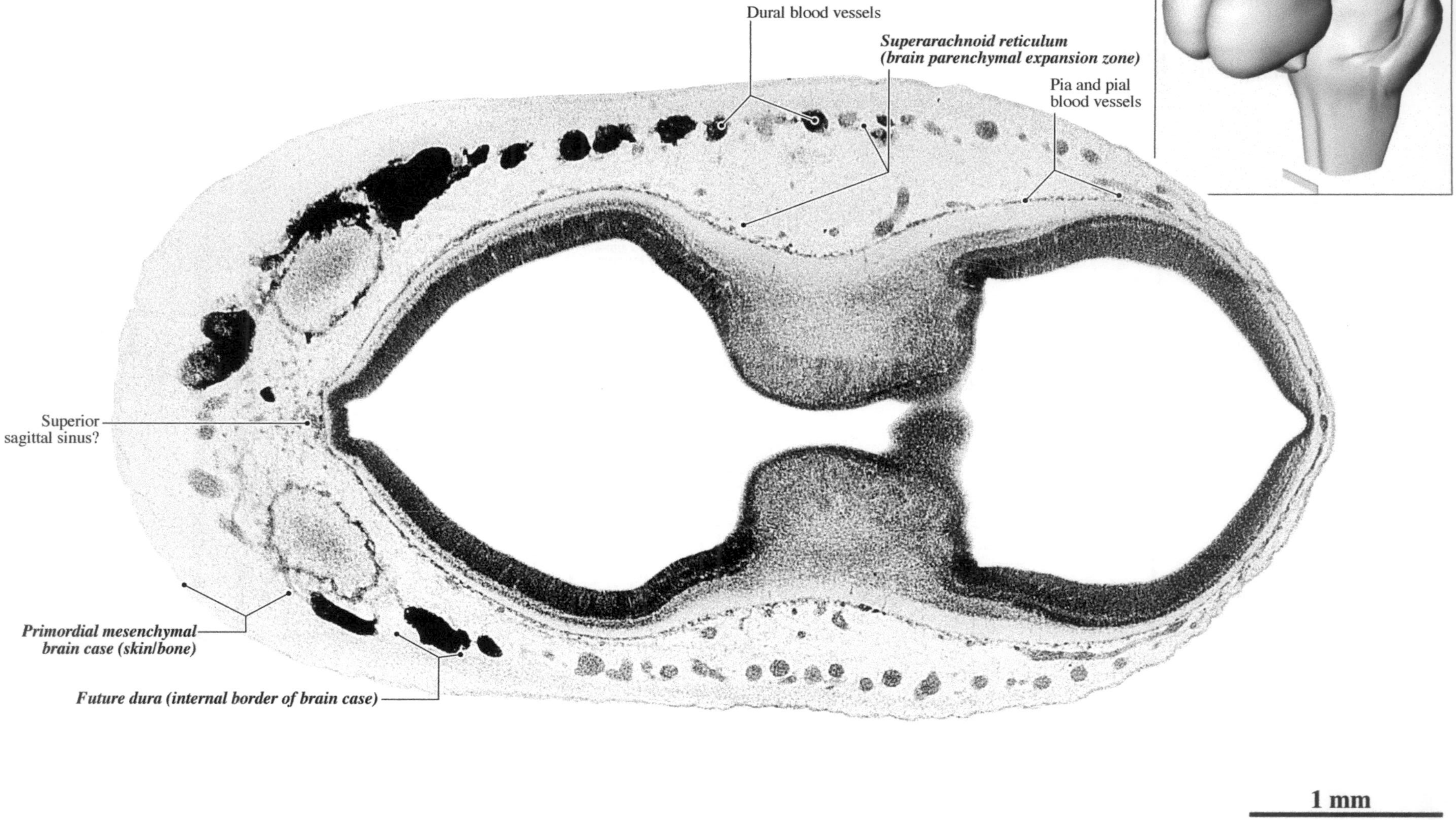

Section 36 Brain *in situ*

Superarachnoid reticulum
(brain parenchymal expansion zone)

Pia and pial
blood vessels

Dural blood vessels

Superior
sagittal sinus?

Primordial mesenchymal
brain case (skin/bone)

Future dura (internal border of brain case)

1 mm

Dark stain in some blood vessels is injected ink.

111

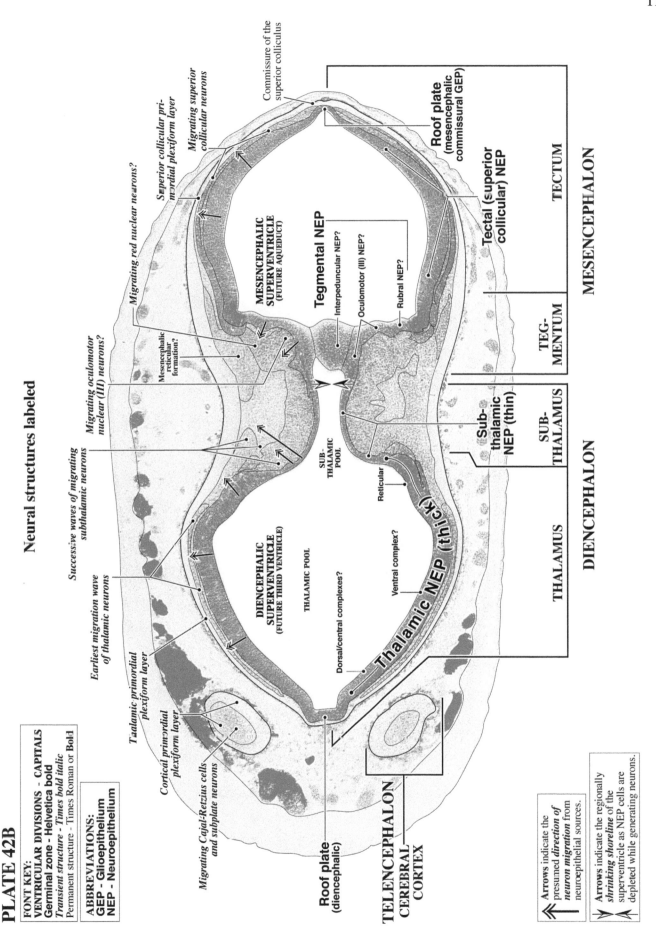

PLATE 42B

FONT KEY:
VENTRICULAR DIVISIONS - CAPITALS
Germinal zone - Helvetica bold
Transient structure - Times bold italic
Permanent structure - Times Roman or Bold

ABBREVIATIONS:
GEP - Glioepithelium
NEP - Neuroepithelium

Neural structures labeled

Commissure of the superior colliculus

Roof plate (mesencephalic commissural GEP)

Tectal (superior collicular) NEP

TECTUM

MESENCEPHALON

TEG-MENTUM

Tegmental NEP

Interpeduncular NEP?

Oculomotor (III) NEP?

Rubral NEP?

Sub-thalamic NEP (thin)

SUB-THALAMUS

Reticular

Ventral complex?

THALAMUS

DIENCEPHALON

Thalamic NEP (thick)

Dorsal/central complexes?

MESENCEPHALIC SUPERVENTRICLE (FUTURE AQUEDUCT)

SUB-THALAMIC POOL

DIENCEPHALIC SUPERVENTRICLE (FUTURE THIRD VENTRICLE)

THALAMIC POOL

Superior collicular pri-mordial plexiform layer

Migrating superior collicular neurons

Migrating red nuclear neurons?

Mesencephalic reticular formation?

Migrating oculomotor nuclear (III) neurons?

Successive waves of migrating subthalamic neurons

Earliest migration wave of thalamic neurons

Thalamic primordial plexiform layer

Cortical primordial plexiform layer

Migrating Cajal-Retzius cells and subplate neurons

Roof plate (diencephalic)

TELENCEPHALON CEREBRAL CORTEX

Arrows indicate the presumed direction of neuron migration from neuroepithelial sources.

Arrows indicate the regionally shrinking shoreline of the superventricle as NEP cells are depleted while generating neurons.

Section 39 Brain *in situ*

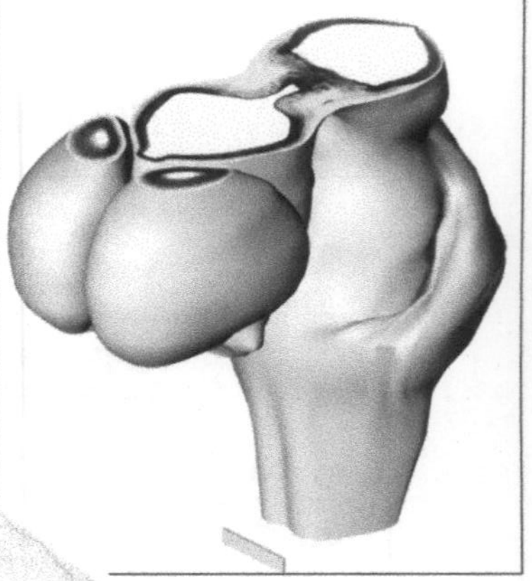

PLATE 43A
CR 15.8 mm, GW 7.5, C492
Horizontal, Section 39

Non-neural structures labeled

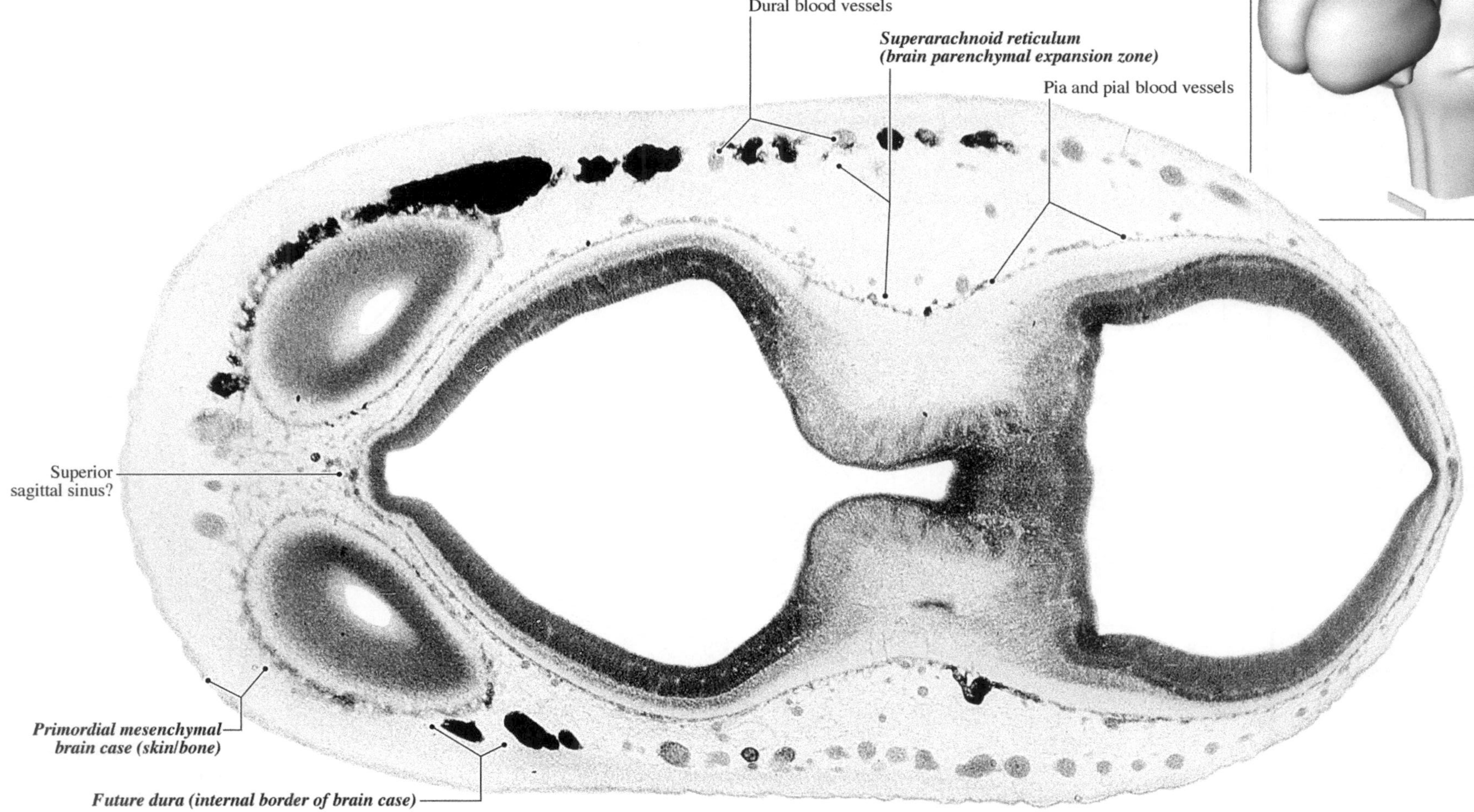

1 mm

Pia and pial blood vessels

Superarachnoid reticulum
(brain parenchymal expansion zone)

Dural blood vessels

Future dura (internal border of brain case)

Primordial mesenchymal
brain case (skin/bone)

Superior
sagittal sinus?

Dark stain in some blood vessels is injected ink.

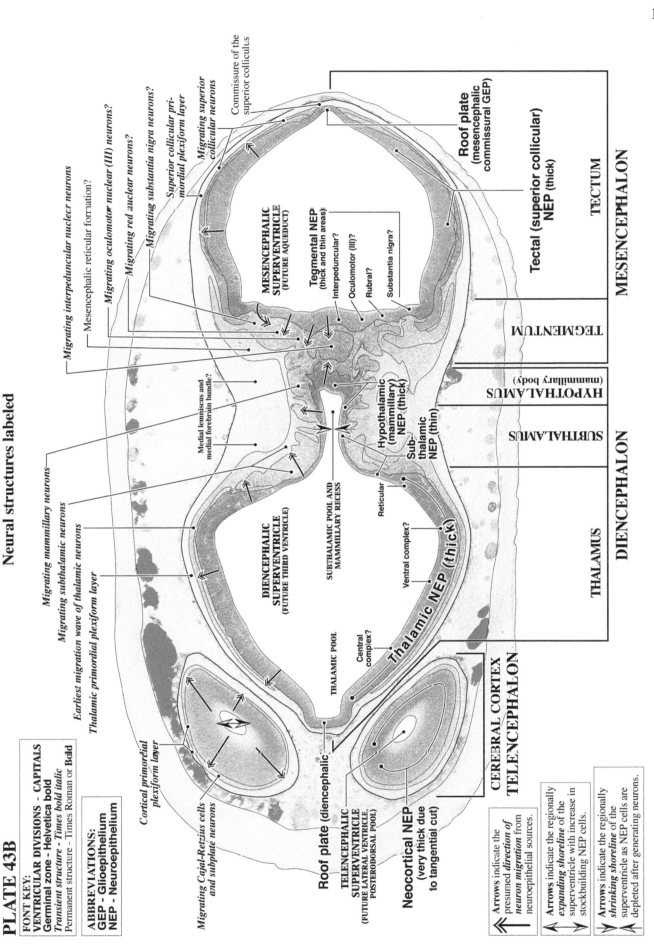

PLATE 43B

FONT KEY:
VENTRICULAR DIVISIONS - CAPITALS
Germinal zone - Helvetica bold
Transient structure - Times bold italic
Permanent structure - Times Roman or **Bold**

ABBREVIATIONS:
GEP - Glioepithelium
NEP - Neuroepithelium

Neural structures labeled

Migrating mammillary neurons

Migrating subthalamic neurons

Earliest migration wave of thalamic neurons

Thalamic primordial plexiform layer

Migrating Cajal-Retzius cells and subplate neurons

Cortical primordial plexiform layer

Migrating interpeduncular nuclear neurons

Mesencephalic reticular formation?

Migrating oculomotor nuclear (III) neurons?

Migrating red nuclear neurons?

Migrating substantia nigra neurons?

Superior collicular primordial plexiform layer

Migrating superior collicular neurons

Commissure of the superior colliculus

Roof plate (mesencephalic commissural GEP)

Tectal (superior collicular) NEP (thick)

TECTUM

MESENCEPHALON

MESENCEPHALIC SUPERVENTRICLE (FUTURE AQUEDUCT)

Tegmental NEP (thick and thin areas)

Interpeduncular?
Oculomotor (III)?
Rubral?
Substantia nigra?

TEGMENTUM

HYPOTHALAMUS (mammillary body)

SUBTHALAMUS

THALAMUS

DIENCEPHALON

Medial lemniscus and medial forebrain bundle?

Hypothalamic (mammillary) NEP (thick)

Subthalamic NEP (thin)

Reticular

Ventral complex?

DIENCEPHALIC SUPERVENTRICLE (FUTURE THIRD VENTRICLE)

SUBTHALAMIC POOL AND MAMMILLARY RECESS

Central complex?

Thalamic NEP (thick)

THALAMIC POOL

CEREBRAL CORTEX TELENCEPHALON

Roof plate (diencephalic)

TELENCEPHALIC SUPERVENTRICLE (FUTURE LATERAL VENTRICLE, POSTERODORSAL POOL)

Neocortical NEP (very thick due to tangential cut)

Arrows indicate the presumed *direction of neuron migration* from neuroepithelial sources.

Arrows indicate the regionally *expanding shoreline* of the superventricle with increase in stockbuilding NEP cells.

Arrows indicate the regionally *shrinking shoreline* of the superventricle as NEP cells are depleted after generating neurons.

114

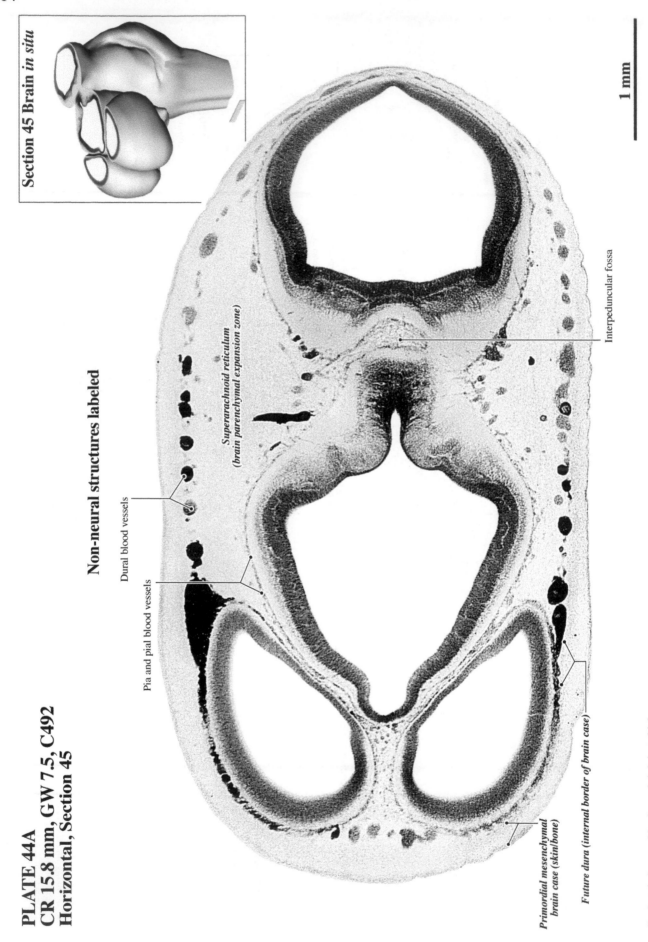

Section 45 Brain *in situ*

1 mm

PLATE 44A
CR 15.8 mm, GW 7.5, C492
Horizontal, Section 45

Non-neural structures labeled

Dural blood vessels

Superarachnoid reticulum
(brain parenchymal expansion zone)

Pia and pial blood vessels

Primordial mesenchymal
brain case (skin/bone)

Future dura (internal border of brain case)

Interpeduncular fossa

Dark stain in some blood vessels is injected ink.

PLATE 44B

Neural structures labeled

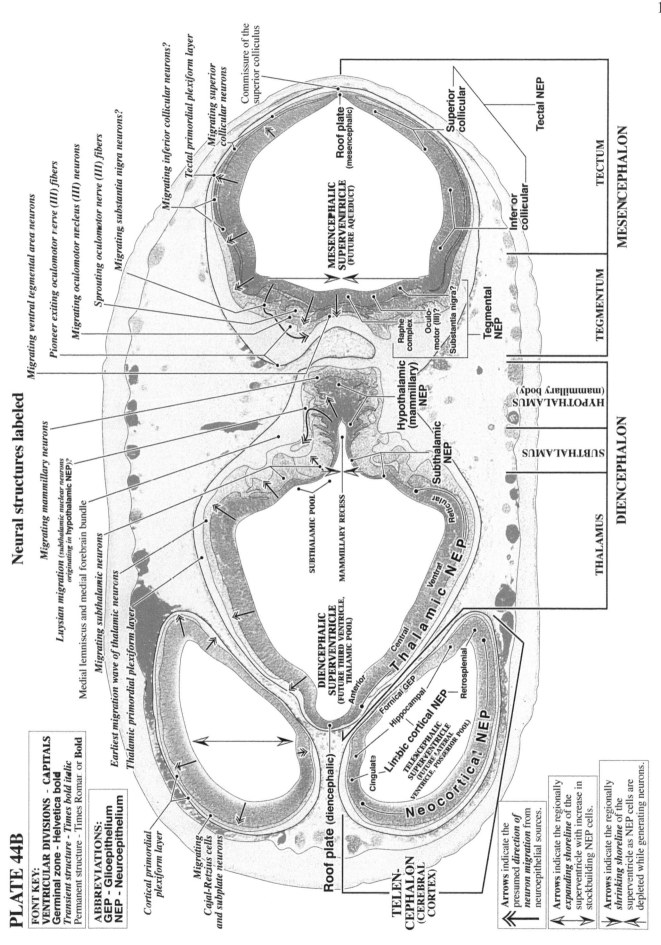

Migrating ventral tegmental area neurons

Pioneer exiting oculomotor nerve (III) fibers

Migrating oculomotor nucleus (III) neurons

Sprouting oculomotor nerve (III) fibers

Migrating substantia nigra neurons?

Migrating inferior collicular neurons?

Tectal primordial plexiform layer

Migrating superior collicular neurons

Commissure of the superior colliculus

Roof plate (mesencephalic)

Superior collicular

Inferior collicular

Tectal NEP

TECTUM

MESENCEPHALON

TEGMENTUM

Tegmental NEP

Raphe complex

Oculo-motor (III)?

Substantia nigra?

MESENCEPHALIC SUPERVENTRICLE (FUTURE AQUEDUCT)

HYPOTHALAMUS (mammillary body)

SUBTHALAMUS

THALAMUS

DIENCEPHALON

Hypothalamic (mammillary) NEP

Subthalamic NEP

Reticular

Central

Ventral

Thalamic NEP

SUBTHALAMIC POOL

MAMMILLARY RECESS

DIENCEPHALIC SUPERVENTRICLE (FUTURE THIRD VENTRICLE, THALAMIC POOL)

Anterior

Migrating mammillary neurons

Luysian migration (subthalamic nuclear neurons originating in hypothalamic NEP)?

Medial lemniscus and medial forebrain bundle

Migrating subthalamic neurons

Earliest migration wave of thalamic neurons

Thalamic primordial plexiform layer

Cortical primordial plexiform layer

Migrating Cajal-Retzius cells and subplate neurons

Roof plate (diencephalic)

Fornical GEP

Hippocampal

Limbic cortical NEP

TELENCEPHALIC SUPERVENTRICLE (FUTURE LATERAL VENTRICLE, POSTERIOR POOL)

Cingulate

Retrosplenial

Neocortical NEP

TELEN-CEPHALON (CEREBRAL CORTEX)

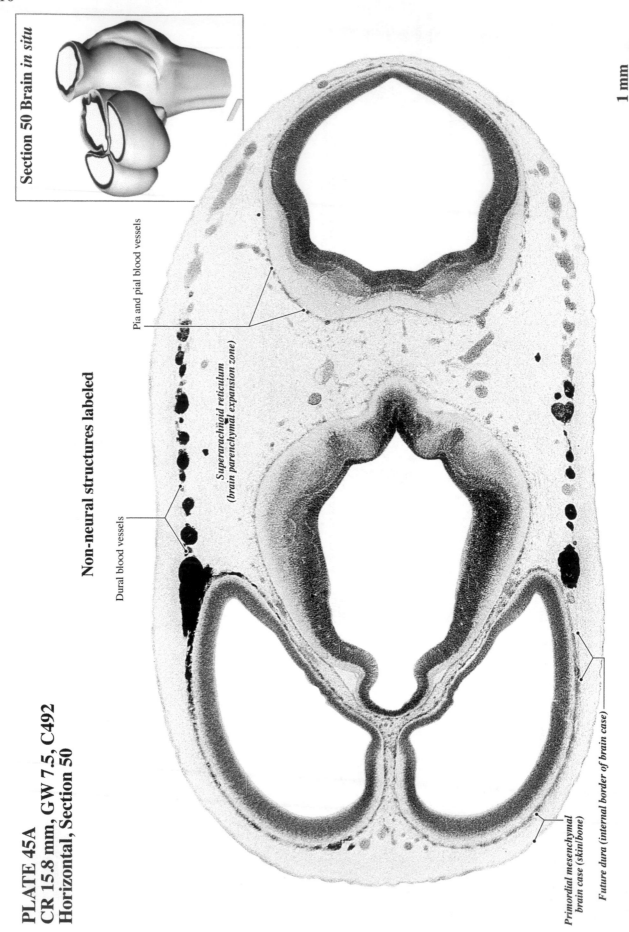

Section 50 Brain *in situ*

1 mm

PLATE 45A
CR 15.8 mm, GW 7.5, C492
Horizontal, Section 50

Non-neural structures labeled

Pia and pial blood vessels

Dural blood vessels

Superarachnoid reticulum
(brain parenchymal expansion zone)

Primordial mesenchymal
brain case (skin/bone)

Future dura (internal border of brain case)

Dark stain in some blood vessels is injected ink.

PLATE 45B

Neural structures labeled

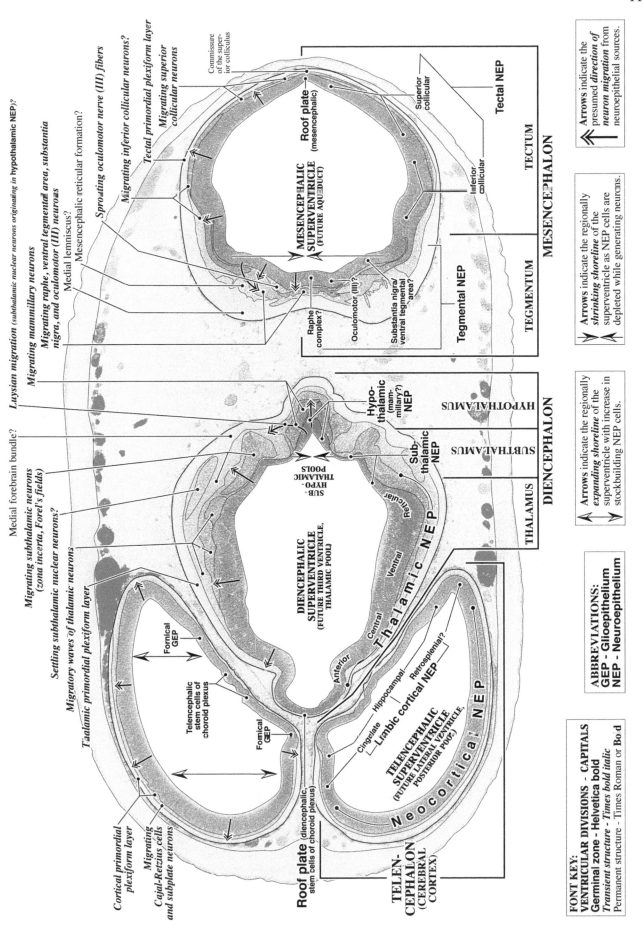

117

Medial forebrain bundle?

Luysian migration *(subthalamic nuclear neurons originating in hypothalamic NEP)?*

Migrating mammillary neurons

Migrating raphe, ventral tegmental area, substantia nigra, and oculomotor (III) neurons

Medial lemniscus?

Mesencephalic reticular formation?

Sprouting oculomotor nerve (III) fibers

Migrating inferior collicular neurons?

Tectal primordial plexiform layer

Migrating superior collicular neurons

Commissure of the superior colliculus

Roof plate (mesencephalic)

Superior collicular

Inferior collicular

Tectal NEP

TECTUM

MESENCEPHALIC SUPERVENTRICLE (FUTURE AQUEDUCT)

MESENCEPHALON

Raphe complex?

Oculomotor (III)?

Substantia nigra/ ventral tegmental area?

Tegmental NEP

TEGMENTUM

Hypo-thalamic (mam-millary?) NEP

HYPOTHALAMUS

Sub-thalamic NEP

SUBTHALAMUS

THALAMUS

DIENCEPHALON

Reticular

Ventral

Central

Anterior

T-h-a-l-a-m-i-c N E P

SUB-HYPO-THALAMIC POOLS

DIENCEPHALIC SUPERVENTRICLE (FUTURE THIRD VENTRICLE, THALAMIC POOL)

Migrating subthalamic neurons (zona incerta, Forel's fields)

Settling subthalamic nuclear neurons?

Migratory waves of thalamic neurons

Thalamic primordial plexiform layer

Cortical primordial plexiform layer

Migrating Cajal-Retzius cells and subplate neurons

Fornical GEP

Telencephalic stem cells of choroid plexus

Fornical GEP

Roof plate (diencephalic, stem cells of choroid plexus)

Cingulate

Hippocampal

Retrosplenial?

Limbic cortical NEP

TELENCEPHALIC SUPERVENTRICLE (FUTURE LATERAL VENTRICLE, POSTERIOR POOL)

N e o c o r t i c a l N E P

TELEN-CEPHALON (CEREBRAL CORTEX)

Arrows indicate the presumed *direction of neuron migration* from neuroepithelial sources.

Arrows indicate the regionally *shrinking shoreline* of the superventricle as NEP cells are depleted while generating neurons.

Arrows indicate the regionally *expanding shoreline* of the superventricle with increase in stockbuilding NEP cells.

Section 55 Brain *in situ*

1 mm

PLATE 46A
CR 15.8 mm, GW 7.5, C492
Horizontal, Section 55

Non-neural and peripheral neural structures labeled

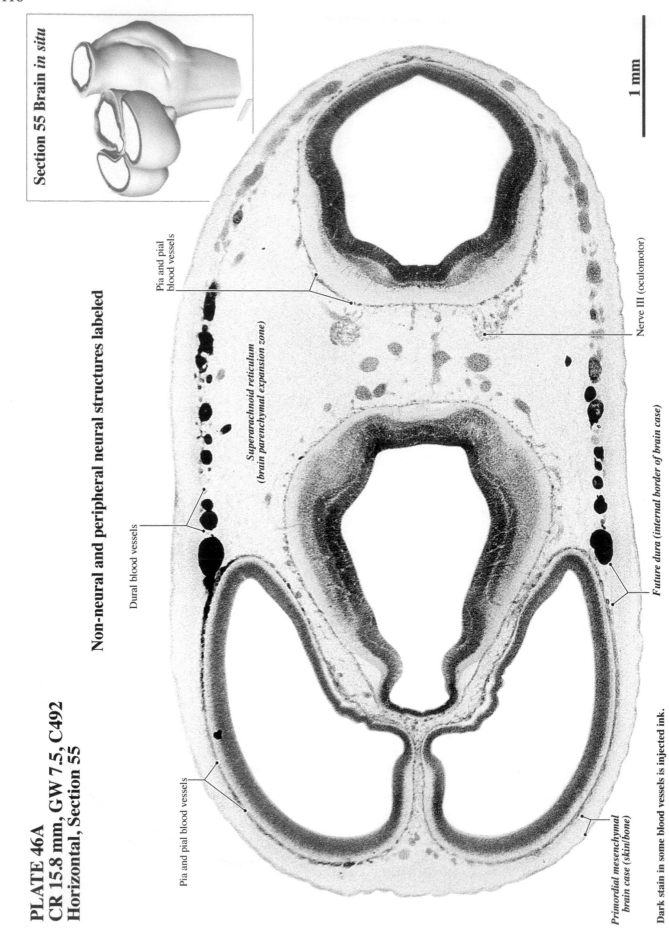

Pia and pial blood vessels

Nerve III (oculomotor)

Superarachnoid reticulum (brain parenchymal expansion zone)

Future dura (internal border of brain case)

Dural blood vessels

Pia and pial blood vessels

Primordial mesenchymal brain case (skin/bone)

Dark stain in some blood vessels is injected ink.

PLATE 46B

Central neural structures labeled

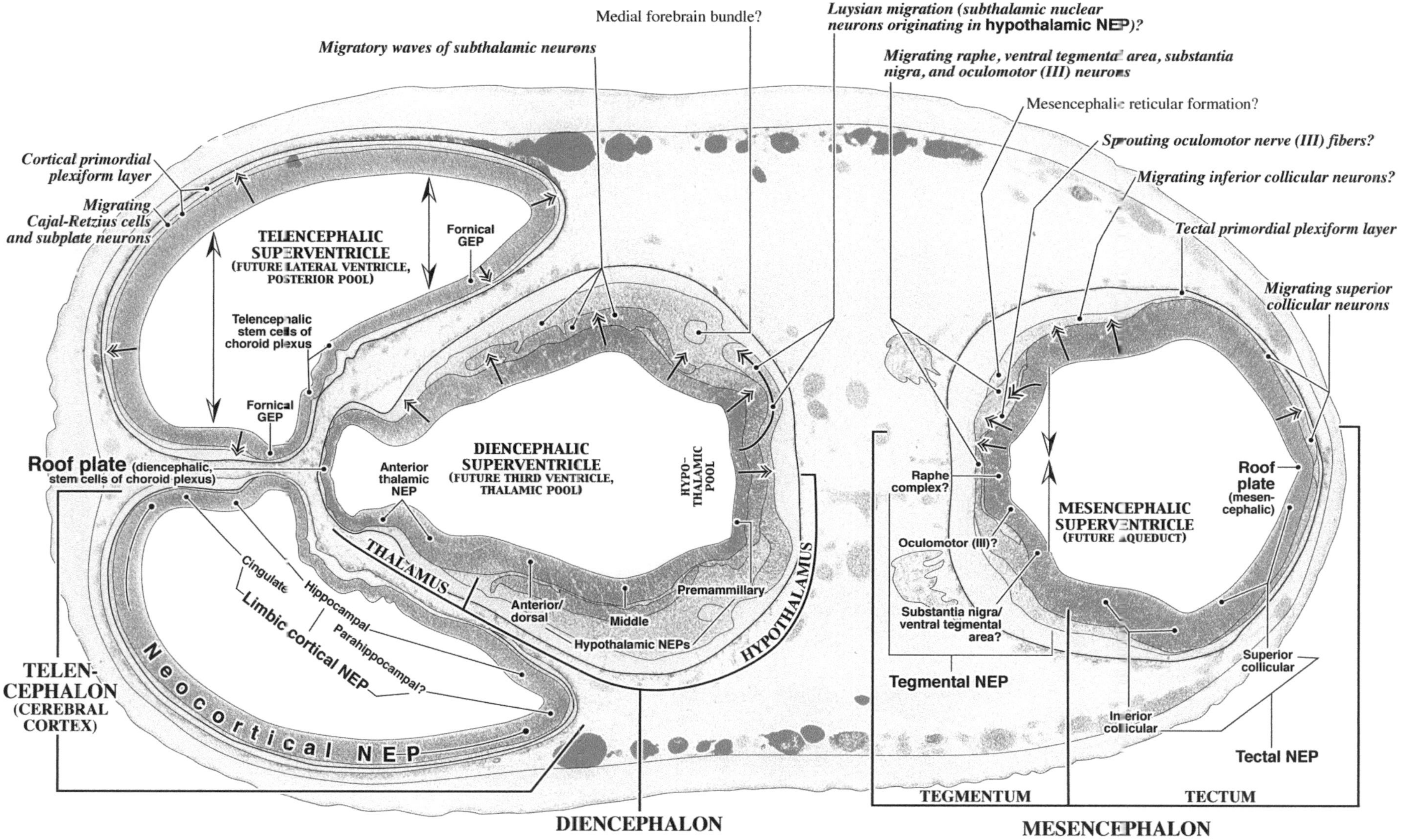

119

120

PLATE 47A
CR 15.8 mm, GW 7.5, C492
Horizontal, Section 65

Non-neural and peripheral neural structures labeled

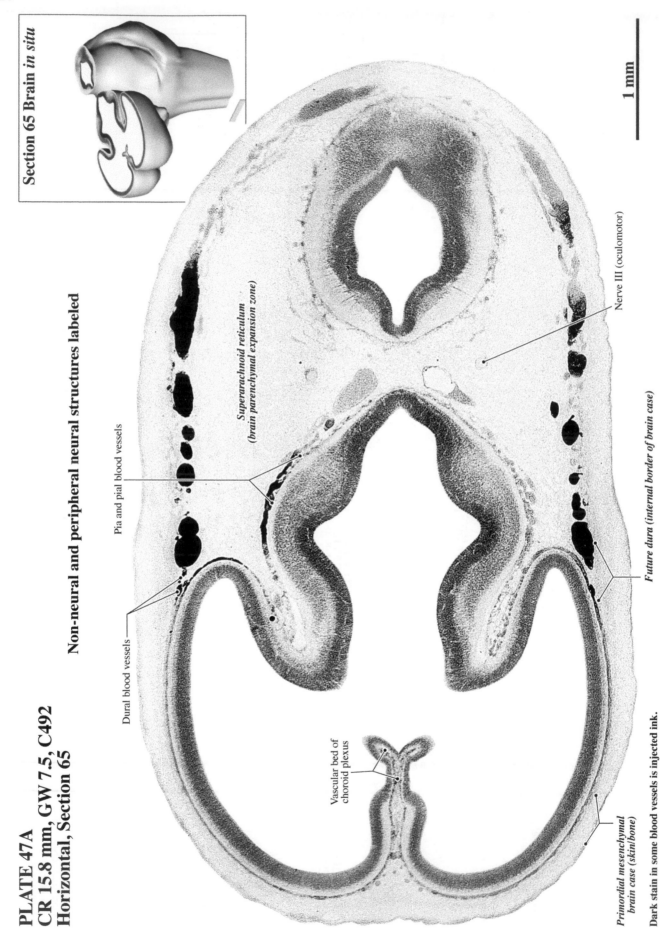

Pia and pial blood vessels

Superarachnoid reticulum (brain parenchymal expansion zone)

Nerve III (oculomotor)

Dural blood vessels

Future dura (internal border of brain case)

Vascular bed of choroid plexus

Primordial mesenchymal brain case (skull bone)

Dark stain in some blood vessels is injected ink.

1 mm

PLATE 47B

Central neural structures labeled

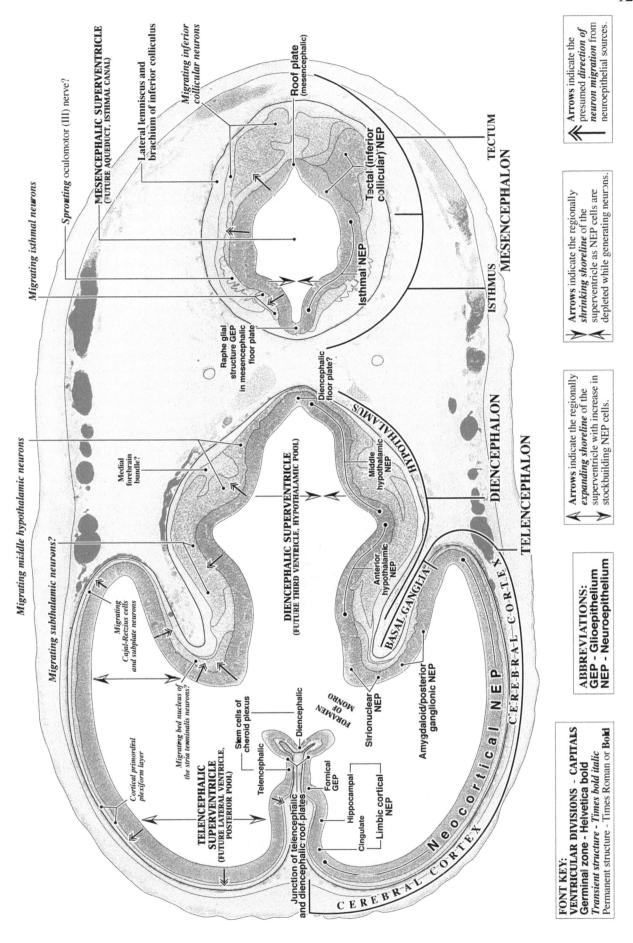

121

Migrating middle hypothalamic neurons

MESENCEPHALIC SUPERVENTRICLE (FUTURE AQUEDUCT, ISTHMAL CANAL)

Sprouting oculomotor (III) nerve?

Lateral lemniscus and brachium of inferior colliculus

Migrating inferior collicular neurons

Migrating isthmal neurons

Roof plate (mesencephalic)

Tectal (inferior collicular) NEP

Isthmal NEP

Raphe glial structure GEP in mesencephalic floor plate

TECTUM

MESENCEPHALON

ISTHMUS

Diencephalic floor plate?

Migrating subthalamic neurons?

Medial forebrain bundle?

Middle hypothalamic NEP

HYPOTHALAMUS

DIENCEPHALON

Anterior hypothalamic NEP

DIENCEPHALIC SUPERVENTRICLE (FUTURE THIRD VENTRICLE, HYPOTHALAMIC POOL)

BASAL GANGLIA

TELENCEPHALON

Migrating Cajal-Retzius cells and subplate neurons

Cortical primordial plexiform layer

Migrating bed nucleus of the stria terminalis neurons?

TELENCEPHALIC SUPERVENTRICLE (FUTURE LATERAL VENTRICLE, POSTERIOR POOL)

Stem cells of choroid plexus

Diencephalic

Telencephalic

FORAMEN OF MONRO

Diencephalic

Fornical GEP

Hippocampal NEP

Cingulate

Limbic cortical NEP

Strionuclear NEP

Amygdaloid/posterior ganglionic NEP

Junction of telencephalic and diencephalic roof plates

Neocortical N E P

C-E-R-E-B-R-A-L C-O-R-T-E-X

CEREBRAL CORTEX

PLATE 48A
CR 15.8 mm, GW 7.5, C492
Horizontal, Section 71

Non-neural and peripheral neural structures labeled

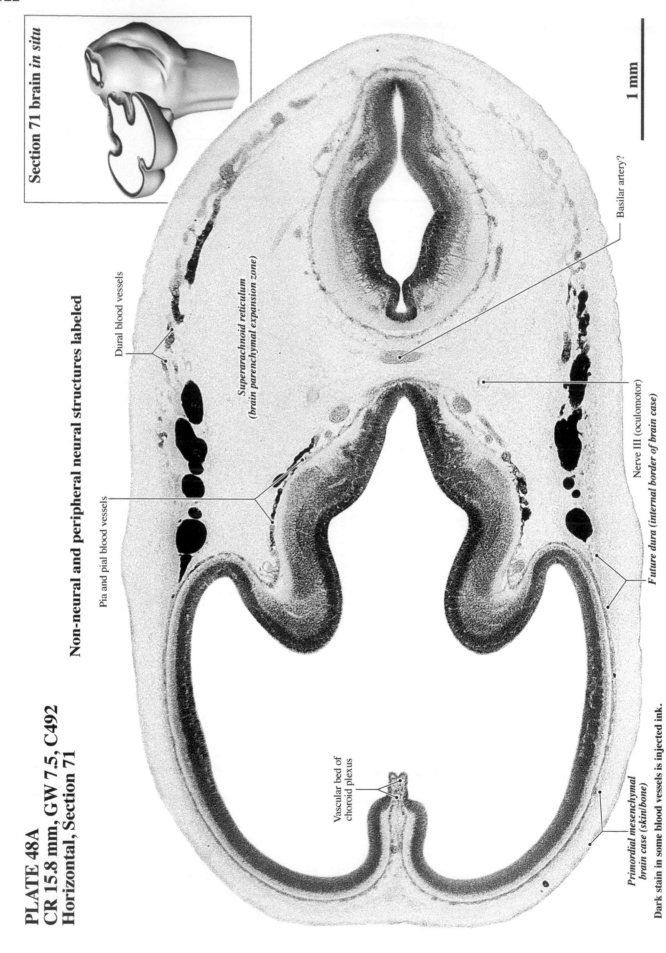

Section 71 brain *in situ*

1 mm

Dural blood vessels

Superarachnoid reticulum (brain parenchymal expansion zone)

Pia and pial blood vessels

Basilar artery?

Nerve III (oculomotor)

Future dura (internal border of brain case)

Vascular bed of choroid plexus

Primordial mesenchymal brain case (skin/bone)
Dark stain in some blood vessels is injected ink.

PLATE 48B

Central neural structures labeled

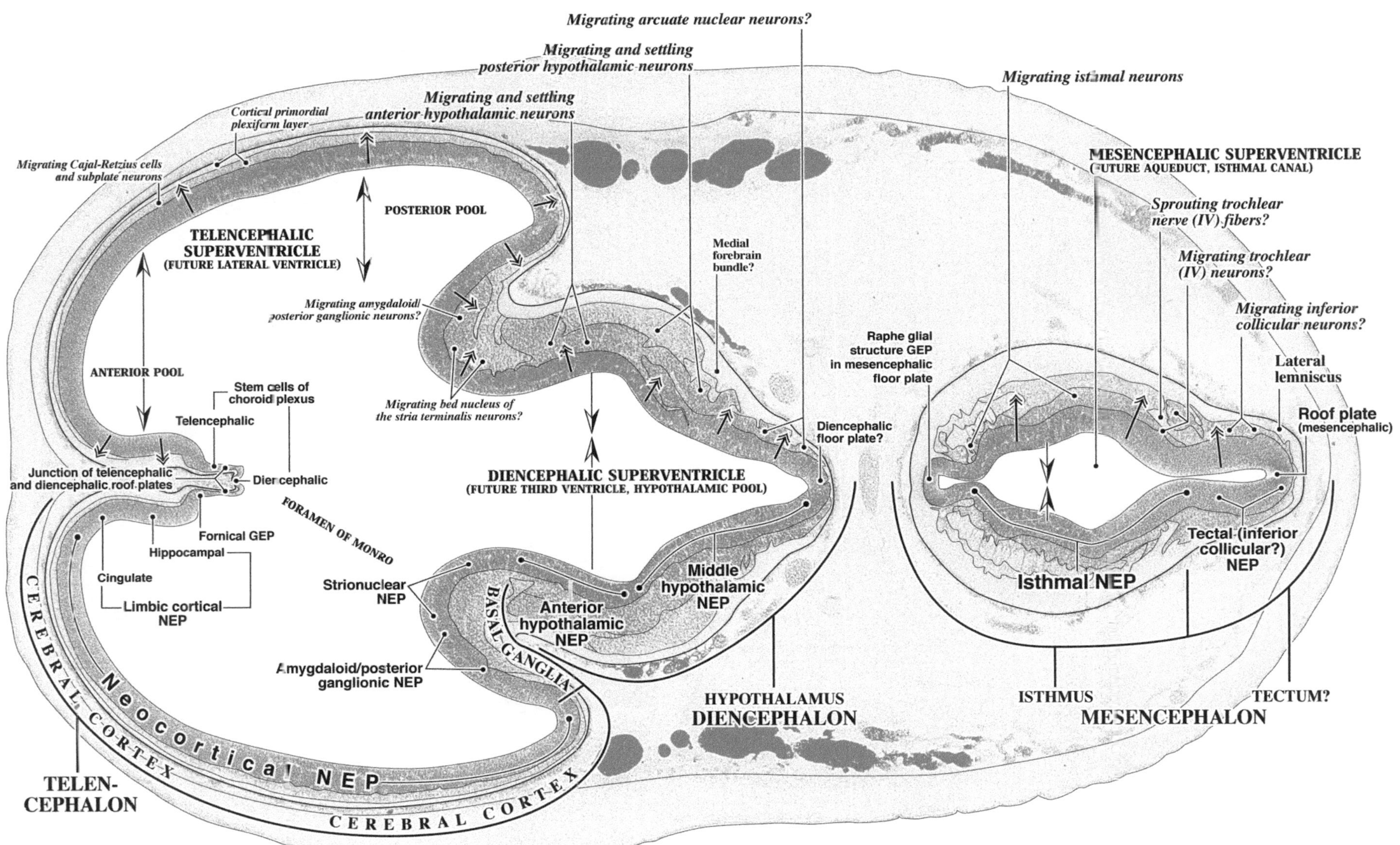

123

FONT KEY:
VENTRICULAR DIVISIONS - CAPITALS
Germinal zone - Helvetica bold
Transient structure - Times bold italic
Permanent structure - Times Roman or Bold

ABBREVIATIONS:
GEP - Glioepithelium
NEP - Neuroepithelium

Arrows indicate the presumed direction of neuron migration from neuroepithelial sources.

Arrows indicate the regionally expanding shoreline of the superventricle with increase in stockbuilding NEP cells.

Arrows indicate the regionally shrinking shoreline of the superventricle as NEP cells are depleted while generating neurons.

Labels on figure:

Migrating arcuate nuclear neurons?

Migrating and settling posterior hypothalamic neurons

Migrating and settling anterior-hypothalamic neurons

Migrating Cajal-Retzius cells and subplate neurons

Cortical primordial plexiform layer

POSTERIOR POOL

TELENCEPHALIC SUPERVENTRICLE (FUTURE LATERAL VENTRICLE)

ANTERIOR POOL

Stem cells of choroid plexus

Diencephalic

Telencephalic

Junction of telencephalic and diencephalic roof plates

Fornical GEP

Hippocampal

Cingulate

Limbic cortical NEP

FORAMEN OF MONRO

Migrating bed nucleus of the stria terminalis neurons?

Migrating amygdaloid/ posterior ganglionic neurons?

Medial forebrain bundle?

Diencephalic floor plate?

Raphe glial structure GEP in mesencephalic floor plate

MESENCEPHALIC SUPERVENTRICLE (FUTURE AQUEDUCT, ISTHMAL CANAL)

Sprouting trochlear nerve (IV) fibers?

Migrating trochlear (IV) neurons?

Migrating inferior collicular neurons?

Lateral lemniscus

Roof plate (mesencephalic)

Tectal- (inferior collicular?) NEP

Isthmal NEP

TECTUM?

ISTHMUS

MESENCEPHALON

Migrating isthmal neurons

DIENCEPHALIC SUPERVENTRICLE (FUTURE THIRD VENTRICLE, HYPOTHALAMIC POOL)

Middle hypothalamic NEP

Anterior hypothalamic NEP

HYPOTHALAMUS

DIENCEPHALON

BASAL GANGLIA

Strionuclear NEP

Amygdaloid/posterior ganglionic NEP

CEREBRAL CORTEX

Neocortical NEP

CEREBRAL CORTEX

TELEN-CEPHALON

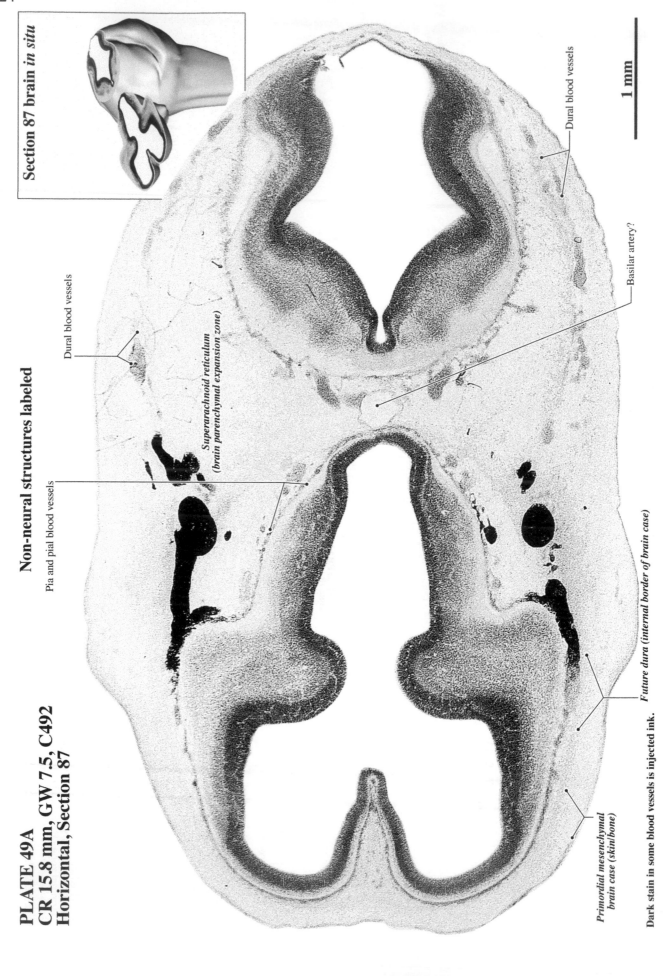

Section 87 brain *in situ*

PLATE 49A
CR 15.8 mm, GW 7.5, C492
Horizontal, Section 87

Non-neural structures labeled

Dural blood vessels

Dural blood vessels

Basilar artery?

Superarachnoid reticulum
(brain parenchymal expansion zone)

Pia and pial blood vessels

Future dura (internal border of brain case)

Primordial mesenchymal
brain case (skinbone)

Dark stain in some blood vessels is injected ink.

1 mm

PLATE 49B

Neural structures labeled

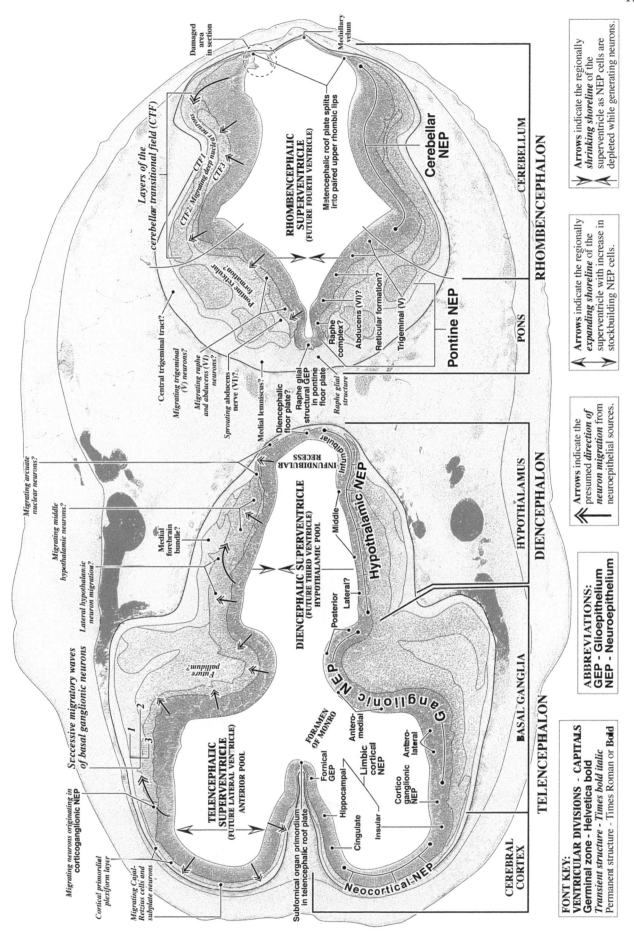

Migrating neurons originating in corticoganglionic NEP

Migrating arcuate nuclear neurons?

Migrating middle hypothalamic neurons?

Lateral hypothalamic neuron migration?

Medial forebrain bundle?

Cortical primordial plexiform layer

Migrating Cajal-Retzius cells and subplate neurons

Successive migratory waves of basal ganglionic neurons

Central trigeminal tract?

Migrating trigeminal (V) neurons?

Migrating raphe and abducens neurons?

Sprouting abducens nerve (VI)?

Medial lemniscus?

Diencephalic floor plate?

Raphe glial structural GEP in pontine floor plate

Raphe glial structure

Damaged area in section

Layers of the cerebellar transitional field (CTF)

gr neurons

CTF1
CTF2: Migrating deep nuclei
CTF3

Pontine reticular formation?

RHOMBENCEPHALIC SUPERVENTRICLE (FUTURE FOURTH VENTRICLE)

Metencephalic roof plate splits into paired upper rhombic lips

Medullary velum

Cerebellar NEP

Raphe complex?

Abducens (VI)?

Reticular formation?

Trigeminal (V)

Pontine NEP

DIENCEPHALIC SUPERVENTRICLE (FUTURE THIRD VENTRICLE) HYPOTHALAMIC POOL

INFUNDIBULAR RECESS

Infundibular

Hypothalamic NEP

Middle

Posterior

Lateral?

Future pallidum?

Ganglionic NEP

TELENCEPHALIC SUPERVENTRICLE (FUTURE LATERAL VENTRICLE) ANTERIOR POOL

1
2
3

FORAMEN OF MONRO

Fornical GEP

Hippocampal

Cingulate

Insular

Antero-medial

Limbic cortical NEP

Antero-lateral

Cortico ganglionic NEP

Subfornical organ primordium in telencephalic roof plate

Neocortical NEP

CEREBELLUM

RHOMBENCEPHALON

PONS

HYPOTHALAMUS

DIENCEPHALON

BASAL GANGLIA

TELENCEPHALON

CEREBRAL CORTEX

Arrows indicate the regionally *shrinking shoreline* of the superventricle as NEP cells are depleted while generating neurons.

Arrows indicate the regionally *expanding shoreline* of the superventricle with increase in stockbuilding NEP cells.

Arrows indicate the presumed *direction of neuron migration* from neuroepithelial sources.

ABBREVIATIONS:
GEP - Glioepithelium
NEP - Neuroepithelium

FONT KEY:
VENTRICULAR DIVISIONS - CAPITALS
Germinal zone - Helvetica bold
Transient structure - Times bold italic
Permanent structure - Times Roman or **Bold**

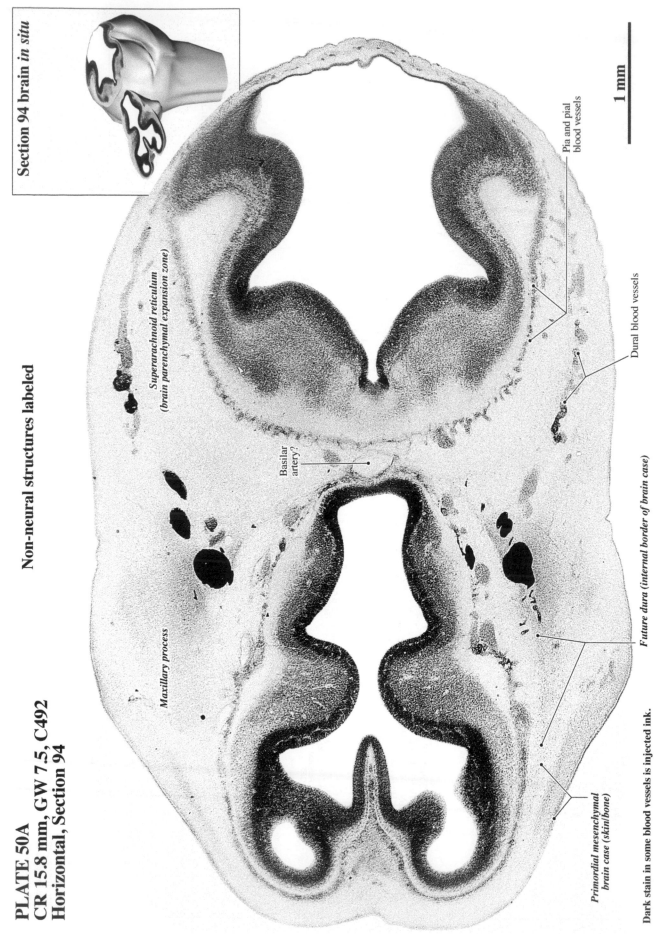

126

PLATE 50A
CR 15.8 mm, GW 7.5, C492
Horizontal, Section 94

Non-neural structures labeled

Section 94 brain *in situ*

1 mm

Superarachnoid reticulum
(brain parenchymal expansion zone)

Pia and pial
blood vessels

Dural blood vessels

Basilar
artery?

Maxillary process

Future dura (internal border of brain case)

Primordial mesenchymal
brain case (skin/bone)

Dark stain in some blood vessels is injected ink.

PLATE 50B

Neural structures labeled

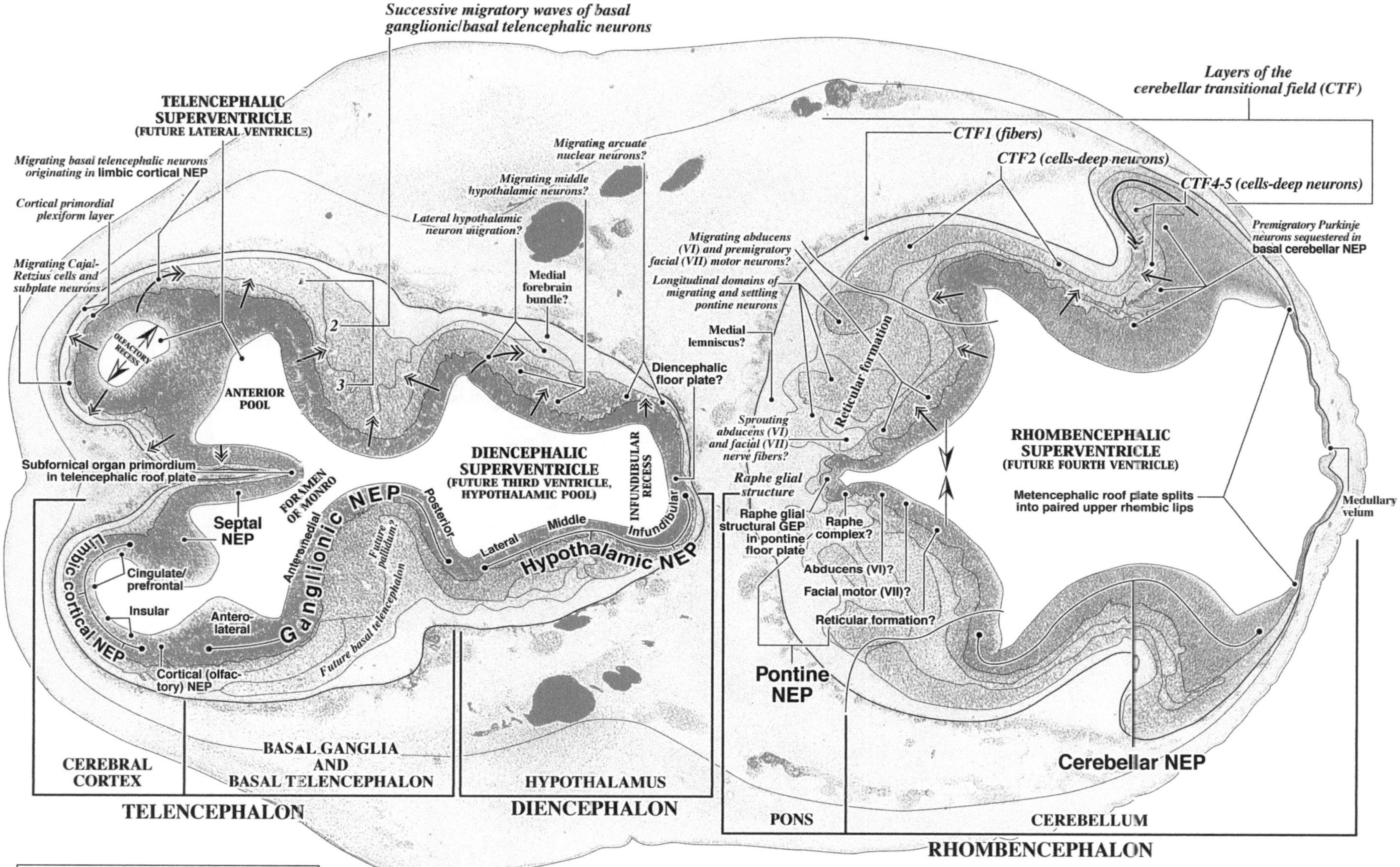

Layers of the cerebellar transitional field (CTF)

Successive migratory waves of basal ganglionic/basal telencephalic neurons

CTF-4-5 (cells-deep neurons)

Premigratory Purkinje neurons sequestered in basal cerebellar NEP

CTF2 (cells-deep neurons)

CTF1 (fibers)

TELENCEPHALIC SUPERVENTRICLE (FUTURE LATERAL VENTRICLE)

Migrating basal telencephalic neurons originating in limbic cortical NEP

Cortical primordial plexiform layer

Migrating Cajal-Retzius cells and subplate neurons?

OLFACTORY RECESS

ANTERIOR POOL

FORAMEN OF MONRO

Subfornical organ primordium in telencephalic roof plate

Septal NEP

Ganglionic NEP

Future basal telencephalon

Anteromedial

Anterolateral

Cingulate/ prefrontal

Insular

Cortical (olfactory) NEP

Limbic cortical NEP

Migrating arcuate nuclear neurons?

Migrating middle hypothalamic neurons?

Lateral hypothalamic neuron migration?

Medial forebrain bundle?

Diencephalic floor plate?

DIENCEPHALIC SUPERVENTRICLE (FUTURE THIRD VENTRICLE, HYPOTHALAMIC POOL)

INFUNDIBULAR RECESS

Hypothalamic NEP

Posterior

Lateral

Middle

Infundibular

Medial lemniscus?

Reticular formation

Migrating abducens (VI) and premigratory facial (VII) motor neurons?

Longitudinal domains of migrating and settling pontine neurons

Sprouting abducens (VI) and facial (VII) nerve fibers?

Raphe glial structure

Raphe glial structural GEP in pontine floor plate

Raphe complex?

Abducens (VI)?

Facial motor (VII)?

Reticular formation?

Pontine NEP

RHOMBENCEPHALIC SUPERVENTRICLE (FUTURE FOURTH VENTRICLE)

Metencephalic roof plate splits into paired upper rhembic lips

Medullary velum

Cerebellar NEP

CEREBELLUM

PONS

CEREBRAL CORTEX

BASAL GANGLIA AND BASAL TELENCEPHALON

TELENCEPHALON

HYPOTHALAMUS

DIENCEPHALON

RHOMBENCEPHALON

Arrows indicate the presumed *direction of neuron migration* from neuroepithelial sources.

Arrows indicate the regionally *expanding shoreline* of the superventricle with increase in stockbuilding NEP cells.

Arrows indicate the regionally *shrinking shoreline* of the superventricle as NEP cells are depleted while generating neurons.

ABBREVIATIONS:
GEP - **Glioepithelium**
NEP - **Neuroepithelium**

FONT KEY:
VENTRICULAR DIVISIONS - CAPITALS
Germinal zone - **Helvetica bold**
Transient structure - Times bold italic
Permanent structure - Times Roman or **Bold**

1 2 3

128

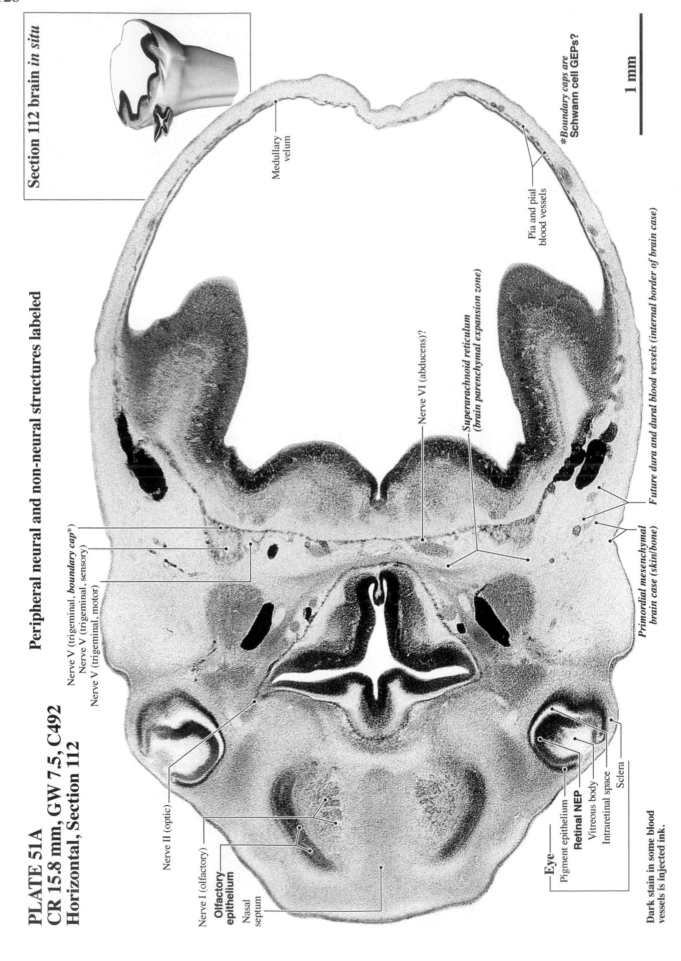

Section 112 brain *in situ*

PLATE 51A
CR 15.8 mm, GW 7.5, C492
Horizontal, Section 112

Peripheral neural and non-neural structures labeled

1 mm

Medullary velum

Pia and pial blood vessels

*Boundary caps are Schwann cell GEPs?

*Superarachnoid reticulum (brain parenchymal expansion zone)

Nerve VI (abducens)?

Nerve V (trigeminal, *boundary cap**)
Nerve V (trigeminal, sensory)
Nerve V (trigeminal, motor)

Future dura and dural blood vessels (internal border of brain case)

Primordial mesenchymal brain case (skin/bone)

Nerve II (optic)

Nerve I (olfactory)

Olfactory epithelium

Nasal septum

Eye
Pigment epithelium
Retinal NEP
Vitreous body
Intraretinal space
Sclera

Dark stain in some blood vessels is injected ink.

PLATE 51B

Central neural structures labeled

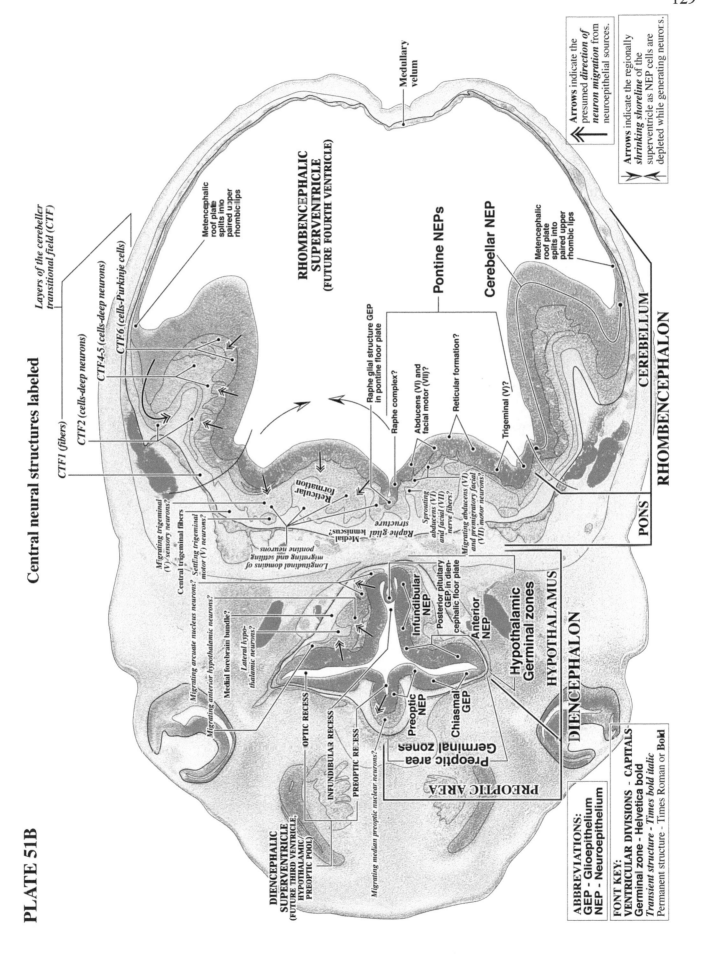

129

Medullary velum

Layers of the cerebellar transitional field (CTF)

CTF1 (fibers)

CTF2 (cells-deep neurons)

CTF4-5 (cells-deep neurons)

CTF6 (cells-Purkinje cells)

Metencephalic roof plate splits into paired upper rhombic lips

RHOMBENCEPHALIC SUPERVENTRICLE (FUTURE FOURTH VENTRICLE)

Raphe glial structure GEP in pontine floor plate

Raphe complex?

Pontine NEPs

Abducens (VI) and facial motor (VII)?

Reticular formation?

Cerebellar NEP

Trigeminal (V)?

Metencephalic roof plate splits into paired upper rhombic lips

CEREBELLUM

RHOMBENCEPHALON

PONS

Migrating trigeminal (V) sensory neurons?

Central trigeminal fibers

Settling trigeminal motor (V) neurons?

Reticular formation

Medial glial lemniscus?

Raphe glial structure

Sprouting abducens (VI) and facial (VII) nerve fibers?

Migrating abducens (VI) and premigratory facial (VII) motor neurons?

Longitudinal domains of migrating and settling pontine neurons

Migrating arcuate nucleus neurons?

Migrating anterior hypothalamic neurons?

Medial forebrain bundle?

Lateral hypothalamic neurons?

Posterior pituitary GEP in diencephalic floor plate

Infundibular NEP

Anterior NEP

Hypothalamic Germinal zones

HYPOTHALAMUS

DIENCEPHALON

OPTIC RECESS

INFUNDIBULAR RECESS

PREOPTIC RECESS

Migrating median preoptic nuclear neurons?

Preoptic NEP

Chiasmal GEP

Preoptic area Germinal zones

PREOPTIC AREA

DIENCEPHALIC SUPERVENTRICLE (FUTURE THIRD VENTRICLE, HYPOTHALAMIC/ PREOPTIC POOL)

ABBREVIATIONS:
GEP - Glioepithelium
NEP - Neuroepithelium

FONT KEY:
VENTRICULAR DIVISIONS - CAPITALS
Germinal zone - Helvetica bold
Transient structure - Times bold italic
Permanent structure - Times Roman or **Bold**

Arrows indicate the presumed *direction of neuron migration* from neuroepithelial sources.

Arrows indicate the regionally *shrinking shoreline* of the superventricle as NEP cells are depleted while generating neurons.

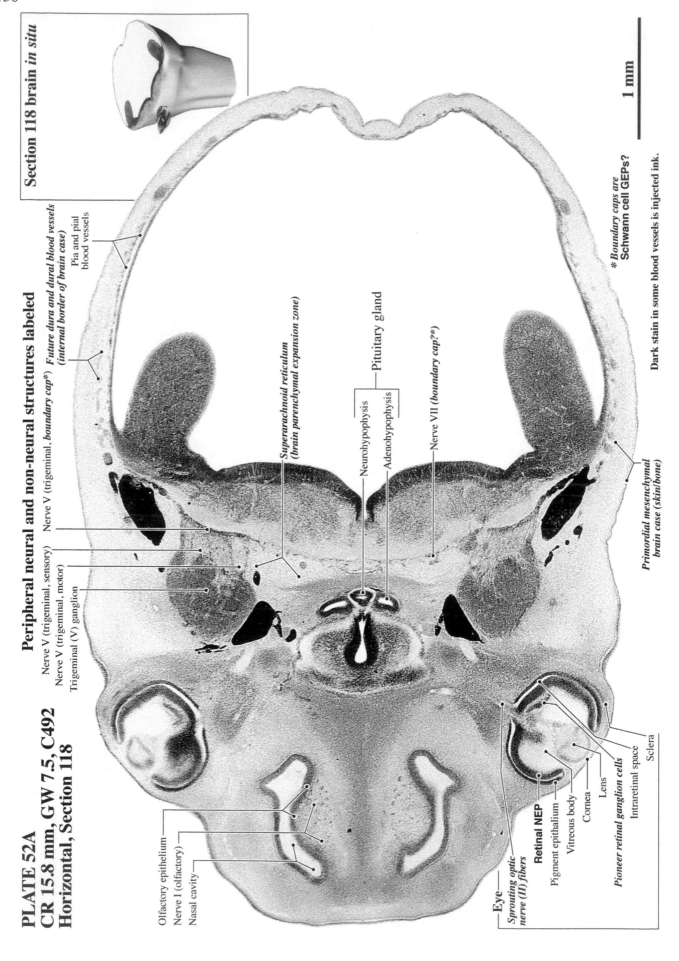

130

PLATE 52A
CR 15.8 mm, GW 7.5, C492
Horizontal, Section 118

Peripheral neural and non-neural structures labeled

Section 118 brain *in situ*

1 mm

*Future dura and dural blood vessels
(internal border of brain case)*

Nerve V (trigeminal, *boundary cap**)

Pia and pial
blood vessels

Nerve V (trigeminal, sensory)
Nerve V (trigeminal, motor)
Trigeminal (V) ganglion

*Superarachnoid reticulum
(brain parenchymal expansion zone)*

Neurohypophysis
Adenohypophysis

Pituitary gland

Nerve VII (*boundary cap?**)

*Primordial mesenchymal
brain case (skin/bone)*

** Boundary caps are
Schwann cell GEPs?*

Dark stain in some blood vessels is injected ink.

Olfactory epithelium
Nerve I (olfactory)
Nasal cavity

Eye
*Sprouting optic
nerve (II) fibers*

Retinal NEP
Pigment epithalium
Vitreous body
Cornea
Lens
Pioneer retinal ganglion cells
Intraretinal space
Sclera

PLATE 52B

Central neural structures labeled

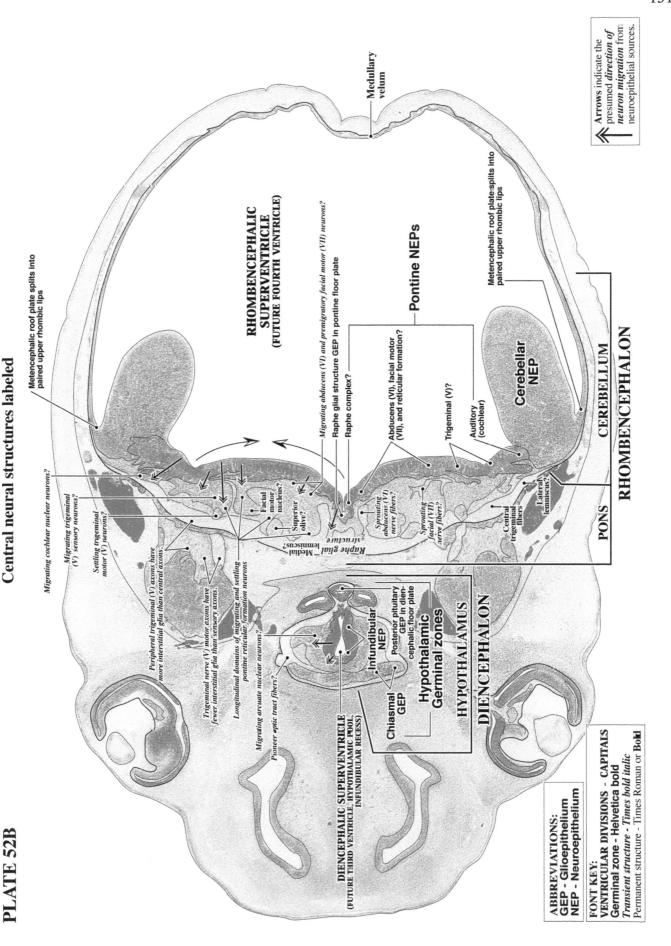

Arrows indicate the presumed *direction of neuron migration* from neuroepithelial sources.

Medullary velum

Metencephalic roof plate splits into paired upper rhombic lips

RHOMBENCEPHALIC SUPERVENTRICLE (FUTURE FOURTH VENTRICLE)

Migrating abducens (VI) and premigratory facial motor (VII) neurons?

Raphe glial structure GEP in pontine floor plate

Raphe complex?

Pontine NEPs

Abducens (VI), facial motor (VII), and reticular formation?

Trigeminal (V)?

Auditory (cochlear)

Metencephalic roof plate splits into paired upper rhombic lips

Cerebellar NEP

Migrating cochlear nuclear neurons?

Migrating trigeminal (V) sensory neurons?

Settling trigeminal motor (V) neurons?

Facial motor nuclei?

Superior olive?

Raphe glial structure

Medial lemniscus?

Sprouting abducens (VI) nerve fibers?

Sprouting facial (VII) nerve fibers?

Central trigeminal fibers

Lateral lemniscus?

CEREBELLUM

RHOMBENCEPHALON

PONS

Peripheral trigeminal (V) axons have more interstitial glia than central axons.

Trigeminal nerve (V) motor axons have fewer interstitial glia than sensory axons.

Longitudinal domains of migrating and settling pontine reticular formation neurons

Migrating arcuate nuclear neurons?

Pioneer optic tract fibers?

Infundibular NEP

Posterior pituitary GEP in diencephalic floor plate

Hypothalamic Germinal zones

HYPOTHALAMUS

Chiasmal GEP

DIENCEPHALON

DIENCEPHALIC SUPERVENTRICLE
(FUTURE THIRD VENTRICLE, HYPOTHALAMIC POOL, INFUNDIBULAR RECESS)

ABBREVIATIONS:
GEP - Glioepithelium
NEP - Neuroepithelium

FONT KEY:
VENTRICULAR DIVISIONS - CAPITALS
Germinal zone - Helvetica bold
Transient structure - Times bold italic
Permanent structure - Times Roman or **Bold**

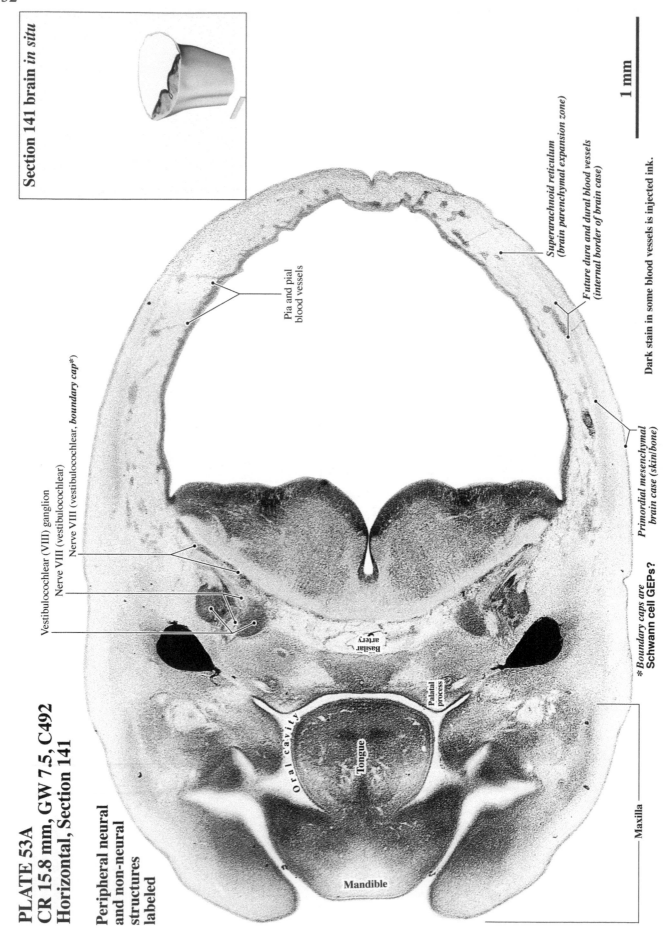

Section 141 brain *in situ*

PLATE 53A
CR 15.8 mm, GW 7.5, C492
Horizontal, Section 141

**Peripheral neural
and non-neural
structures
labeled**

Vestibulocochlear (VIII) ganglion

Nerve VIII (vestibulocochlear)

Nerve VIII (vestibulocochlear, *boundary cap**)

Pia and pial
blood vessels

*Superarachnoid reticulum
(brain parenchymal expansion zone)*

*Future dura and dural blood vessels
(internal border of brain case)*

Dark stain in some blood vessels is injected ink.

*Primordial mesenchymal
brain case (skin/bone)*

* Boundary caps are
Schwann cell GEPs?

1 mm

Basilar
artery

Palatal
process

O r a l   c a v i t y

Tongue

Mandible

Maxilla

PLATE 53B

Central neural structures labeled

133

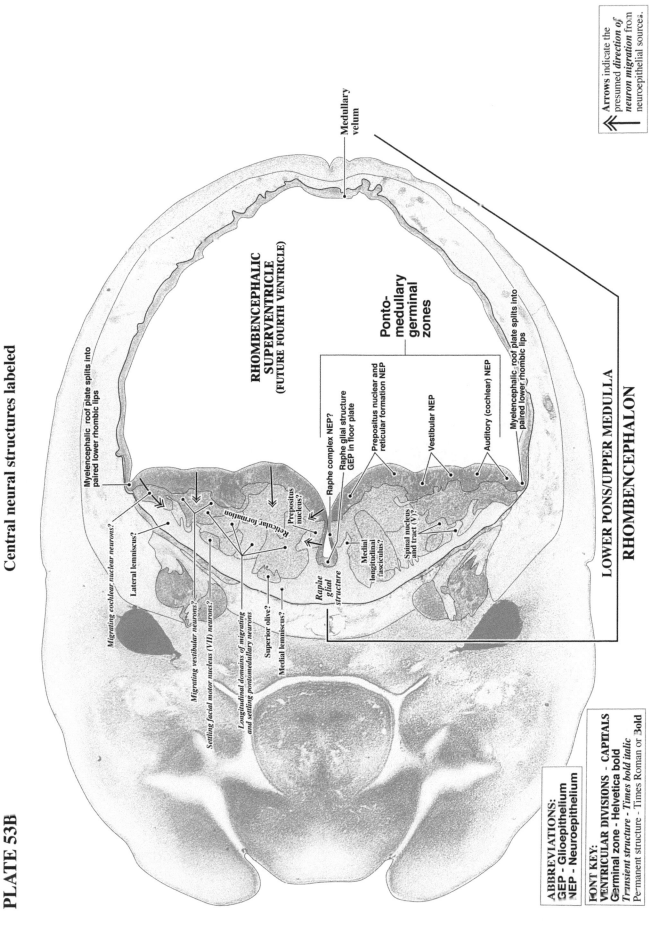

Arrows indicate the presumed *direction of neuron migration* from neuroepithelial sources.

Medullary velum

RHOMBENCEPHALIC SUPERVENTRICLE (FUTURE FOURTH VENTRICLE)

Myelencephalic roof plate splits into paired lower rhombic lips

Ponto-medullary germinal zones

Raphe complex NEP?

Raphe glial structure GEP in floor plate

Prepositus nuclear and reticular formation NEP

Vestibular NEP

Auditory (cochlear) NEP

Myelencephalic roof plate splits into paired lower rhombic lips

Migrating cochlear nuclear neurons?

Lateral lemniscus?

Migrating vestibular neurons?

Settling facial motor nucleus (VII) neurons?

Longitudinal domains of migrating and settling pontomedullary neurons

Superior olive?

Medial lemniscus?

Reticular formation

Prepositus nucleus?

Raphe glial structure

Medial longitudinal fasciculus?

Spinal nucleus and tract (V)?

LOWER PONS/UPPER MEDULLA

RHOMBENCEPHALON

ABBREVIATIONS:
GEP - Glioepithelium
NEP - Neuroepithelium

FONT KEY:
VENTRICULAR DIVISIONS - CAPITALS
Germinal zone - Helvetica bold
Transient structure - Times bold italic
Permanent structure - Times Roman or Bold

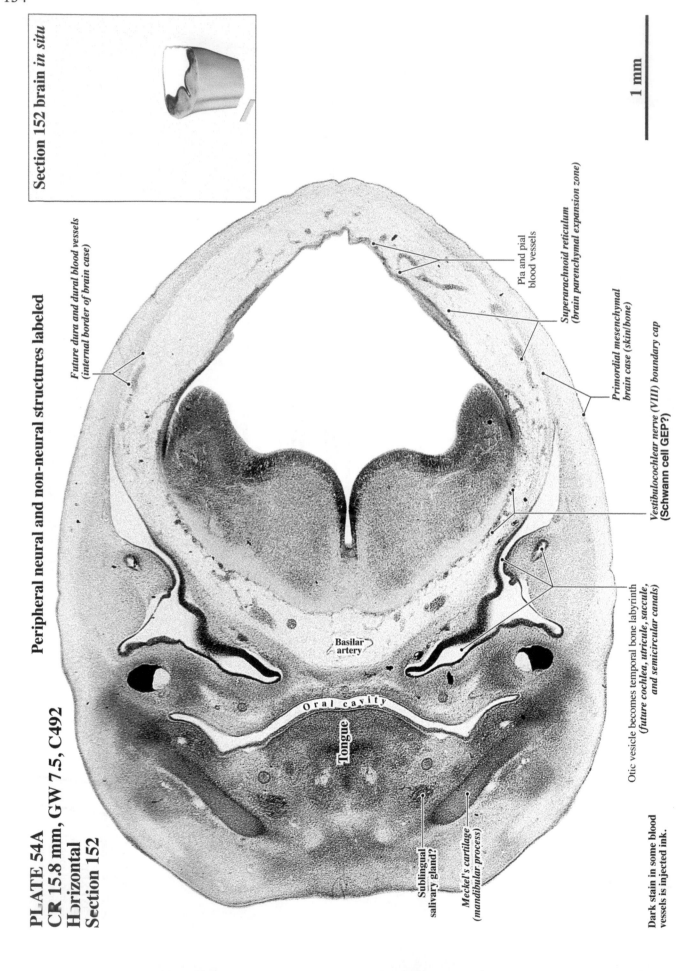

Section 152 brain *in situ*

1 mm

PLATE 54A
CR 15.8 mm, GW 7.5, C492
Horizontal
Section 152

Peripheral neural and non-neural structures labeled

Future dura and dural blood vessels
(internal border of brain case)

Pia and pial
blood vessels

Superarachnoid reticulum
(brain parenchymal expansion zone)

Primordial mesenchymal
brain case (skin/bone)

Vestibulocochlear nerve (VIII) boundary cap
(Schwann cell GEP?)

Basilar
artery

Otic vesicle becomes temporal bone labyrinth
(future cochlea, utricule, saccule,
and semicircular canals)

O r a l   c a v i t y

Tongue

Sublingual
salivary gland?

Meckel's cartilage
(mandibular process)

Dark stain in some blood
vessels is injected ink.

PLATE 54B

Central neural structures labeled

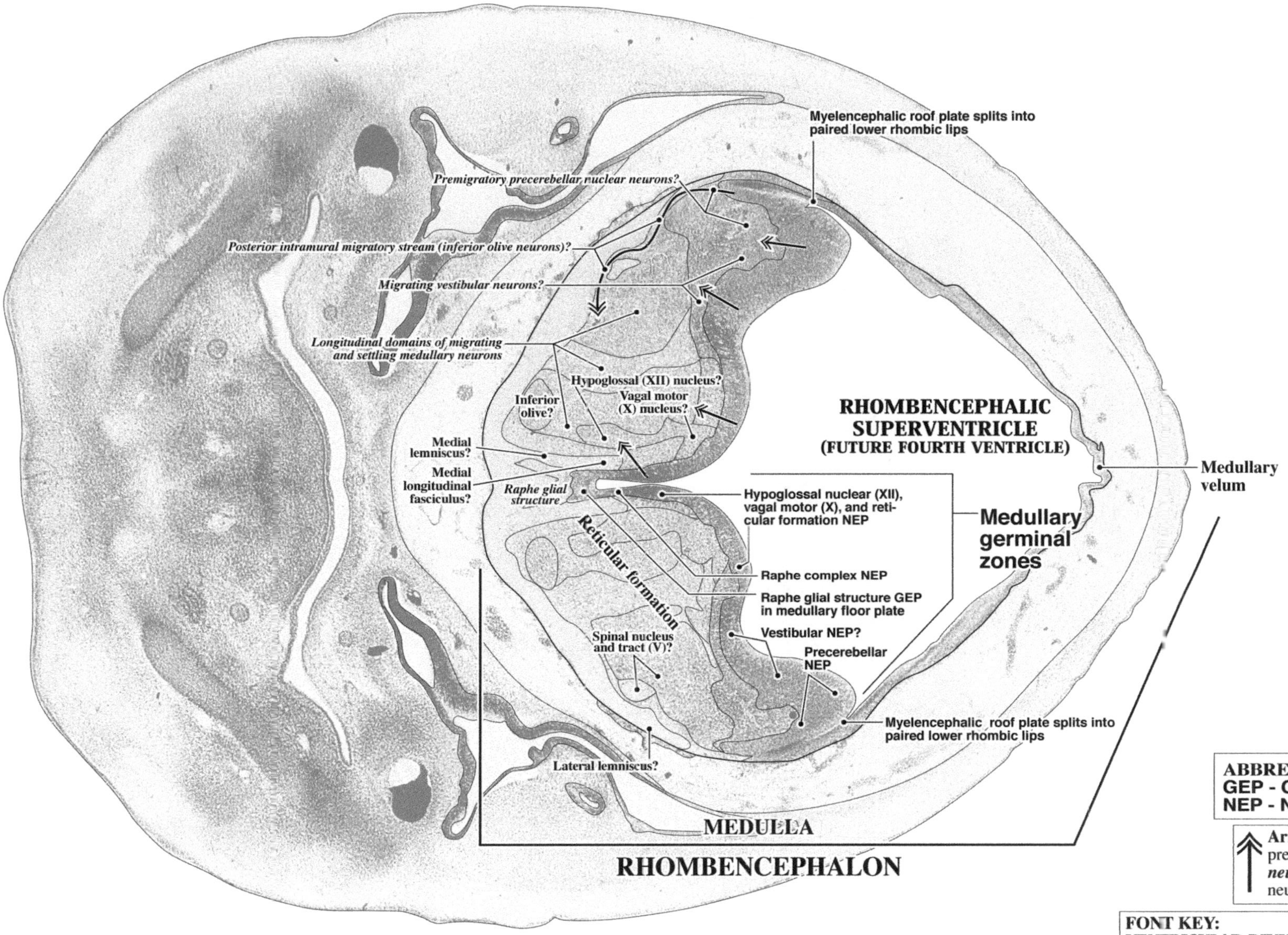

ABBREVIATIONS:
GEP - Glioepithelium
NEP - Neuroepithelium

Arrows indicate the presumed *direction of neuron migration* from neuroepithelial sources.

FONT KEY:
VENTRICULAR DIVISIONS - CAPITALS
Germinal zone - Helvetica bold
Transient structure - Times bold italic
Permanent structure - Times Roman or **Bold**

Medullary velum

Myelencephalic roof plate splits into paired lower rhombic lips

RHOMBENCEPHALIC SUPERVENTRICLE (FUTURE FOURTH VENTRICLE)

Medullary germinal zones

Hypoglossal nuclear (XII), vagal motor (X), and reticular formation NEP

Raphe complex NEP

Raphe glial structure GEP in medullary floor plate

Vestibular NEP?

Precerebellar NEP

Myelencephalic roof plate splits into paired lower rhombic lips

MEDULLA

RHOMBENCEPHALON

Premigratory precerebellar nuclear neurons?

Posterior intramural migratory stream (inferior olive neurons)?

Migrating vestibular neurons?

Longitudinal domains of migrating and settling medullary neurons

Hypoglossal (XII) nucleus?

Vagal motor (X) nucleus?

Inferior olive?

Raphe glial structure

Medial lemniscus?

Medial longitudinal fasciculus?

Reticular formation

Spinal nucleus and tract (V)?

Lateral lemniscus?

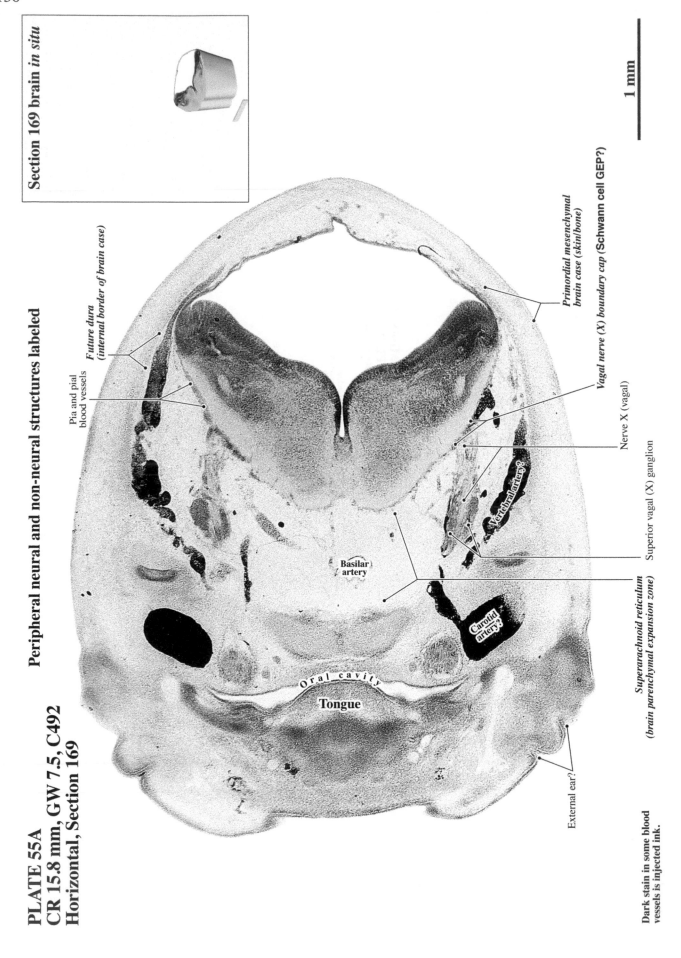

Section 169 brain *in situ*

PLATE 55A
CR 15.8 mm, GW 7.5, C492
Horizontal, Section 169

Peripheral neural and non-neural structures labeled

Future dura (internal border of brain case)

Pia and pial blood vessels

Primordial mesenchymal brain case (skin/bone)

Vagal nerve (X) boundary cap (Schwann cell GEP?)

Nerve X (vagal)

Superior vagal (X) ganglion

Vertebral artery?

Basilar artery

Carotid artery?

Superarachnoid reticulum (brain parenchymal expansion zone)

Oral cavity

Tongue

External ear?

1 mm

Dark stain in some blood vessels is injected ink.

PLATE 55B

Central neural structures labeled

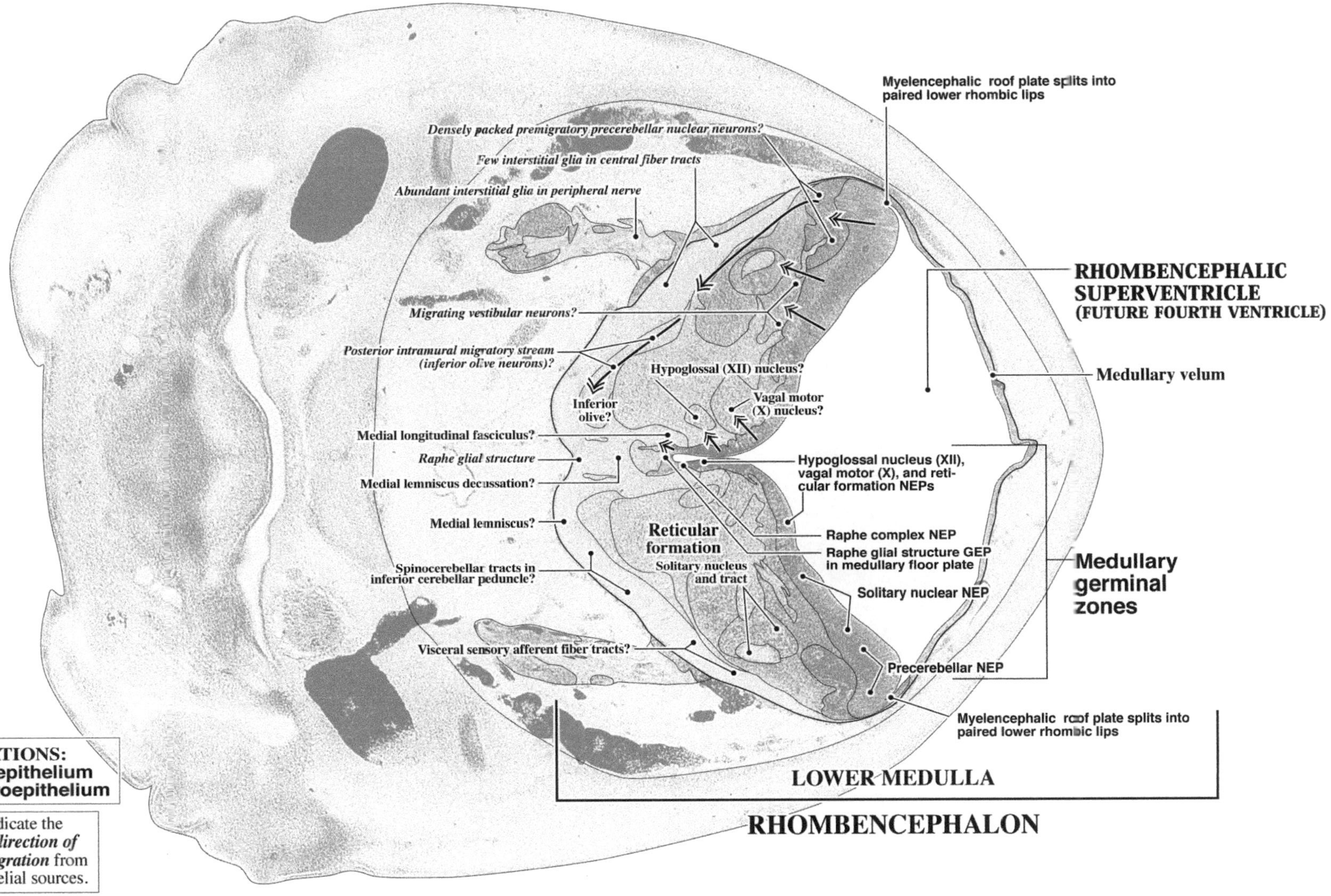

RHOMBENCEPHALIC SUPERVENTRICLE (FUTURE FOURTH VENTRICLE)

Medullary velum

Myelencephalic roof plate splits into paired lower rhombic lips

Medullary germinal zones

Myelencephalic roof plate splits into paired lower rhombic lips

LOWER MEDULLA

RHOMBENCEPHALON

Hypoglossal nucleus (XII), vagal motor (X), and reticular formation NEPs

Raphe complex NEP

Raphe glial structure GEP in medullary floor plate

Solitary nuclear NEP

Precerebellar NEP

Vagal motor (X) nucleus?

Hypoglossal (XII) nucleus?

Reticular formation

Solitary nucleus and tract

Densely packed premigratory precerebellar nuclear neurons?

Few interstitial glia in central fiber tracts

Abundant interstitial glia in peripheral nerve

Migrating vestibular neurons?

Posterior intramural migratory stream (inferior olive neurons)?

Inferior olive?

Medial longitudinal fasciculus?

Raphe glial structure?

Medial lemniscus decussation?

Medial lemniscus?

Spinocerebellar tracts in inferior cerebellar peduncle?

Visceral sensory afferent fiber tracts?

ABBREVIATIONS:
GEP - Glioepithelium
NEP - Neuroepithelium

Arrows indicate the presumed *direction of neuron migration* from neuroepithelial sources.

FONT KEY:
VENTRICULAR DIVISIONS - CAPITALS
Germinal zone - Helvetica bold
Transient structure - Times bold italic
Permanent structure - Times Roman or **Bold**

138

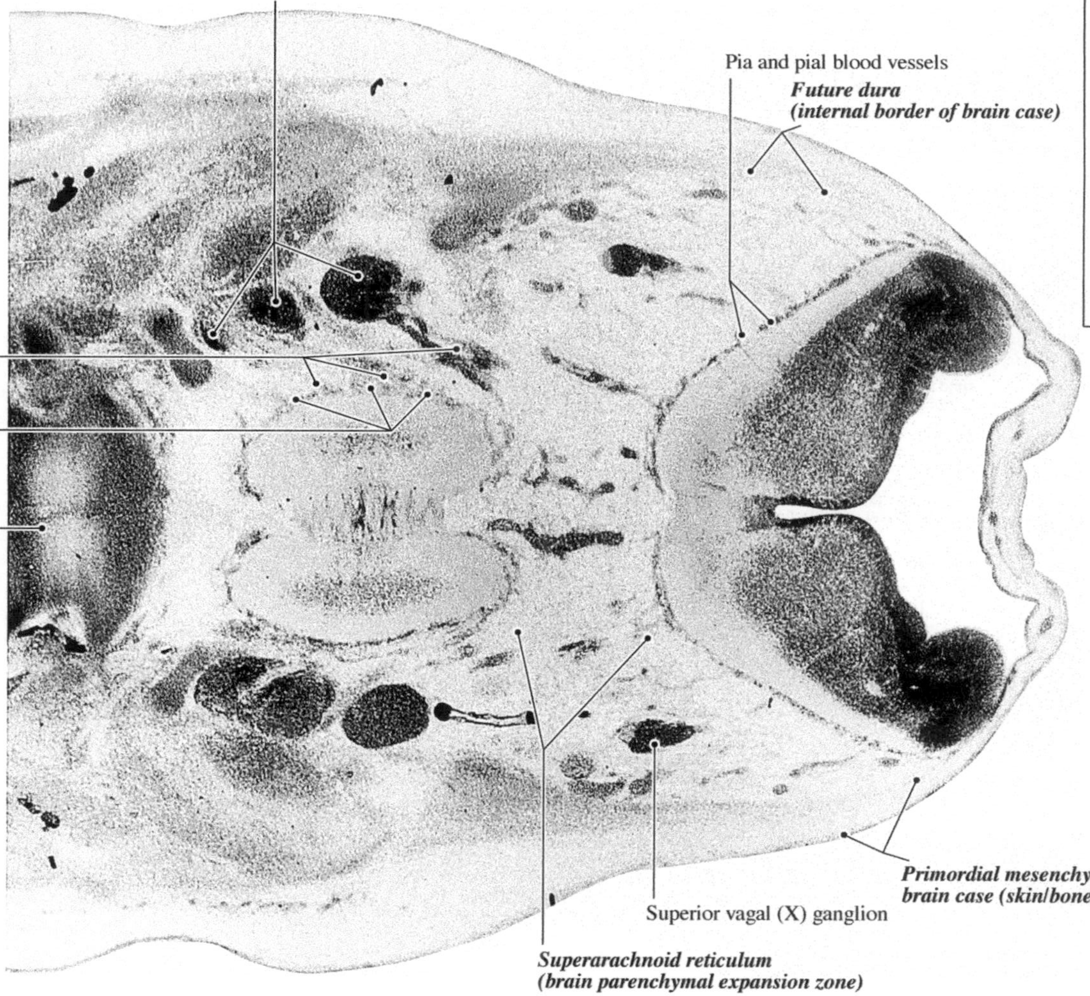

Section 205 brain *in situ*

1 mm

Peripheral neural and non-neural structures labeled

PLATE 56A
CR 15.8 mm, GW 7.5, C492
Horizontal, Section 205

Pia and pial blood vessels

Future dura
(internal border of brain case)

Primordial mesenchymal
brain case (skin/bone)

Superior vagal (X) ganglion

Superarachnoid reticulum
(brain parenchymal expansion zone)

Dorsal root ganglia

Spinal nerve (dorsal roots)

Dorsal root boundary caps

Vertebral column

Dark stain in some blood vessels is injected ink.

PLATE 56B

Central neural structures labeled

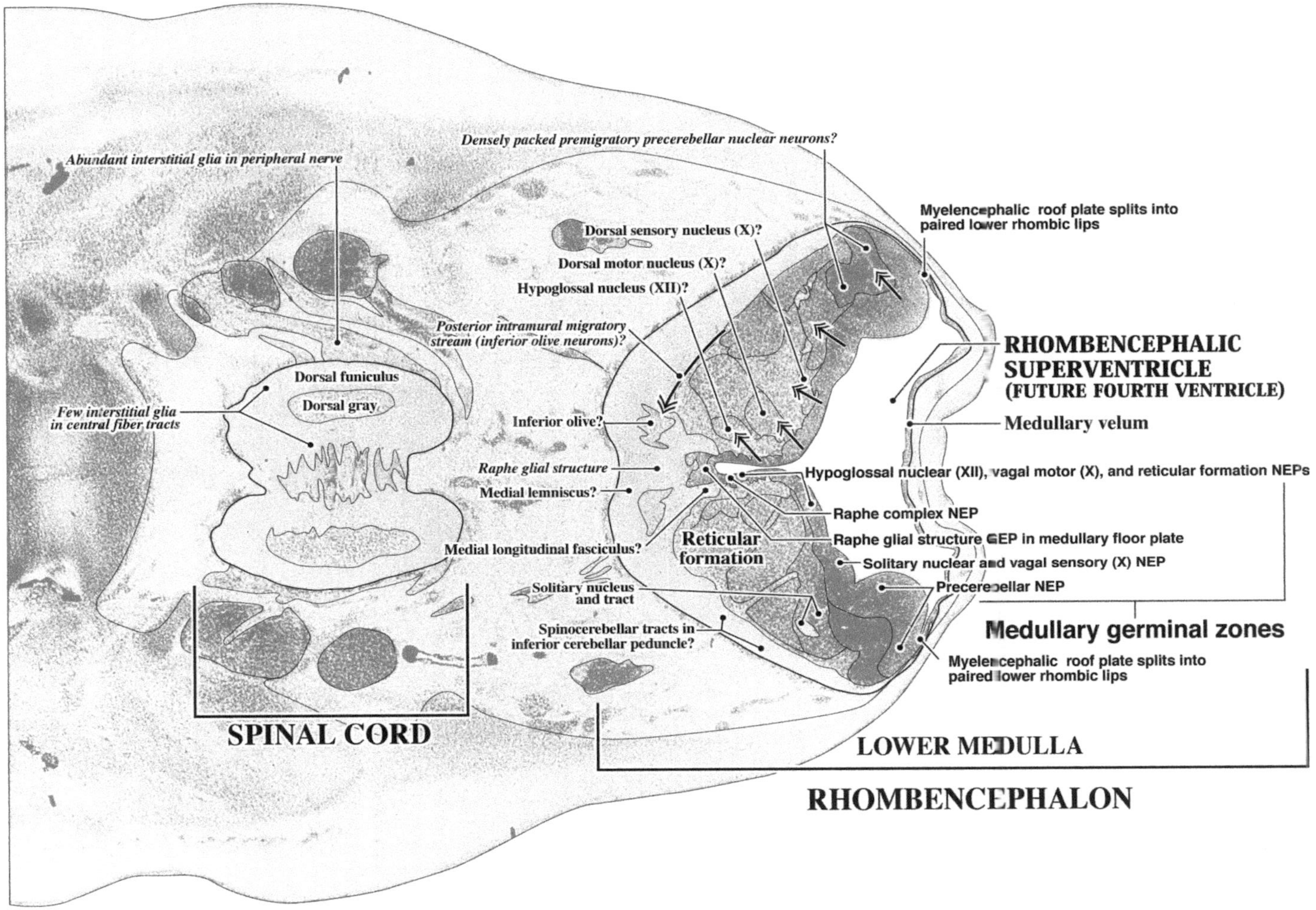

RHOMBENCEPHALIC SUPERVENTRICLE (FUTURE FOURTH VENTRICLE)

Myelencephalic roof plate splits into paired lower rhombic lips

Medullary velum

Hypoglossal nuclear (XII), vagal motor (X), and reticular formation NEPs

Hypoglossal nuclear NEP

Raphe complex NEP

Raphe glial structure GEP in medullary floor plate

Solitary nuclear and vagal sensory (X) NEP

Precerebellar NEP

Medullary germinal zones

Myelencephalic roof plate splits into paired lower rhombic lips

LOWER MEDULLA

RHOMBENCEPHALON

Densely packed premigratory precerebellar nuclear neurons?

Dorsal sensory nucleus (X)?

Dorsal motor nucleus (X)?

Hypoglossal nucleus (XII)?

Posterior intramural migratory stream (inferior olive neurons)?

Inferior olive?

Raphe glial structure?

Medial lemniscus?

Medial longitudinal fasciculus?

Solitary nucleus and tract

Spinocerebellar tracts in inferior cerebellar peduncle?

Reticular formation

Abundant interstitial glia in peripheral nerve

Dorsal funiculus

Dorsal gray

Few interstitial glia in central fiber tracts

SPINAL CORD

ABBREVIATIONS:
GEP - Glioepithelium
NEP - Neuroepithelium

Arrows indicate the presumed *direction of neuron migration* from neuroepithelial sources.

FONT KEY:
VENTRICULAR DIVISIONS - CAPITALS
Germinal zone - Helvetica bold
Transient structure - Times bold italic
Permanent structure - Times Roman or Bold

PART V: C1390
CR 18.0 mm (GW 7.8)
Sagittal

Carnegie Collection specimen #1390 (designated here as C1390) was collected in 1916 from a tubal pregnancy. The crown-rump length (CR) is 18 mm estimated to be at gestational week (GW) 7.8. C1390 was fixed in formalin, embedded in paraffin, and was cut in 20-μm sagittal sections that were stained with aluminum cochineal. Various orientations of the computer-aided 3-D reconstruction of M2155's brain is used to show the gross external features of a GW7.8 brain (**Figure 15**). Like most sagittally cut specimens, C1390's sections are not parallel to the midline; **Figure 15** shows the approximate rotations in top (**B**) and back views (**C**). We photographed 62 sections at low magnification from the left to right sides of the brain. Seven of the sections, mainly from the left side of the brain, are illustrated in **Plates 57AB to 63AB**. Each illustrated section shows the brain with all surrounding tissues. Labels in **A Plates** (normal-contrast images) indicate the approximate location of the midline and identify non-neural structures, peripheral neural structures, and brain ventricular divisions; labels in **B Plates** (low-contrast images) identify central neural structures. **Plates 64AB** show three high-magnification sections in the region of the facial nerve genu. The brain of C1390 is at a similar stage of development as M2155, the next GW7.8 specimen.

Throughout the telencephalon, the cortical neuroepithelium (NEP) surrounds the enlarging roof of the telencephalic superventricle. All telencephalic NEPs are either just entering or are well into their neurogenetic stage, as more basal telencephalic neurons are generated and have migrated into an expanding parenchyma. Very few pioneer Cajal-Retzius neurons have migrated into the primordial plexiform layer adjacent to the cortical NEP, but many pyramidal cells destined to settle in layers VI and V are sequestered in the NEP itself, waiting to migrate later. The sagittal plane is ideal to show the slight evagination of the olfactory NEP in exactly the same region that is contacted by olfactory nerve fibers.

The diencephalic NEP surrounds a large superventricle. It is shrinking in the hypothalamic and subthalamic areas where stem cells are depleted as they generate neurons. Many migrating and settling young neurons are in the parenchyma surrounding these NEPs. In contrast, the superventricular shoreline is expanding in thalamic areas as the thalamic NEP continues to add more postmitotic neurons that sequester inside the thick NEP. The few neurons outside the thalamic NEP are best seen in sections grazing the dorsolateral part of the diencephalon.

The roof (tectum and pretectum) of the mesencephalon contains neurogenetic NEPs adjacent to a very thin layer of pioneer migrating neurons. However, bundles of fibers in the posterior commissure are very distinct, with spike-like arrays of cells extending between these bundles. Possibly premigratory neurons produce axons while they are sequestered in the pretectal NEP. The tegmental and isthmal NEPs are much thinner because most of their neuronal progeny has migrated out. These cells accumulate as inner dense clumps and outer sparse arrays are interspersed among the thick subpial fiber band in the tegmental and isthmal parenchyma.

Both the pons and medulla have NEPs that are shrinking as stem cells unload their neuronal and glial progeny into an expanding parenchyma. The longitudinal arrays at the pontine flexure are easy to see in the sagittal plane. The genu of the facial motor nerve forms fascicles adjacent to the neuroepithelium in medial sections; these fascicles are adjacent to the pial surface in lateral sections (**Plate 64**). What is presumed to be the solitary tract is the most prominent internal fiber tract in the medulla. Both the pons and medulla have a thick subpial fibrous layer. Lateral sections show peripheral sensory nerves contacting the brain. No doubt, many of the superficial fibers are the afferent axons of these ganglia along with ascending fiber tracts from the spinal cord. All of the peripheral nerves (most clearly shown in the trigeminal nerve) have dense glia (Schwann cells), while central fiber tracts are clear. Thus, peripheral gliogenesis precedes the generation of oligodendrocytes in central fiber tracts.

The exceptionally thick cerebellar NEP, in comparison with the thin adjacent pontine NEP, is most easily seen in lateral sections where it sharply juts into the rhombencephalic superventricle. In spite of its thickness, the cerebellar NEP has already generated most of the deep nuclear neurons and is nearly finished generating Purkinje cells that are sojourning in the basal NEP. Most deep nuclear neurons are sojourning in the superficial layers of the cerebellar transitional field, while the Purkinje cells pile up below them.

EXTERNAL FEATURES OF THE GW7.8 BRAIN

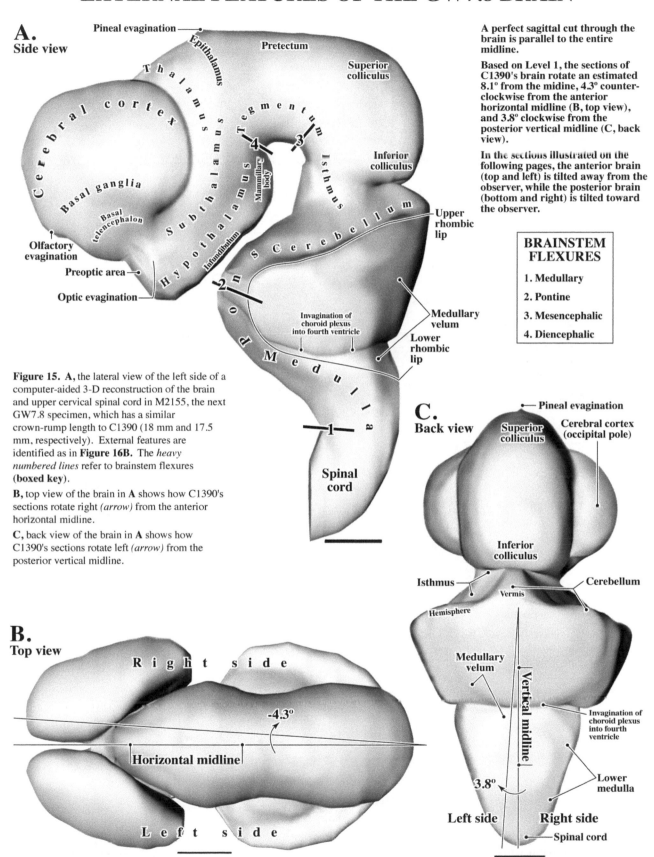

A. Side view

Pineal evagination

Epithalamus

Pretectum

Superior colliculus

Thalamus

Cerebral cortex

Tegmentum

Basal ganglia

Subthalamus

Basal telencephalon

Inferior colliculus

Isthmus

Mammillary body

Olfactory evagination

Preoptic area

Hypothalamus

Infundibulum

Optic evagination

Pons

Cerebellum

Upper rhombic lip

Medulla

Invagination of choroid plexus into fourth ventricle

Medullary velum

Lower rhombic lip

Spinal cord

A perfect sagittal cut through the brain is parallel to the entire midline.

Based on Level 1, the sections of C1390's brain rotate an estimated 8.1° from the midine, 4.3° counterclockwise from the anterior horizontal midline (B, top view), and 3.8° clockwise from the posterior vertical midline (C, back view).

In the sections illustrated on the following pages, the anterior brain (top and left) is tilted away from the observer, while the posterior brain (bottom and right) is tilted toward the observer.

BRAINSTEM FLEXURES

1. Medullary
2. Pontine
3. Mesencephalic
4. Diencephalic

Figure 15. A, the lateral view of the left side of a computer-aided 3-D reconstruction of the brain and upper cervical spinal cord in M2155, the next GW7.8 specimen, which has a similar crown-rump length to C1390 (18 mm and 17.5 mm, respectively). External features are identified as in **Figure 16B.** The *heavy numbered lines* refer to brainstem flexures **(boxed key)**.

B, top view of the brain in **A** shows how C1390's sections rotate right *(arrow)* from the anterior horizontal midline.

C, back view of the brain in **A** shows how C1390's sections rotate left *(arrow)* from the posterior vertical midline.

C. Back view

Pineal evagination

Cerebral cortex (occipital pole)

Superior colliculus

Inferior colliculus

Isthmus

Vermis

Cerebellum

Hemisphere

Medullary velum

Vertical midline

Invagination of choroid plexus into fourth ventricle

3.8°

Lower medulla

Left side

Right side

Spinal cord

B. Top view

Right side

-4.3°

Horizontal midline

Left side

Scale bars = 1 mm

PLATE 57A

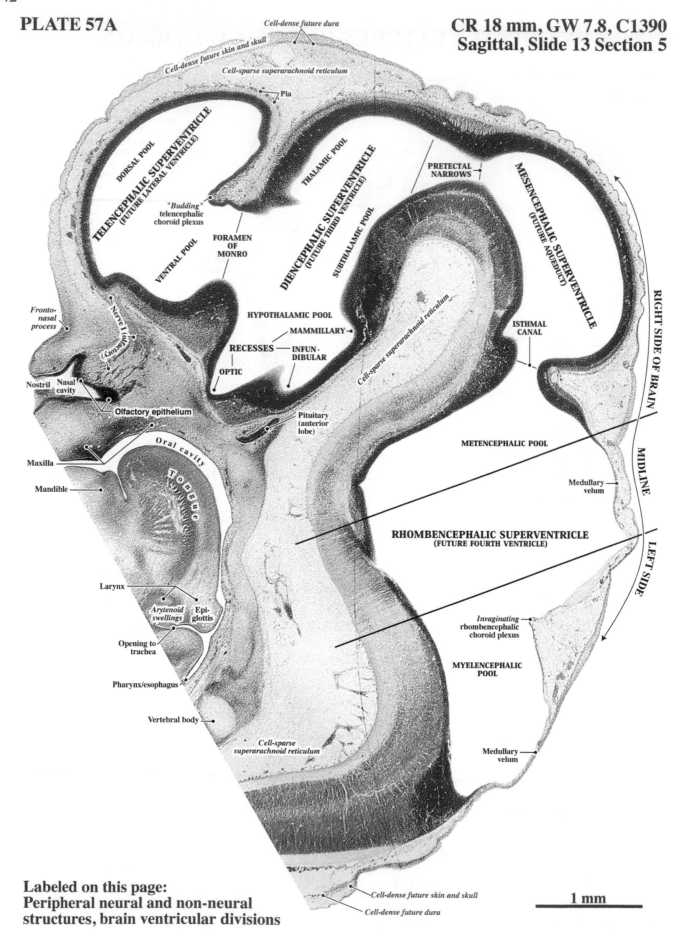

Cell-dense future dura

Cell-dense future skin and skull

Cell-sparse superarachnoid reticulum

Pia

DORSAL POOL

TELENCEPHALIC SUPERVENTRICLE
(FUTURE LATERAL VENTRICLE)

THALAMIC POOL

DIENCEPHALIC SUPERVENTRICLE
(FUTURE THIRD VENTRICLE)

PRETECTAL
NARROWS

MESENCEPHALIC SUPERVENTRICLE
(FUTURE AQUEDUCT)

"Budding"
telencephalic
choroid plexus

FORAMEN
OF
MONRO

SUBTHALAMIC POOL

VENTRAL POOL

*Fronto-
nasal
process*

Nerve I (olfactory)

HYPOTHALAMIC POOL

MAMMILLARY

RECESSES — INFUN-
DIBULAR

ISTHMAL
CANAL

RIGHT SIDE OF BRAIN

Cell-sparse superarachnoid reticulum

OPTIC

Nostril Nasal
cavity

Olfactory epithelium

Pituitary
(anterior
lobe)

METENCEPHALIC POOL

MIDLINE

Oral cavity

Maxilla

Mandible

Tongue

Medullary
velum

RHOMBENCEPHALIC SUPERVENTRICLE
(FUTURE FOURTH VENTRICLE)

LEFT SIDE

Larynx

*Arytenoid
swellings* Epi-
glottis

Opening to
trachea

Invaginating
rhombencephalic
choroid plexus

Pharynx/esophagus

MYELENCEPHALIC
POOL

Vertebral body

*Cell-sparse
superarachnoid reticulum*

Medullary
velum

**Labeled on this page:
Peripheral neural and non-neural
structures, brain ventricular divisions**

Cell-dense future skin and skull

Cell-dense future dura

1 mm

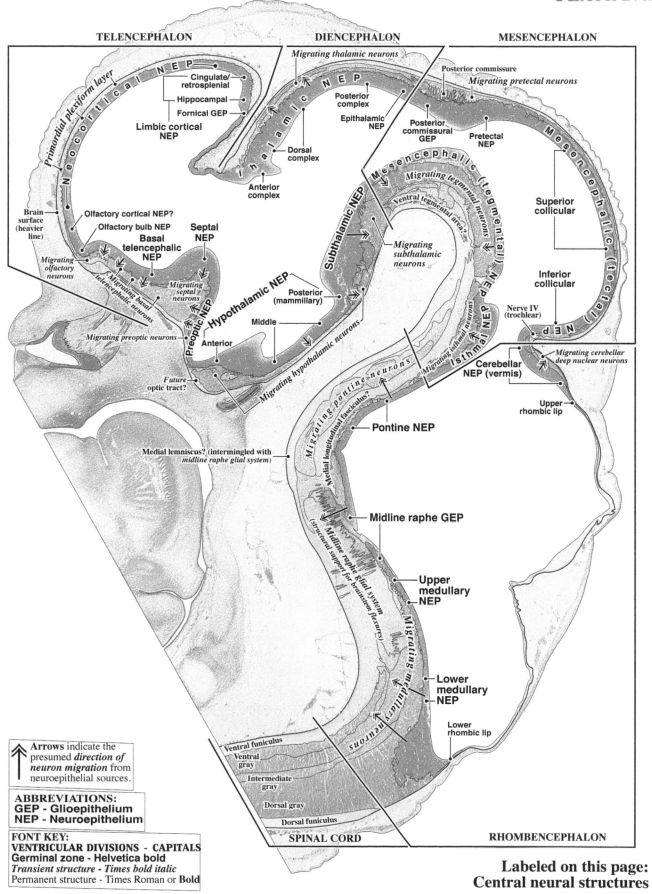

TELENCEPHALON

DIENCEPHALON

MESENCEPHALON

Primordial plexiform layer

Neocortical N E P

Cingulate/
retrosplenial

Hippocampal

Fornical GEP

Limbic cortical
NEP

Migrating thalamic neurons

Thalamic N E P

Posterior
complex

Epithalamic
NEP

Dorsal
complex

Anterior
complex

Posterior commissure

Migrating pretectal neurons

Posterior
commissural
GEP

*Posterior
commissural
GEP*

Pretectal
NEP

Mesencephalic (tectal) N E P

Brain
surface
(heavier
line)

Olfactory cortical NEP?

Olfactory bulb NEP

Basal
telencephalic
NEP

Septal
NEP

*Migrating
olfactory
neurons*

*Migrating basal
telencephalic neurons*

*Migrating
septal
neurons*

Preoptic NEP

Migrating preoptic neurons

Anterior

*Future
optic tract?*

Hypothalamic NEP

Posterior
(mammillary)

Middle

Subthalamic NEP

Mesencephalic (tegmental) NEP

Migrating tegmental neurons

Ventral tegmental area?

*Migrating
subthalamic
neurons*

Superior
collicular

Inferior
collicular

Nerve IV
(trochlear)

Mesencephalic (tectal) NEP

Migrating isthmal neurons

Isthmal NEP

Cerebellar
NEP (vermis)

*Migrating cerebellar
deep nuclear neurons*

Upper
rhombic lip

Migrating hypothalamic neurons

Migrating pontine neurons

Medial longitudinal fasciculus?

Pontine NEP

Medial lemniscus? (intermingled with
midline raphe glial system)

*Midline raphe glial system
(structural support for brainstem flexures)*

Midline raphe GEP

**Upper
medullary
NEP**

Migrating medullary neurons

**Lower
medullary
NEP**

Lower
rhombic lip

Ventral funiculus

Ventral
gray

Intermediate
gray

Dorsal gray

Dorsal funiculus

SPINAL CORD

RHOMBENCEPHALON

↑ **Arrows** indicate the
presumed *direction of
neuron migration* from
neuroepithelial sources.

ABBREVIATIONS:
GEP - Glioepithelium
NEP - Neuroepithelium

FONT KEY:
VENTRICULAR DIVISIONS - CAPITALS
Germinal zone - Helvetica bold
Transient structure - Times bold italic
Permanent structure - Times Roman or **Bold**

Labeled on this page:
Central neural structures

PLATE 58A

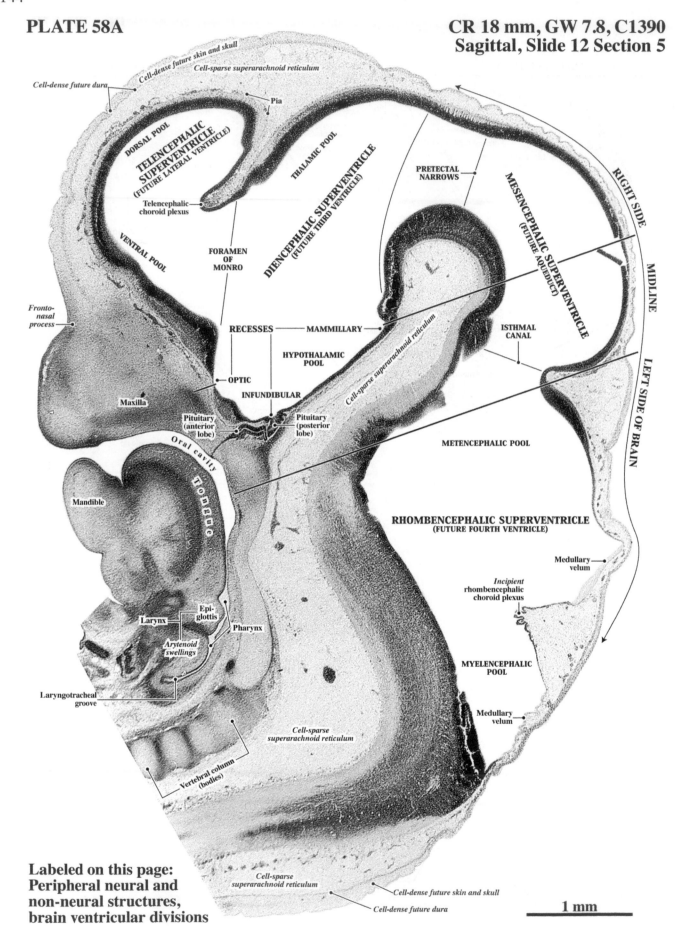

Cell-dense future dura

Cell-dense future skin and skull

Cell-sparse superarachnoid reticulum

Pia

DORSAL POOL

TELENCEPHALIC SUPERVENTRICLE (FUTURE LATERAL VENTRICLE)

THALAMIC POOL

DIENCEPHALIC SUPERVENTRICLE (FUTURE THIRD VENTRICLE)

PRETECTAL NARROWS

MESENCEPHALIC SUPERVENTRICLE (FUTURE AQUEDUCT)

RIGHT SIDE

MIDLINE

Telencephalic choroid plexus

VENTRAL POOL

FORAMEN OF MONRO

Fronto-nasal process

RECESSES — MAMMILLARY

HYPOTHALAMIC POOL

ISTHMAL CANAL

LEFT SIDE OF BRAIN

OPTIC

INFUNDIBULAR

Cell-sparse superarachnoid reticulum

Maxilla

Pituitary (anterior lobe)

Pituitary (posterior lobe)

METENCEPHALIC POOL

Oral cavity

Tongue

Mandible

RHOMBENCEPHALIC SUPERVENTRICLE (FUTURE FOURTH VENTRICLE)

Medullary velum

Incipient rhombencephalic choroid plexus

Epi-glottis

Larynx

Pharynx

Arytenoid swellings

Laryngotracheal groove

MYELENCEPHALIC POOL

Medullary velum

Cell-sparse superarachnoid reticulum

Vertebral column (bodies)

Cell-sparse superarachnoid reticulum

Cell-dense future skin and skull

Cell-dense future dura

**Labeled on this page:
Peripheral neural and
non-neural structures,
brain ventricular divisions**

1 mm

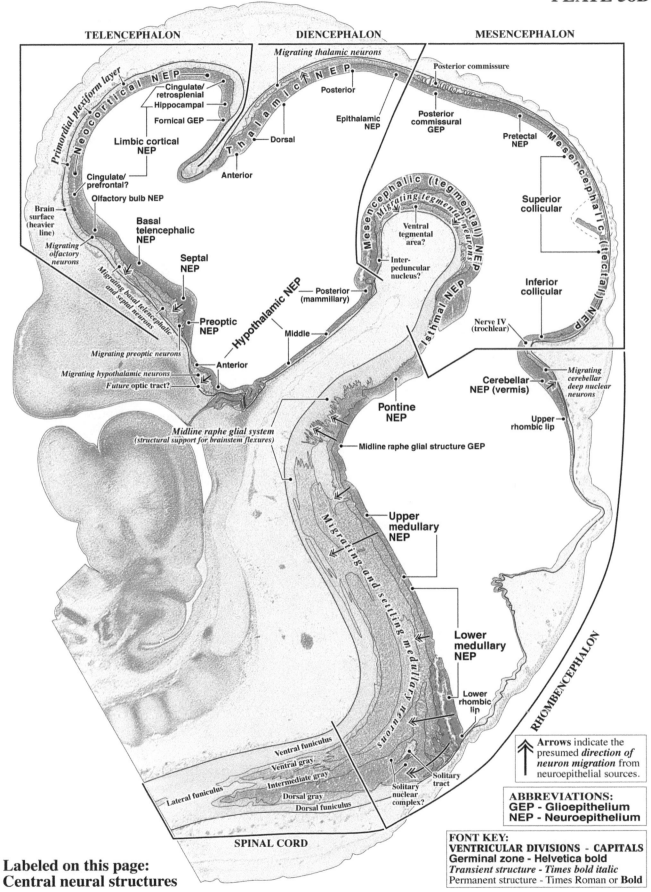

TELENCEPHALON

DIENCEPHALON

MESENCEPHALON

Primordial plexiform layer

Neocortical NEP

Cingulate/retrosplenial

Hippocampal

Fornical GEP

Limbic cortical NEP

Cingulate/prefrontal?

Olfactory bulb NEP

Brain surface (heavier line)

Migrating olfactory neurons

Basal telencephalic NEP

Migrating basal telencephalic and septal neurons

Septal NEP

Preoptic NEP

Migrating preoptic neurons

Hypothalamic NEP

Thalamic NEP

Migrating thalamic neurons

Posterior

Dorsal

Anterior

Epithalamic NEP

Posterior commissure

Posterior commissural GEP

Pretectal NEP

Mesencephalic (tectal) NEP

Superior collicular

Inferior collicular

Mesencephalic (tegmental) NEP

Migrating tegmental neurons

Ventral tegmental area?

Interpeduncular nucleus?

Posterior (mammillary)

Middle

Anterior

Migrating hypothalamic neurons

Future optic tract?

Midline raphe glial system (structural support for brainstem flexures)

Isthmal NEP

Nerve IV (trochlear)

Cerebellar NEP (vermis)

Migrating cerebellar deep nuclear neurons

Upper rhombic lip

Pontine NEP

Midline raphe glial structure GEP

Migrating and settling medullary neurons

Upper medullary NEP

Lower medullary NEP

Lower rhombic lip

RHOMBENCEPHALON

Ventral funiculus

Ventral gray

Intermediate gray

Lateral funiculus

Dorsal gray

Dorsal funiculus

Solitary tract

Solitary nuclear complex?

SPINAL CORD

Arrows indicate the presumed *direction of neuron migration* from neuroepithelial sources.

ABBREVIATIONS:
GEP - **Glioepithelium**
NEP - **Neuroepithelium**

FONT KEY:
VENTRICULAR DIVISIONS - CAPITALS
Germinal zone - Helvetica bold
Transient structure - Times bold italic
Permanent structure - Times Roman or **Bold**

Labeled on this page:
Central neural structures

146

CR 18 mm, GW 7.8, C1390
Sagittal, Slide 11 Section 5

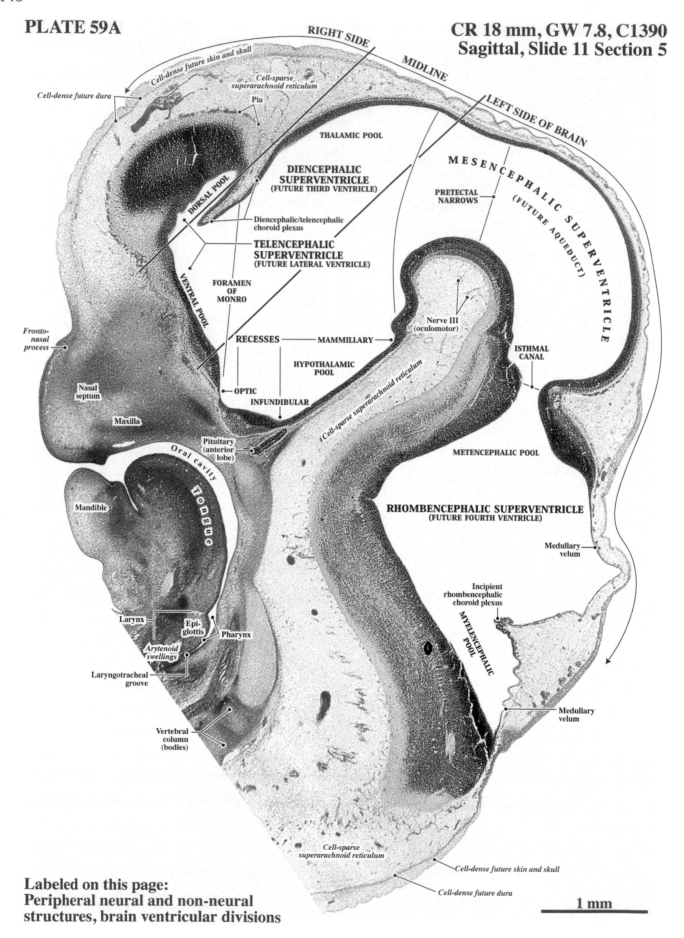

Cell-dense future skin and skull

Cell-dense future dura

Cell-sparse, superarachnoid reticulum

Pia

RIGHT SIDE

MIDLINE

LEFT SIDE OF BRAIN

THALAMIC POOL

MESENCEPHALIC SUPERVENTRICLE
(FUTURE AQUEDUCT)

DIENCEPHALIC
SUPERVENTRICLE
(FUTURE THIRD VENTRICLE)

PRETECTAL
NARROWS

DORSAL POOL

Diencephalic/telencephalic
choroid plexus

TELENCEPHALIC
SUPERVENTRICLE
(FUTURE LATERAL VENTRICLE)

Nerve III
(oculomotor)

VENTRAL POOL

FORAMEN
OF
MONRO

ISTHMAL
CANAL

*Fronto-
nasal
process*

RECESSES — MAMMILLARY

HYPOTHALAMIC
POOL

Nasal
septum

OPTIC

Maxilla

INFUNDIBULAR

Cell-sparse superarachnoid reticulum

METENCEPHALIC POOL

Pituitary
(anterior
lobe)

Oral cavity

Mandible

Tongue

RHOMBENCEPHALIC SUPERVENTRICLE
(FUTURE FOURTH VENTRICLE)

Medullary
velum

Incipient
rhombencephalic
choroid plexus

Larynx

Epi-
glottis

Pharynx

*Arytenoid
swellings*

MYELENCEPHALIC
POOL

Laryngotracheal
groove

Medullary
velum

Vertebral
column
(bodies)

*Cell-sparse
superarachnoid reticulum*

Cell-dense future skin and skull

Cell-dense future dura

Labeled on this page:
Peripheral neural and non-neural
structures, brain ventricular divisions

1 mm

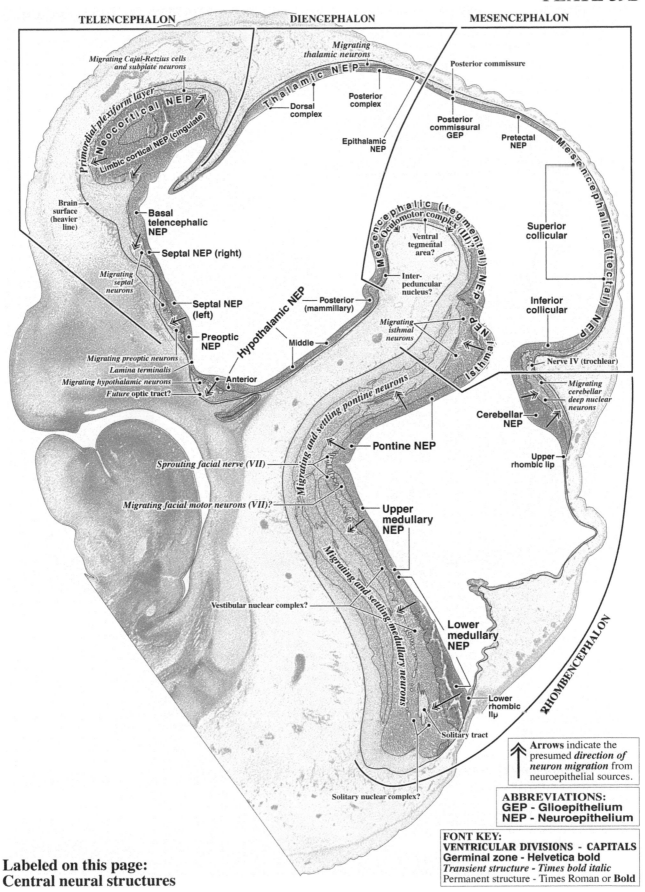

TELENCEPHALON DIENCEPHALON MESENCEPHALON

Migrating Cajal-Retzius cells and subplate neurons

Migrating thalamic neurons

Posterior commissure

Primordial plexiform layer

Thalamic NEP

Neocortical NEP

Dorsal complex

Posterior complex

Posterior commissural GEP

Limbic cortical NEP (cingulate)

Epithalamic NEP

Pretectal NEP

Brain surface (heavier line)

Basal telencephalic NEP

Mesencephalic (tegmental) NEP
Oculomotor complex (III)?

Ventral tegmental area?

Superior collicular

Mesencephalic (tectal) NEP

Septal NEP (right)

Migrating septal neurons

Inter-peduncular nucleus?

Inferior collicular

Septal NEP (left)

Hypothalamic NEP

Posterior (mammillary)

Migrating isthmal neurons

Isthmal NEP

Preoptic NEP

Middle

Nerve IV (trochlear)

Migrating preoptic neurons
Lamina terminalis
Migrating hypothalamic neurons
Future optic tract?

Anterior

Migrating and settling pontine neurons

Migrating cerebellar deep nuclear neurons

Cerebellar NEP

Pontine NEP

Upper rhombic lip

Sprouting facial nerve (VII)

Migrating facial motor neurons (VII)?

Migrating and settling medullary neurons

Upper medullary NEP

Vestibular nuclear complex?

Lower medullary NEP

Lower rhombic lip

Solitary tract

RHOMBENCEPHALON

Solitary nuclear complex?

Arrows indicate the presumed *direction of neuron migration* from neuroepithelial sources.

ABBREVIATIONS:
GEP - Glioepithelium
NEP - Neuroepithelium

FONT KEY:
VENTRICULAR DIVISIONS - CAPITALS
Germinal zone - Helvetica bold
Transient structure - Times bold italic
Permanent structure - Times Roman or **Bold**

**Labeled on this page:
Central neural structures**

PLATE 60A

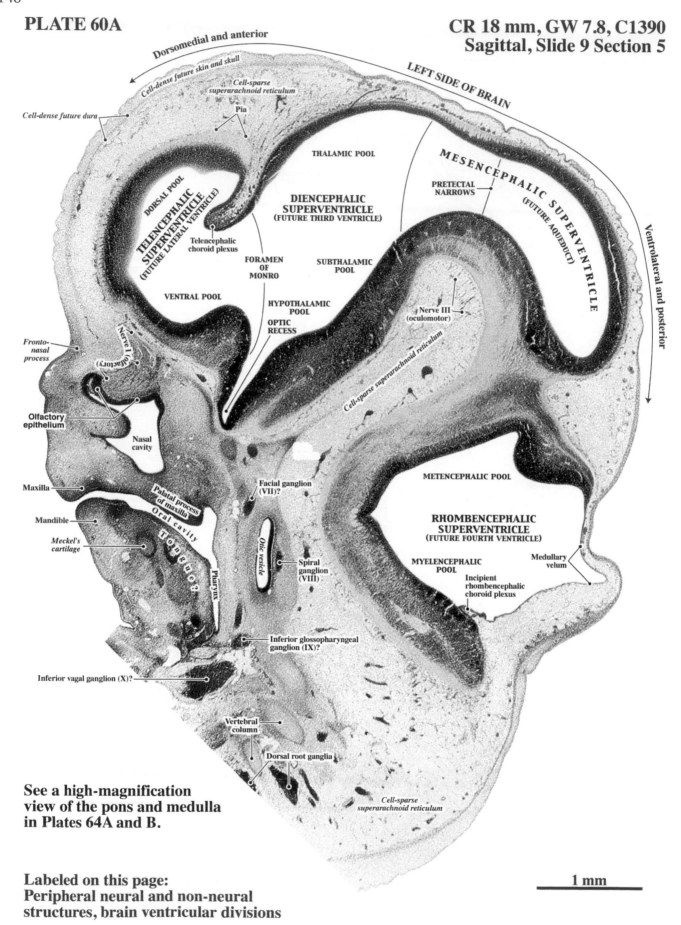

Dorsomedial and anterior

LEFT SIDE OF BRAIN

Cell-dense future skin and skull

Cell-sparse superarachnoid reticulum

Pia

Cell-dense future dura

THALAMIC POOL

MESENCEPHALIC SUPERVENTRICLE
(FUTURE AQUEDUCT)

PRETECTAL NARROWS

DORSAL POOL

TELENCEPHALIC SUPERVENTRICLE
(FUTURE LATERAL VENTRICLE)

DIENCEPHALIC SUPERVENTRICLE
(FUTURE THIRD VENTRICLE)

Telencephalic choroid plexus

Ventrolateral and posterior

SUBTHALAMIC POOL

FORAMEN OF MONRO

VENTRAL POOL

HYPOTHALAMIC POOL

Nerve III (oculomotor)

OPTIC RECESS

Fronto-nasal process

Nerve I (olfactory)

Cell-sparse superarachnoid reticulum

Olfactory epithelium

Nasal cavity

Maxilla

Palatal process of maxilla

Facial ganglion (VII)?

METENCEPHALIC POOL

Oral cavity

Mandible

RHOMBENCEPHALIC SUPERVENTRICLE
(FUTURE FOURTH VENTRICLE)

Meckel's cartilage

Tongue?

Otic vesicle

Pharynx

Spiral ganglion (VIII)

MYELENCEPHALIC POOL

Medullary velum

Incipient rhombencephalic choroid plexus

Inferior glossopharyngeal ganglion (IX)?

Inferior vagal ganglion (X)?

Vertebral column

Dorsal root ganglia

See a high-magnification view of the pons and medulla in Plates 64A and B.

Cell-sparse superarachnoid reticulum

1 mm

Labeled on this page: Peripheral neural and non-neural structures, brain ventricular divisions

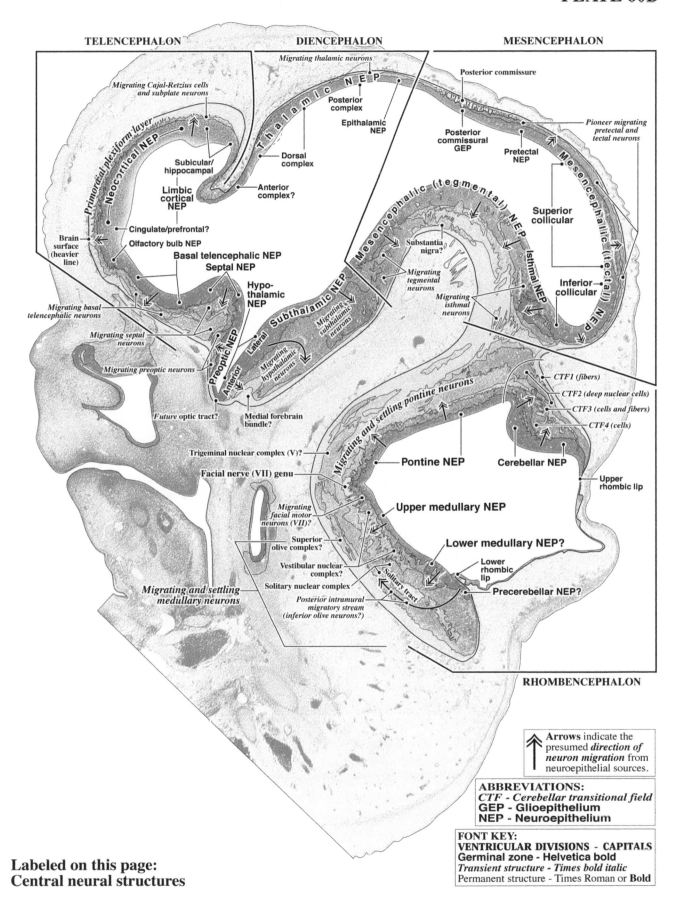

TELENCEPHALON

Migrating Cajal-Retzius cells and subplate neurons

Primordial plexiform layer

Neocortical NEP

Subicular/hippocampal

Limbic cortical NEP

Brain surface (heavier line)

Cingulate/prefrontal?

Olfactory bulb NEP

Basal telencephalic NEP
Septal NEP

Migrating basal telencephalic neurons

Migrating septal neurons

Migrating preoptic neurons

Hypothalamic NEP

Preoptic NEP

Anterior

Future optic tract?

Medial forebrain bundle?

DIENCEPHALON

Migrating thalamic neurons

Thalamic NEP

Posterior complex

Epithalamic NEP

Dorsal complex

Anterior complex?

Lateral Subthalamic NEP

Migrating subthalamic neurons

Migrating hypothalamic neurons

MESENCEPHALON

Posterior commissure

Posterior commissural GEP

Pretectal NEP

Mesencephalic (tegmental) NEP

Superior collicular

Pioneer migrating pretectal and tectal neurons

Mesencephalic (tectal) NEP

Isthmal NEP

Substantia nigra?

Migrating tegmental neurons

Migrating isthmal neurons

Inferior collicular

Trigeminal nuclear complex (V)?

Facial nerve (VII) genu

Migrating and settling pontine neurons

Migrating facial motor neurons (VII)?

Superior olive complex?

Vestibular nuclear complex?

Solitary nuclear complex

Migrating and settling medullary neurons

Posterior intramural migratory stream (inferior olive neurons?)

Pontine NEP

Solitary tract

CTF1 (fibers)
CTF2 (deep nuclear cells)
CTF3 (cells and fibers)
CTF4 (cells)

Cerebellar NEP

Upper rhombic lip

Upper medullary NEP

Lower medullary NEP?

Lower rhombic lip

Precerebellar NEP?

RHOMBENCEPHALON

Arrows indicate the presumed *direction of neuron migration* from neuroepithelial sources.

ABBREVIATIONS:
CTF - Cerebellar transitional field
GEP - Glioepithelium
NEP - Neuroepithelium

FONT KEY:
VENTRICULAR DIVISIONS - CAPITALS
Germinal zone - Helvetica bold
Transient structure - Times bold italic
Permanent structure - Times Roman or **Bold**

Labeled on this page:
Central neural structures

PLATE 61A

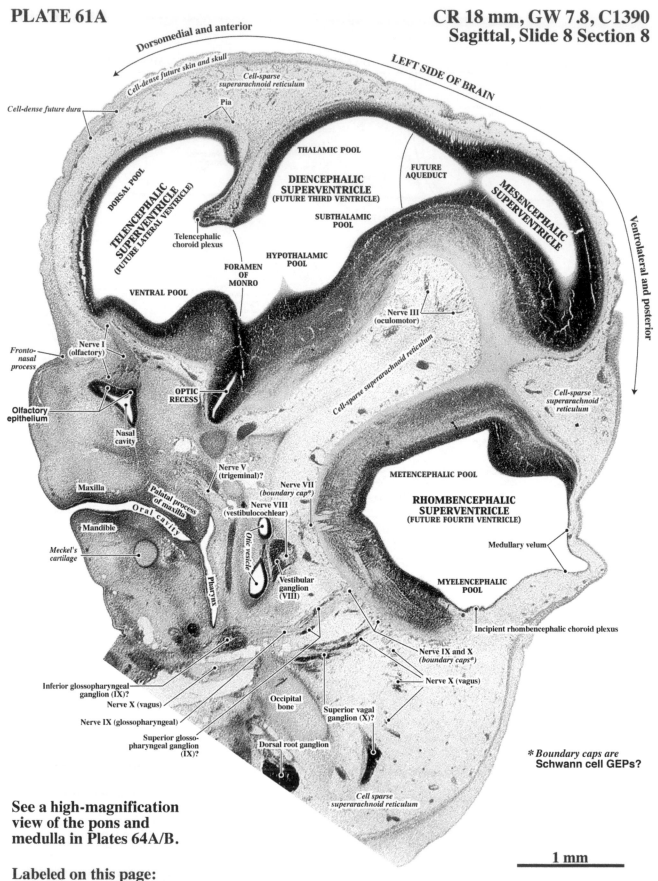

Dorsomedial and anterior

LEFT SIDE OF BRAIN

Cell-dense future skin and skull

Cell-sparse superarachnoid reticulum

Cell-dense future dura

Pia

THALAMIC POOL

DORSAL POOL

FUTURE AQUEDUCT

DIENCEPHALIC SUPERVENTRICLE (FUTURE THIRD VENTRICLE)

MESENCEPHALIC SUPERVENTRICLE

TELENCEPHALIC SUPERVENTRICLE (FUTURE LATERAL VENTRICLE)

SUBTHALAMIC POOL

Telencephalic choroid plexus

HYPOTHALAMIC POOL

FORAMEN OF MONRO

VENTRAL POOL

Nerve III (oculomotor)

Ventrolateral and posterior

Nerve I (olfactory)

Fronto-nasal process

OPTIC RECESS

Cell-sparse superarachnoid reticulum

Cell-sparse superarachnoid reticulum

Olfactory epithelium

Nasal cavity

Nerve V (trigeminal)?

Nerve VII (*boundary cap**)

Maxilla

Palatal process of maxilla

Nerve VIII (vestibulocochlear)

METENCEPHALIC POOL

RHOMBENCEPHALIC SUPERVENTRICLE (FUTURE FOURTH VENTRICLE)

Mandible

Oral cavity

Otic vesicle

Medullary velum

Meckel's cartilage

Vestibular ganglion (VIII)

MYELENCEPHALIC POOL

Pharynx

Incipient rhombencephalic choroid plexus

Inferior glossopharyngeal ganglion (IX)?

Nerve IX and X (*boundary caps**)

Nerve X (vagus)

Nerve X (vagus)

Occipital bone

Superior vagal ganglion (X)?

Nerve IX (glossopharyngeal)

Superior glosso-pharyngeal ganglion (IX)?

Dorsal root ganglion

Cell sparse superarachnoid reticulum

**Boundary caps are* Schwann cell GEPs?

See a high-magnification view of the pons and medulla in Plates 64A/B.

Labeled on this page: Peripheral neural and non-neural structures, brain ventricular divisions

1 mm

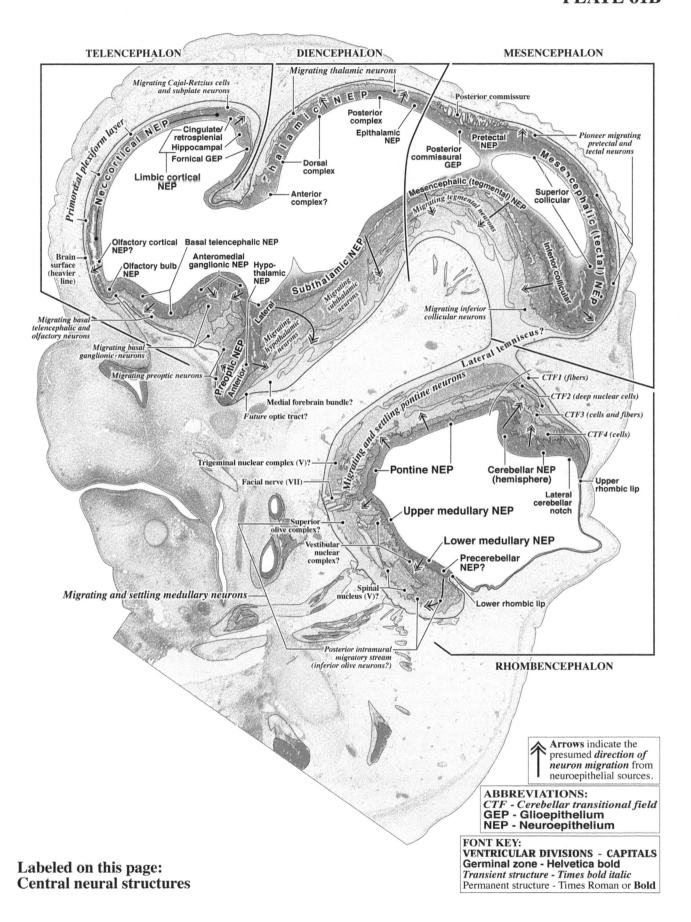

TELENCEPHALON   DIENCEPHALON   MESENCEPHALON

Migrating Cajal-Retzius cells and subplate neurons

Migrating thalamic neurons

Posterior commissure

Primordial plexiform layer

Neocortical NEP

Cingulate/ retrosplenial

Hippocampal

Fornical GEP

Limbic cortical NEP

Posterior complex

Epithalamic NEP

Pretectal NEP

Pioneer migrating pretectal and tectal neurons

Dorsal complex

Posterior commissural GEP

Mesencephalic (tectal) NEP

Anterior complex?

Mesencephalic (tegmental) NEP

Superior collicular

Migrating tegmental neurons

Brain surface (heavier line)

Olfactory cortical NEP?

Basal telencephalic NEP

Inferior collicular

Olfactory bulb NEP

Anteromedial ganglionic NEP

Hypo-thalamic NEP

Subthalamic NEP

Migrating subthalamic neurons

Migrating inferior collicular neurons

Migrating basal telencephalic and olfactory neurons

Lateral

Migrating hypothalamic neurons

Migrating basal ganglionic neurons

Preoptic NEP

Lateral lemniscus?

Migrating preoptic neurons

Anterior

CTF1 (fibers)

Medial forebrain bundle?

Migrating and settling pontine neurons

CTF2 (deep nuclear cells)

Future optic tract?

CTF3 (cells and fibers)

CTF4 (cells)

Trigeminal nuclear complex (V)?

Pontine NEP

Cerebellar NEP (hemisphere)

Upper rhombic lip

Facial nerve (VII)

Lateral cerebellar notch

Superior olive complex?

Upper medullary NEP

Vestibular nuclear complex?

Lower medullary NEP

Precerebellar NEP?

Migrating and settling medullary neurons

Spinal nucleus (V)?

Lower rhombic lip

Posterior intramural migratory stream (inferior olive neurons?)

RHOMBENCEPHALON

Arrows indicate the presumed *direction of neuron migration* from neuroepithelial sources.

ABBREVIATIONS:
CTF - Cerebellar transitional field
GEP - Glioepithelium
NEP - Neuroepithelium

FONT KEY:
VENTRICULAR DIVISIONS - CAPITALS
Germinal zone - Helvetica bold
Transient structure - Times bold italic
Permanent structure - Times Roman or **Bold**

Labeled on this page:
Central neural structures

PLATE 62A

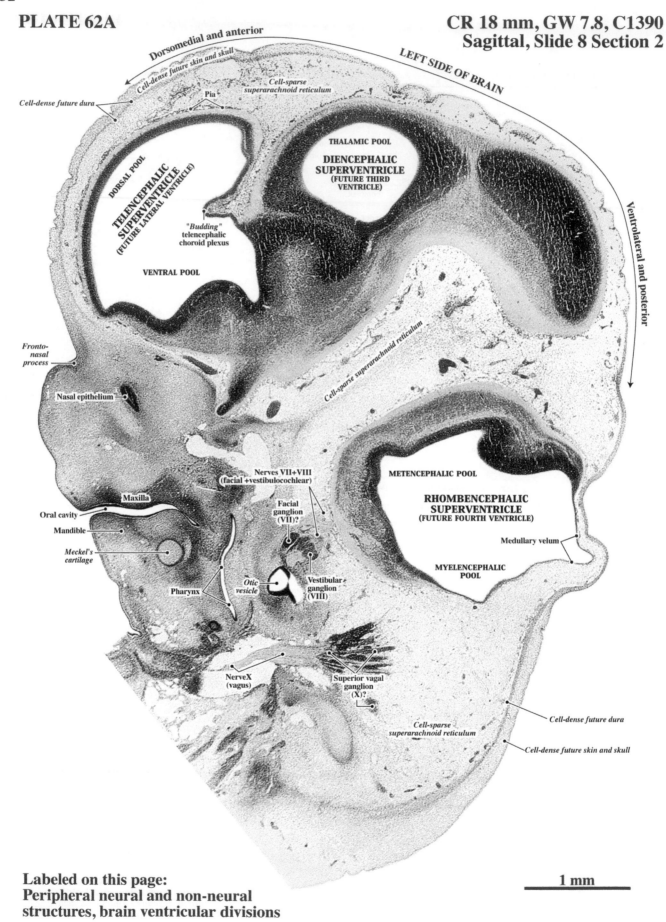

Dorsomedial and anterior

Cell-dense future skin and skull

LEFT SIDE OF BRAIN

Pia

Cell-sparse superarachnoid reticulum

Cell-dense future dura

THALAMIC POOL

DIENCEPHALIC SUPERVENTRICLE (FUTURE THIRD VENTRICLE)

DORSAL POOL

TELENCEPHALIC SUPERVENTRICLE (FUTURE LATERAL VENTRICLE)

Ventrolateral and posterior

"Budding" telencephalic choroid plexus

VENTRAL POOL

Fronto-nasal process

Cell-sparse superarachnoid reticulum

Nasal epithelium

Nerves VII+VIII (facial +vestibulocochlear)

METENCEPHALIC POOL

RHOMBENCEPHALIC SUPERVENTRICLE (FUTURE FOURTH VENTRICLE)

Maxilla

Facial ganglion (VII)?

Oral cavity

Mandible

Medullary velum

Meckel's cartilage

MYELENCEPHALIC POOL

Otic vesicle

Vestibular ganglion (VIII)

Pharynx

Nerve X (vagus)

Superior vagal ganglion (X)?

Cell-dense future dura

Cell-sparse superarachnoid reticulum

Cell-dense future skin and skull

1 mm

**Labeled on this page:
Peripheral neural and non-neural
structures, brain ventricular divisions**

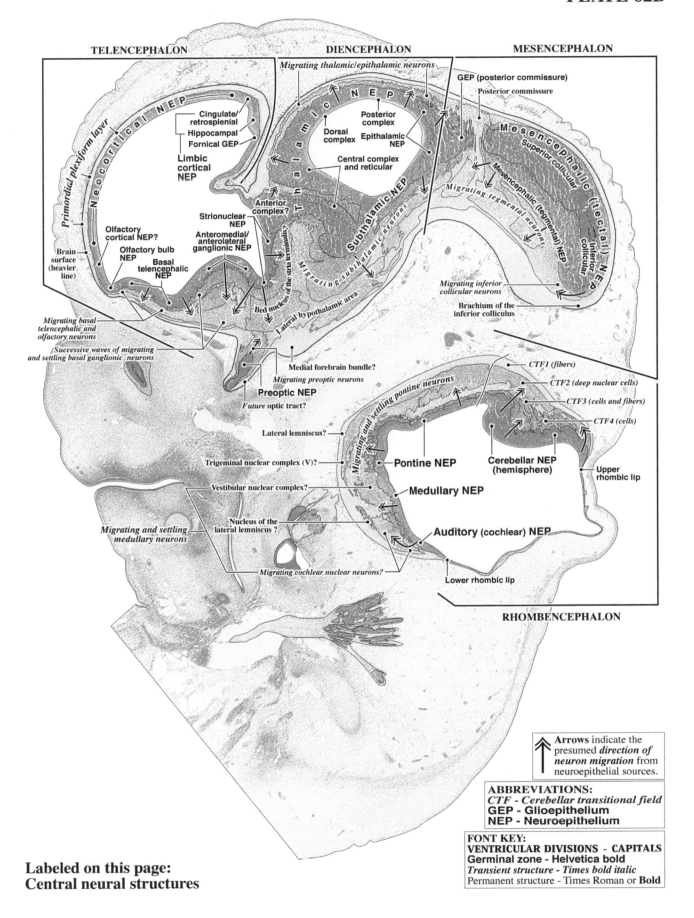

TELENCEPHALON · DIENCEPHALON · MESENCEPHALON

Migrating thalamic/epithalamic neurons

GEP (posterior commissure)

Posterior commissure

Migrating basal telencephalic and olfactory neurons

Successive waves of migrating and settling basal ganglionic neurons

Primordial plexiform layer

Neocortical NEP

Cingulate/retrosplenial
Hippocampal
Fornical GEP

Limbic cortical NEP

Brain surface (heavier line)

Olfactory cortical NEP?
Olfactory bulb NEP

Basal telencephalic NEP

Strionuclear NEP
Anteromedial/anterolateral ganglionic NEP

Anterior complex?

Thalamic NEP

Dorsal complex

Posterior complex
Epithalamic NEP

Central complex and reticular

Subthalamic NEP

Migrating subthalamic neurons

Migrating the stria terminalis?

Bed nuclei of the stria terminalis

Lateral hypothalamic area

Medial forebrain bundle?

Migrating preoptic neurons

Preoptic NEP

Future optic tract?

Mesencephalic Superior collicular

Migrating tegmental neurons

Mesencephalic (tegmental) NEP

Mesencephalic (tectal) NEP
Inferior collicular

Migrating inferior collicular neurons

Brachium of the inferior colliculus

Lateral lemniscus?

Trigeminal nuclear complex (V)?

Vestibular nuclear complex?

Migrating and settling medullary neurons

Nucleus of the lateral lemniscus ?

Migrating cochlear nuclear neurons?

Migrating and settling pontine neurons

CTF1 (fibers)
CTF2 (deep nuclear cells)
CTF3 (cells and fibers)
CTF4 (cells)

Pontine NEP

Cerebellar NEP (hemisphere)

Upper rhombic lip

Medullary NEP

Auditory (cochlear) NEP

Lower rhombic lip

RHOMBENCEPHALON

Arrows indicate the presumed *direction of neuron migration* from neuroepithelial sources.

ABBREVIATIONS:
CTF - Cerebellar transitional field
GEP - Glioepithelium
NEP - Neuroepithelium

FONT KEY:
VENTRICULAR DIVISIONS - CAPITALS
Germinal zone - Helvetica bold
Transient structure - Times bold italic
Permanent structure - Times Roman or **Bold**

**Labeled on this page:
Central neural structures**

PLATE 63A

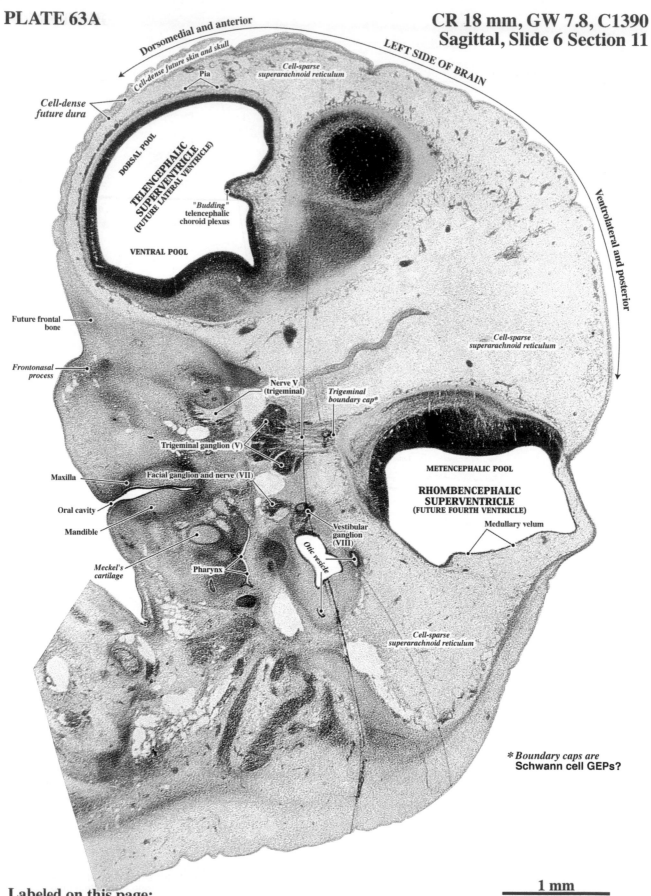

Dorsomedial and anterior

Cell-dense future skin and skull

Pia

Cell-sparse superarachnoid reticulum

LEFT SIDE OF BRAIN

Cell-dense future dura

DORSAL POOL

TELENCEPHALIC SUPERVENTRICLE (FUTURE LATERAL VENTRICLE)

"Budding" telencephalic choroid plexus

VENTRAL POOL

Ventrolateral and posterior

Future frontal bone

Frontonasal process

Cell-sparse superarachnoid reticulum

Nerve V (trigeminal)

Trigeminal boundary cap*

Trigeminal ganglion (V)

Maxilla

Facial ganglion and nerve (VII)

Oral cavity

METENCEPHALIC POOL

RHOMBENCEPHALIC SUPERVENTRICLE (FUTURE FOURTH VENTRICLE)

Medullary velum

Mandible

Vestibular ganglion (VIII)

Meckel's cartilage

Otic vesicle

Pharynx

Cell-sparse superarachnoid reticulum

*Boundary caps are Schwann cell GEPs?

1 mm

Labeled on this page:
Peripheral neural and non-neural structures, brain ventricular divisions

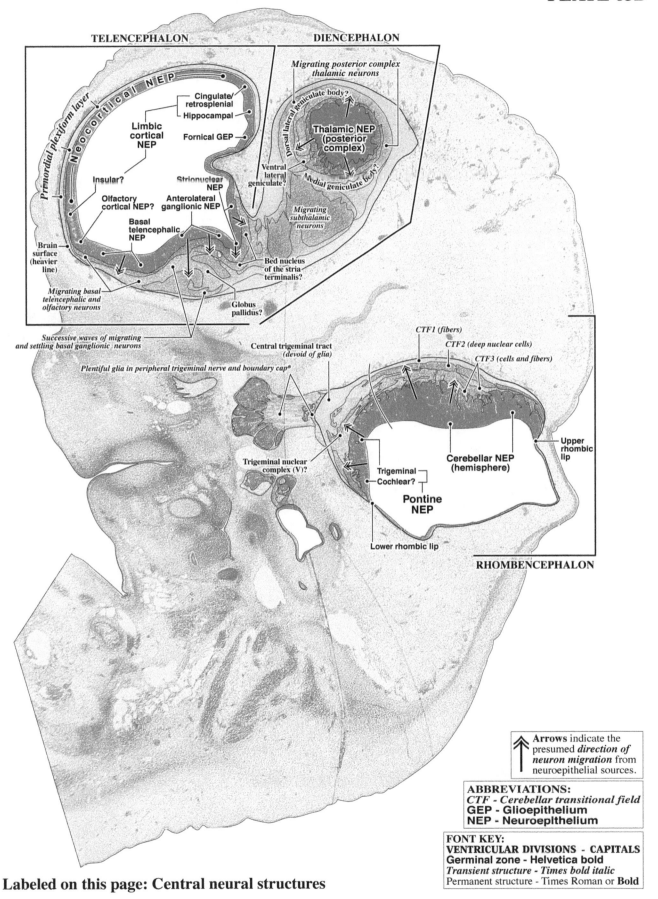

TELENCEPHALON

DIENCEPHALON

Primordial plexiform layer

Neocortical NEP

Cingulate/
retrosplenial

Hippocampal

Limbic
cortical
NEP

Fornical GEP

*Migrating posterior complex
thalamic neurons*

Dorsal lateral geniculate body?

**Thalamic NEP
(posterior
complex)**

Insular?

**Strionuclear
NEP**

Ventral
lateral
geniculate?

Medial geniculate body?

Olfactory
cortical NEP?

**Anterolateral
ganglionic NEP**

*Migrating
subthalamic
neurons*

Basal
telencephalic
NEP

Brain
surface
(heavier
line)

Bed nucleus
of the stria
terminalis?

*Migrating basal
telencephalic and
olfactory neurons*

Globus
pallidus?

*Successive waves of migrating
and settling basal ganglionic neurons*

Central trigeminal tract
(devoid of glia)

Plentiful glia in peripheral trigeminal nerve and boundary cap*

CTF1 (fibers)

CTF2 (deep nuclear cells)

CTF3 (cells and fibers)

**Cerebellar NEP
(hemisphere)**

Upper
rhombic
lip

Trigeminal nuclear
complex (V)?

Trigeminal

Cochlear?

**Pontine
NEP**

Lower rhombic lip

RHOMBENCEPHALON

Arrows indicate the
presumed *direction of
neuron migration* from
neuroepithelial sources.

ABBREVIATIONS:
CTF - Cerebellar transitional field
GEP - Glioepithelium
NEP - Neuroepithelium

FONT KEY:
VENTRICULAR DIVISIONS - CAPITALS
Germinal zone - Helvetica bold
Transient structure - Times bold italic
Permanent structure - Times Roman or **Bold**

Labeled on this page: Central neural structures

PLATE 64A   **CR 18 mm, GW 7.8, C1390, Sagittal**   **PONS/MEDULLA**

MEDIAL
Slide 9,
Section 5

See the entire
section in
Plates 60A/B.

INTERMEDIATE
Slide 9,
Section 2

LATERAL
Slide 8,
Section 8

See the entire
section in
Plates 61A/B.

0.1 mm

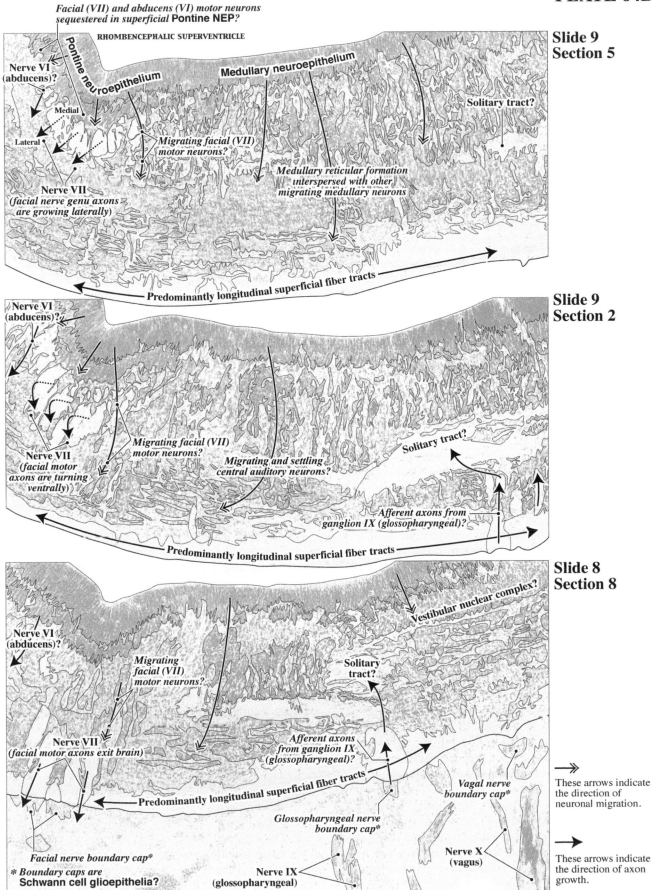

Facial (VII) and abducens (VI) motor neurons sequestered in superficial **Pontine NEP***?*

**Slide 9
Section 5**

RHOMBENCEPHALIC SUPERVENTRICLE

Pontine neuroepithelium

Medullary neuroepithelium

Nerve VI
(abducens)?

Medial

Lateral

Solitary tract?

*Migrating facial (VII)
motor neurons?*

Nerve VII
*(facial nerve genu axons
are growing laterally)*

*Medullary reticular formation
interspersed with other
migrating medullary neurons*

Predominantly longitudinal superficial fiber tracts

**Slide 9
Section 2**

Nerve VI
(abducens)?

*Migrating facial (VII)
motor neurons?*

Solitary tract?

Nerve VII
*(facial motor
axons are turning
ventrally)*

*Migrating and settling
central auditory neurons?*

*Afferent axons from
ganglion IX (glossopharyngeal)?*

Predominantly longitudinal superficial fiber tracts

**Slide 8
Section 8**

Vestibular nuclear complex?

Nerve VI
(abducens)?

*Migrating
facial (VII)
motor neurons?*

Solitary
tract?

Nerve VII
(facial motor axons exit brain)

*Afferent axons
from ganglion IX
(glossopharyngeal)?*

Predominantly longitudinal superficial fiber tracts

Vagal nerve
boundary cap*

These arrows indicate
the direction of
neuronal migration.

*Glossopharyngeal nerve
boundary cap**

Facial nerve boundary cap*

*** Boundary caps are
Schwann cell glioepithelia?**

Nerve IX
(glossopharyngeal)

Nerve X
(vagus)

These arrows indicate
the direction of axon
growth.

PART VI: M2155
CR 17.5 mm (GW 7.8)
Frontal/Horizontal

This specimen is embryo #2155 in the Minot Collection, designated here as M2155. The crown-rump length (CR) is 17.5-mm estimated to be at gestational week (GW) 7.8. Most of M2155's brain sections are cut (10 μm) in the coronal plane, but the plane becomes oblique in the posterior medulla. We photographed 71 sections at low magnification from the frontal prominence to the posterior tips of the mesencephalon and cerebellum. Seventeen of these sections are illustrated in **Plates 65AB to 81AB**. All photographs were used to produce computer-aided 3-D reconstructions of the external features of M2155's brain (**Figure 16**), and to show each illustrated section *in situ* (*insets,* **Plates 65-81A**). Consequently, each illustrated section shows the brain with all surrounding tissues. Labels in **A Plates** (normal-contrast images) identify non-neural and peripheral neural structures; labels in **B Plates** (low-contrast images) identify central neural structures. **Plates 82-84** show high-magnification views of the cerebral cortex, diencephalon, and mesencephalon. The brain of M2155 has considerable variation in the thickness of the neuroepithelium (NEP) and in the number of migrating neurons in various parts of the brain parenchyma.

Throughout the telencephalon, the NEP is the most prominent structure surrounding the enlarging roof of the telencephalic superventricle. A cell-sparse primordial plexiform layer is adjacent to the cerebral cortical neuroepithelium (NEP). A few pioneer Cajal-Retzius neurons have migrated into this layer dorsomedially, more ventrolaterally, and neurons in layers VI and V have had substantial neurogenesis and are sequestered in the NEP. The basal ganglionic and basal telencephalic NEPs are well into neurogenetic stage and are flanked by migrating neurons. In some areas, these neurons appear to migrate together in early (outermost and less dense) to late (innermost and most dense) waves. There is only the slightest indication of an olfactory bulb evagination in spite of the fact that a fully invaginated olfactory epithelium is in the nasal cavity and olfactory nerve fibers already contact the brain just anterior to the basal telencephalon, possibly inducing the cortical neuroepithelium to evaginate into an olfactory bulb.

The diencephalic NEP surrounds the future third ventricle. It is thinnest in the hypothalamic and subthalamic areas, where it is surrounded by densely packed waves of migrating neurons unloading into the parenchyma. In contrast, the thalamic NEP sequesters more postmitotic neurons basally. The few neurons outside the thalamic NEP are postulated to be the oldest neurons in the ventral complex, posterior complex, and the reticular nucleus.

The mesencephalon contains neurogenetic NEPs throughout its extent. The pretectum and tectum are flanked by few migrating neurons. On the other hand, the tegmental and isthmal NEPS are much thinner because most of their neuronal progeny has migrated out in waves of inner dense clumps and outer sparse arrays interspersing among the thick subpial fiber bands in the tegmental and isthmal parenchyma.

Both the pons and medulla have NEPs that are shrinking as they have already unloaded their neuronal precursors into an expanding parenchyma. Cells are migrating and settling in longitudinal arrays at the pontine flexure. A few cells are settling in the superior olive complex and many are settling in the reticular formation throughout the pons and medulla. Facial motor neurons are migrating from medial to lateral, leaving behind their axons in the genu of the facial nerve. Migrating cochlear nuclear neurons are outside the NEP in the anterior part of the lower rhombic lip, while migrating inferior olive neurons are in the posterior intramural migratory stream outside the precerebellar NEP in the posterior lower rhombic lip; some neurons have already settled in the inferior olive. Many neurons have settled in the solitary nucleus surrounding the solitary tract. The hypoglossal nucleus is also distinguishable in the lower medulla. The cerebellar NEP is exceptional in the rhombencephalon because it is still appears to be in the stockbuilding phase. But many Purkinje cells are sequestered in its basal part. Deep nuclear neurons are sojourning in the superficial layers of the cerebellar transitional field, interacting with the fibrous layers that may contain afferents from the spinal cord and the vestibular ganglion.

M2155 Computer-aided 3-D Brain Reconstructions

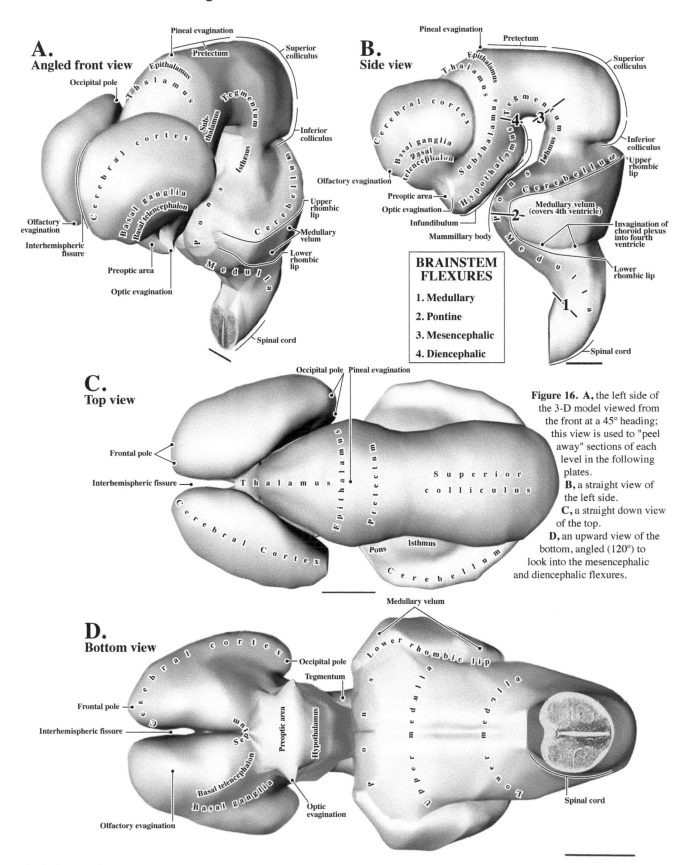

A. Angled front view

Pineal evagination
Pretectum
Superior colliculus
Epithalamus
Occipital pole
Thalamus
Tegmentum
Sub-thalamus
Cerebral cortex
Isthmus
Inferior colliculus
Basal ganglia
Basal telencephalon
Pons
Cerebellum
Upper rhombic lip
Olfactory evagination
Medullary velum
Interhemispheric fissure
Lower rhombic lip
Preoptic area
Medulla
Optic evagination
Spinal cord

B. Side view

Pineal evagination
Pretectum
Epithalamus
Thalamus
Superior colliculus
Cerebral cortex
Tegmentum
Basal ganglia
Basal telencephalon
Subthalamus
Isthmus
Inferior colliculus
Hypothalamus
Pons
Cerebellum
Upper rhombic lip
Olfactory evagination
Preoptic area
Optic evagination
Medullary velum (covers 4th ventricle)
Infundibulum
Mammillary body
Invagination of choroid plexus into fourth ventricle
Medulla
Lower rhombic lip
Spinal cord

BRAINSTEM FLEXURES

1. Medullary
2. Pontine
3. Mesencephalic
4. Diencephalic

C. Top view

Occipital pole   Pineal evagination
Frontal pole
Interhemispheric fissure
Thalamus
Epithalamus
Pretectum
Superior colliculus
Cerebral Cortex
Pons   Isthmus
Cerebellum

Figure 16. A, the left side of the 3-D model viewed from the front at a 45° heading; this view is used to "peel away" sections of each level in the following plates.
  B, a straight view of the left side.
  C, a straight down view of the top.
  D, an upward view of the bottom, angled (120°) to look into the mesencephalic and diencephalic flexures.

D. Bottom view

Medullary velum
Cerebral cortex
Lower rhombic lip
Occipital pole
Tegmentum
Frontal pole
Preoptic area
Hypothalamus
Pons
Upper medulla
Lower medulla
Interhemispheric fissure
Septum
Basal telencephalon
Basal ganglia
Optic evagination
Olfactory evagination
Spinal cord

Scale bars = 1 mm

PLATE 65A
CR 17.5 mm, GW 7.8, M2155
Frontal/Horizontal, Section 50

Non-neural structures labeled

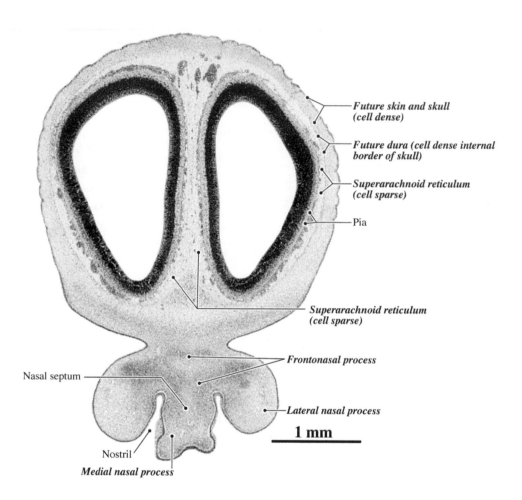

Future skin and skull (cell dense)

Future dura (cell dense internal border of skull)

Superarachnoid reticulum (cell sparse)

Pia

Superarachnoid reticulum (cell sparse)

Frontonasal process

Nasal septum

Lateral nasal process

1 mm

Nostril

Medial nasal process

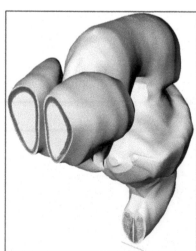

Section 50 brain *in situ*

Neural structures labeled

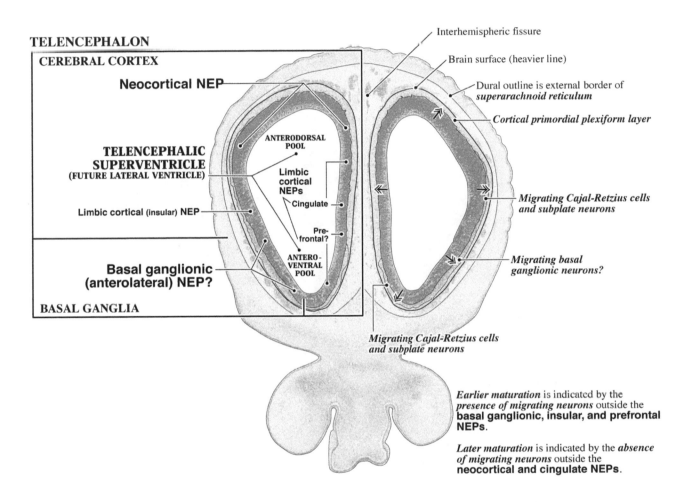

TELENCEPHALON

CEREBRAL CORTEX

Neocortical NEP

TELENCEPHALIC
SUPERVENTRICLE
(FUTURE LATERAL VENTRICLE)

Limbic cortical (insular) NEP

Basal ganglionic
(anterolateral) NEP?

BASAL GANGLIA

ANTERODORSAL
POOL

Limbic
cortical
NEPs

Cingulate

Pre-
frontal?

ANTERO-
VENTRAL
POOL

Interhemispheric fissure

Brain surface (heavier line)

Dural outline is external border of
superarachnoid reticulum

Cortical primordial plexiform layer

*Migrating Cajal-Retzius cells
and subplate neurons*

*Migrating basal
ganglionic neurons?*

*Migrating Cajal-Retzius cells
and subplate neurons*

Earlier maturation is indicated by the
presence of migrating neurons outside the
basal ganglionic, insular, and prefrontal
NEPs.

Later maturation is indicated by the *absence
of migrating neurons* outside the
neocortical and cingulate NEPs.

NEP - Neuroepithelium

Arrows indicate the
presumed *direction of
neuron migration* from
neuroepithelial sources.

FONT KEY:
VENTRICULAR DIVISIONS - CAPITALS
Germinal zone - Helvetica bold
Transient structure - Times bold italic
Permanent structure - Times Roman or **Bold**

PLATE 66A
CR 17.5 mm, GW 7.8, M2155
Frontal/Horizontal, Section 116

Peripheral neural and non-neural structures labeled

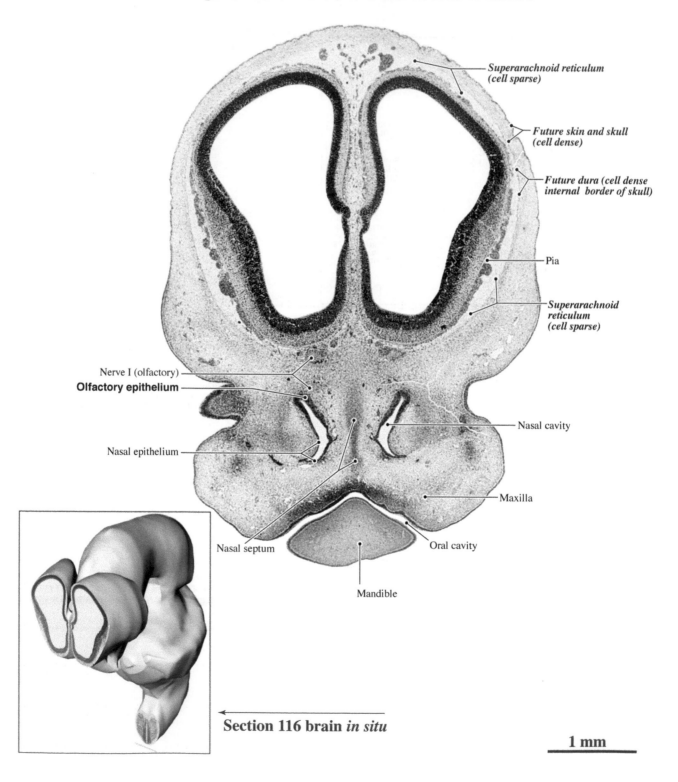

*Superarachnoid reticulum
(cell sparse)*

*Future skin and skull
(cell dense)*

*Future dura (cell dense
internal border of skull)*

Pia

*Superarachnoid
reticulum
(cell sparse)*

Nerve I (olfactory)

Olfactory epithelium

Nasal epithelium

Nasal cavity

Nasal septum

Maxilla

Oral cavity

Mandible

Section 116 brain *in situ*

1 mm

Central neural structures labeled

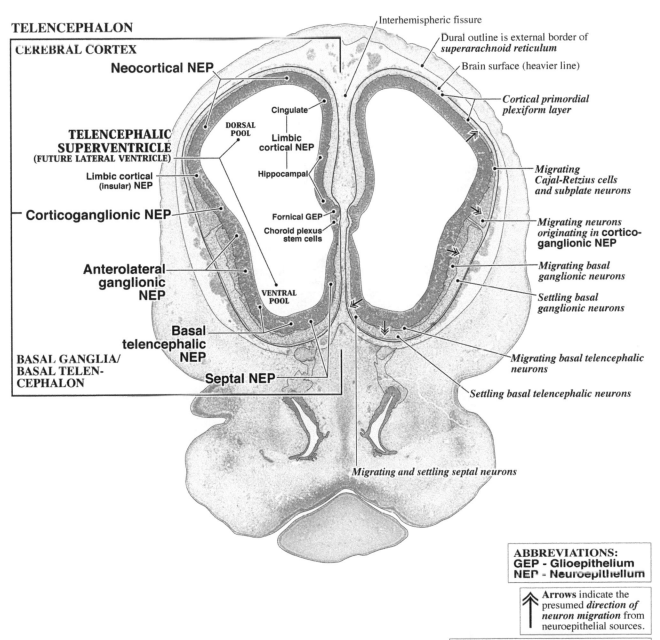

TELENCEPHALON

CEREBRAL CORTEX

Neocortical NEP

TELENCEPHALIC
SUPERVENTRICLE
(FUTURE LATERAL VENTRICLE)

Limbic cortical
(insular) NEP

Corticoganglionic NEP

Anterolateral
ganglionic
NEP

Basal
telencephalic
NEP

Septal NEP

BASAL GANGLIA/
BASAL TELEN-
CEPHALON

DORSAL
POOL

Cingulate

Limbic
cortical NEP

Hippocampal

Fornical GEP

Choroid plexus
stem cells

VENTRAL
POOL

Interhemispheric fissure

Dural outline is external border of
superarachnoid reticulum

Brain surface (heavier line)

*Cortical primordial
plexiform layer*

*Migrating
Cajal-Retzius cells
and subplate neurons*

*Migrating neurons
originating in* cortico-
ganglionic *NEP*

*Migrating basal
ganglionic neurons*

*Settling basal
ganglionic neurons*

*Migrating basal telencephalic
neurons*

Settling basal telencephalic neurons

Migrating and settling septal neurons

ABBREVIATIONS:
GEP - Glioepithelium
NEP - Neuroepithelium

Arrows indicate the
presumed *direction of
neuron migration* from
neuroepithelial sources.

FONT KEY:
VENTRICULAR DIVISIONS - CAPITALS
Germinal zone - Helvetica bold
Transient structure - Times bold italic
Permanent structure - Times Roman or **Bold**

PLATE 67A
CR 17.5 mm, GW 7.8, M2155
Frontal/Horizontal, Section 164

Peripheral neural and non-neural structures labeled

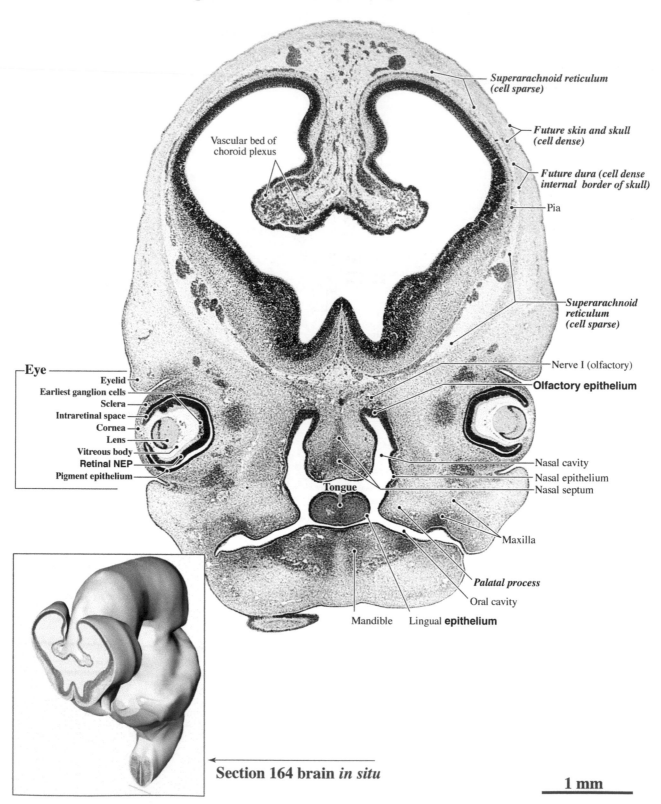

Superarachnoid reticulum (cell sparse)

Future skin and skull (cell dense)

Future dura (cell dense internal border of skull)

Pia

Vascular bed of choroid plexus

Superarachnoid reticulum (cell sparse)

Nerve I (olfactory)

Olfactory epithelium

Eye
Eyelid
Earliest ganglion cells
Sclera
Intraretinal space
Cornea
Lens
Vitreous body
Retinal NEP
Pigment epithelium

Nasal cavity
Nasal epithelium
Nasal septum

Tongue

Maxilla

Palatal process

Oral cavity

Mandible   Lingual **epithelium**

Section 164 brain *in situ*

1 mm

Central neural structures labeled

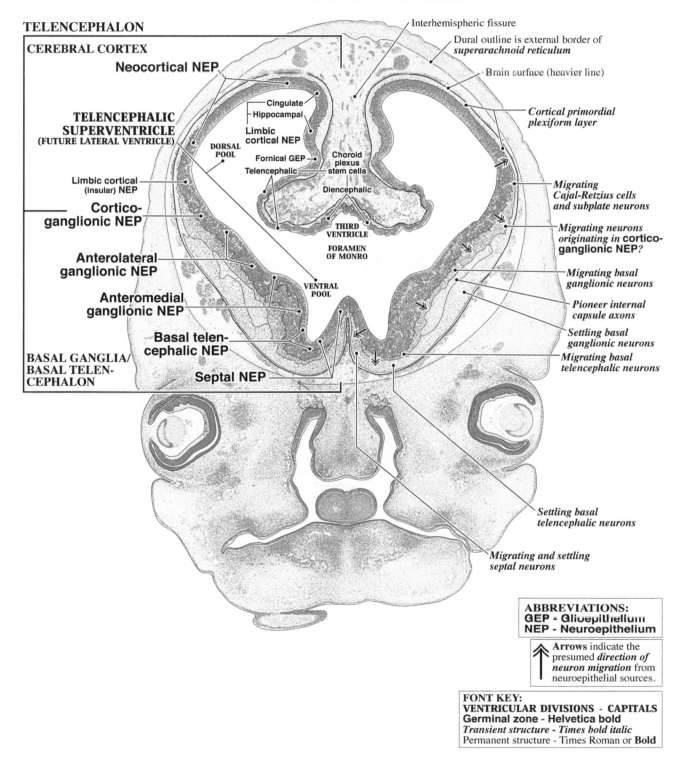

TELENCEPHALON

CEREBRAL CORTEX

Neocortical NEP

TELENCEPHALIC SUPERVENTRICLE
(FUTURE LATERAL VENTRICLE)

DORSAL POOL

Limbic cortical (insular) NEP

Cortico-ganglionic NEP

Anterolateral ganglionic NEP

Anteromedial ganglionic NEP

Basal telen-cephalic NEP

Septal NEP

BASAL GANGLIA/BASAL TELEN-CEPHALON

Cingulate
Hippocampal
Limbic cortical NEP
Fornical GEP
Telencephalic
Choroid plexus stem cells
Diencephalic
THIRD VENTRICLE
FORAMEN OF MONRO
VENTRAL POOL

Interhemispheric fissure

Dural outline is external border of *superarachnoid reticulum*

Brain surface (heavier line)

Cortical primordial plexiform layer

Migrating Cajal-Retzius cells and subplate neurons

Migrating neurons originating in **cortico-ganglionic NEP**?

Migrating basal ganglionic neurons

Pioneer internal capsule axons

Settling basal ganglionic neurons

Migrating basal telencephalic neurons

Settling basal telencephalic neurons

Migrating and settling septal neurons

ABBREVIATIONS:
GEP - Glioepithelium
NEP - Neuroepithelium

Arrows indicate the presumed *direction of neuron migration* from neuroepithelial sources.

FONT KEY:
VENTRICULAR DIVISIONS - CAPITALS
Germinal zone - Helvetica bold
Transient structure - Times bold italic
Permanent structure - Times Roman or **Bold**

PLATE 68A
CR 17.5 mm, GW 7.8, M2155
Frontal/Horizontal, Section 201

See a high-magnification view
of the thalamus and cerebral cortex
in Plates 82A/B.

Peripheral neural and non-neural structures labeled

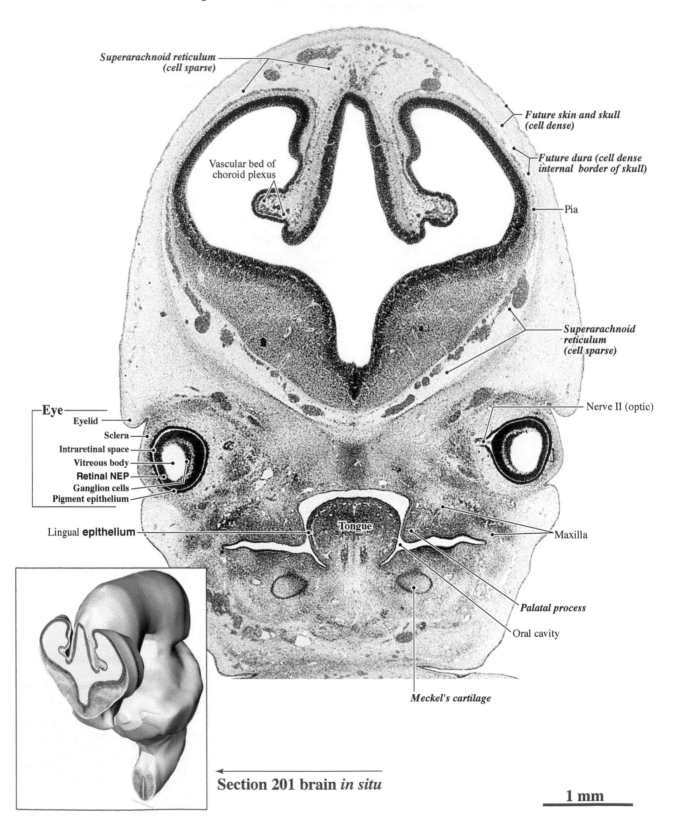

Superarachnoid reticulum
(cell sparse)

Future skin and skull
(cell dense)

Future dura (cell dense
internal border of skull)

Pia

Vascular bed of
choroid plexus

Superarachnoid
reticulum
(cell sparse)

Nerve II (optic)

Eye
— Eyelid
— Sclera
— Intraretinal space
— Vitreous body
— Retinal NEP
— Ganglion cells
— Pigment epithelium

Lingual epithelium

Tongue

Maxilla

Palatal process

Oral cavity

Meckel's cartilage

Section 201 brain *in situ*

1 mm

Central neural structures labeled

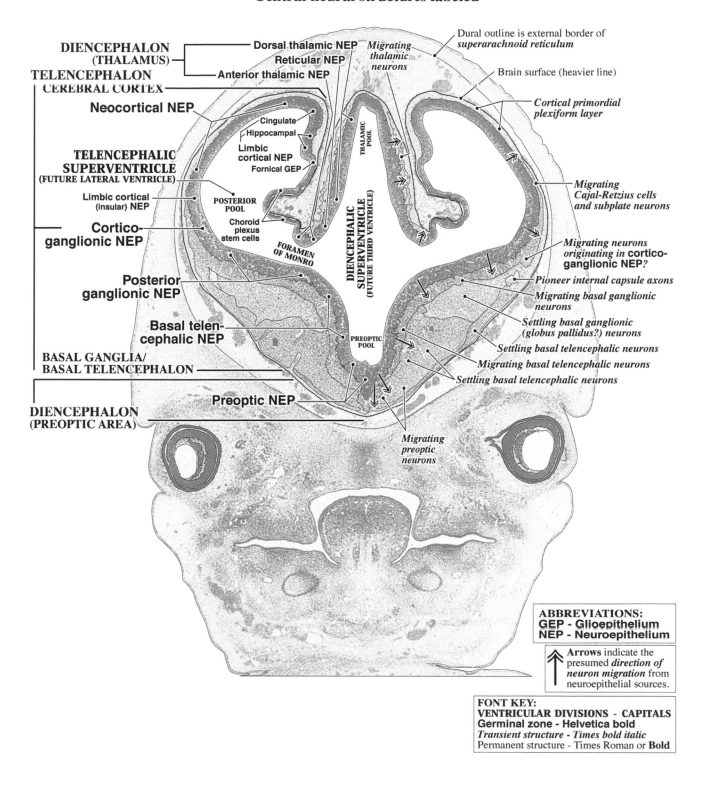

DIENCEPHALON
(THALAMUS)

TELENCEPHALON
CEREBRAL CORTEX

Neocortical NEP

TELENCEPHALIC
SUPERVENTRICLE
(FUTURE LATERAL VENTRICLE)

Limbic cortical
(insular) NEP

Cortico-
ganglionic NEP

Posterior
ganglionic NEP

Basal telen-
cephalic NEP

BASAL GANGLIA/
BASAL TELENCEPHALON

DIENCEPHALON
(PREOPTIC AREA)

Preoptic NEP

Dorsal thalamic NEP
Reticular NEP
Anterior thalamic NEP

Cingulate
Hippocampal
Limbic
cortical NEP
Fornical GEP

POSTERIOR
POOL

Choroid
plexus
stem cells

FORAMEN
OF MONRO

THALAMIC
POOL

DIENCEPHALIC
SUPERVENTRICLE
(FUTURE THIRD VENTRICLE)

PREOPTIC
POOL

*Migrating
thalamic
neurons*

Dural outline is external border of
superarachnoid reticulum

Brain surface (heavier line)

*Cortical primordial
plexiform layer*

*Migrating
Cajal-Retzius cells
and subplate neurons*

*Migrating neurons
originating in* **cortico-
ganglionic NEP?**

Pioneer internal capsule axons

*Migrating basal ganglionic
neurons*

*Settling basal ganglionic
(globus pallidus?) neurons*

Settling basal telencephalic neurons

Migrating basal telencephalic neurons

Settling basal telencephalic neurons

*Migrating
preoptic
neurons*

ABBREVIATIONS:
GEP - Glioepithelium
NEP - Neuroepithelium

Arrows indicate the
presumed *direction of
neuron migration* from
neuroepithelial sources.

FONT KEY:
VENTRICULAR DIVISIONS - CAPITALS
Germinal zone - Helvetica bold
Transient structure - Times bold italic
Permanent structure - Times Roman or **Bold**

PLATE 69A
CR 17.5 mm, GW 7.8, M2155
Frontal/Horizontal, Section 242

See a high-magnification view
of the thalamus and cerebral
cortex from Section 236
in Plates 83A/B.

Peripheral neural and non-neural structures labeled

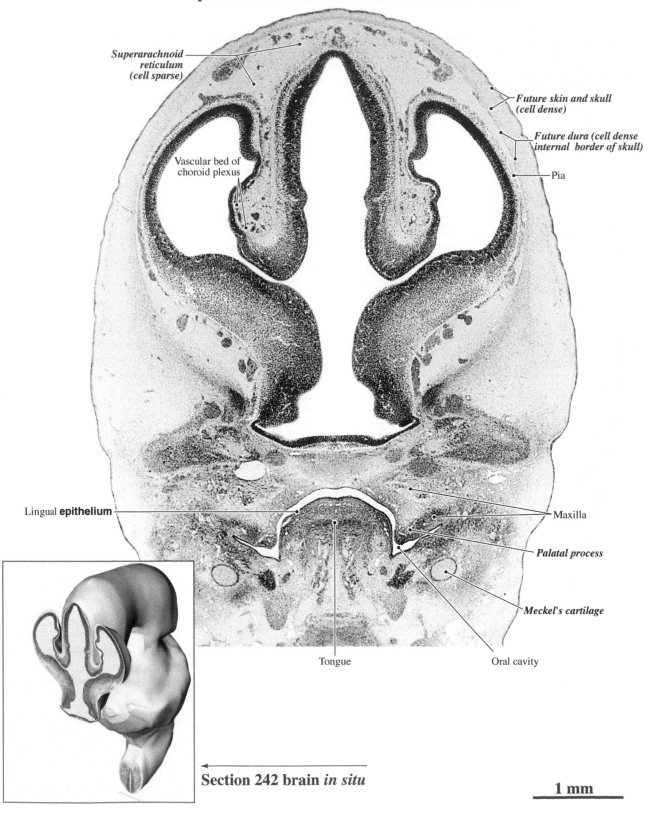

*Superarachnoid
reticulum
(cell sparse)*

*Future skin and skull
(cell dense)*

*Future dura (cell dense
internal border of skull)*

Pia

Vascular bed of
choroid plexus

Lingual **epithelium**

Maxilla

Palatal process

Meckel's cartilage

Tongue

Oral cavity

Section 242 brain *in situ*

1 mm

169

PLATE 69B

Central neural structures labeled

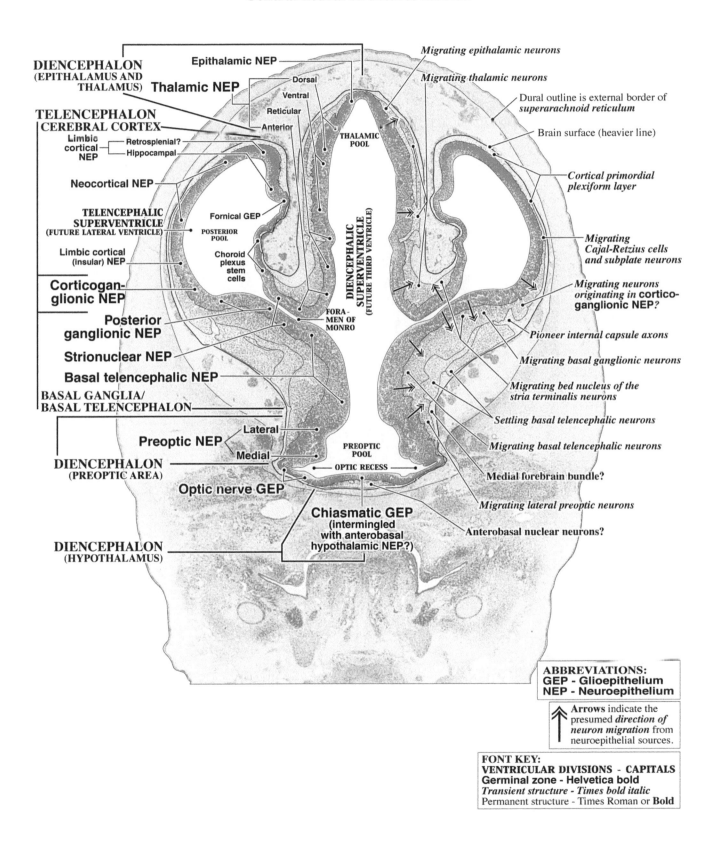

DIENCEPHALON (EPITHALAMUS AND THALAMUS)

Epithalamic NEP

Thalamic NEP

Dorsal
Ventral
Reticular
Anterior

THALAMIC POOL

Migrating epithalamic neurons

Migrating thalamic neurons

Dural outline is external border of *superarachnoid reticulum*

Brain surface (heavier line)

TELENCEPHALON CEREBRAL CORTEX

Limbic cortical NEP

Retrosplenial?
Hippocampal

Neocortical NEP

Cortical primordial plexiform layer

TELENCEPHALIC SUPERVENTRICLE (FUTURE LATERAL VENTRICLE)

Fornical GEP

POSTERIOR POOL

DIENCEPHALIC SUPERVENTRICLE (FUTURE THIRD VENTRICLE)

Migrating Cajal-Retzius cells and subplate neurons

Limbic cortical (insular) NEP

Choroid plexus stem cells

Migrating neurons originating in cortico-ganglionic NEP?

Corticogan-glionic NEP

Posterior ganglionic NEP

FORA-MEN OF MONRO

Pioneer internal capsule axons

Migrating basal ganglionic neurons

Strionuclear NEP

Basal telencephalic NEP

Migrating bed nucleus of the stria terminalis neurons

BASAL GANGLIA/ BASAL TELENCEPHALON

Settling basal telencephalic neurons

Preoptic NEP

Lateral
Medial

PREOPTIC POOL

Migrating basal telencephalic neurons

DIENCEPHALON (PREOPTIC AREA)

OPTIC RECESS

Medial forebrain bundle?

Optic nerve GEP

Migrating lateral preoptic neurons

Chiasmatic GEP (intermingled with anterobasal hypothalamic NEP?)

Anterobasal nuclear neurons?

DIENCEPHALON (HYPOTHALAMUS)

ABBREVIATIONS:
GEP - Glioepithelium
NEP - Neuroepithelium

Arrows indicate the presumed *direction of neuron migration* from neuroepithelial sources.

FONT KEY:
VENTRICULAR DIVISIONS - CAPITALS
Germinal zone - Helvetica bold
Transient structure - Times bold italic
Permanent structure - Times Roman or **Bold**

170

**PLATE 70A
CR 17.5 mm, GW 7.8
M2155
Frontal/Horizontal
Section 283**

**Peripheral neural and
non-neural structures labeled**

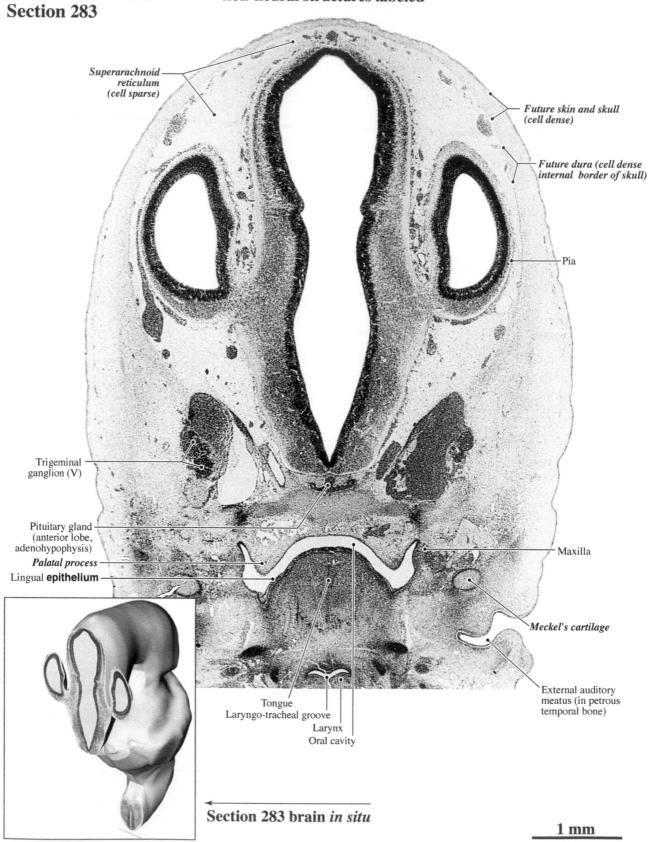

*Superarachnoid
reticulum
(cell sparse)*

*Future skin and skull
(cell dense)*

*Future dura (cell dense
internal border of skull)*

Pia

Trigeminal
ganglion (V)

Pituitary gland
(anterior lobe,
adenohypophysis)

Palatal process

Lingual **epithelium**

Maxilla

Meckel's cartilage

External auditory
meatus (in petrous
temporal bone)

Tongue
Laryngo-tracheal groove
Larynx
Oral cavity

Section 283 brain *in situ*

1 mm

Central neural structures labeled

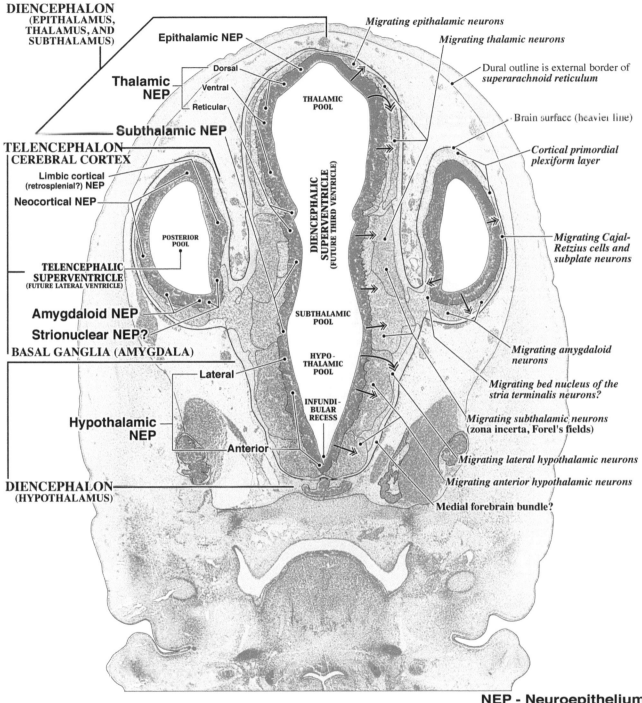

DIENCEPHALON
(EPITHALAMUS,
THALAMUS, AND
SUBTHALAMUS)

Epithalamic NEP

**Thalamic
NEP**

Dorsal

Ventral

Reticular

Subthalamic NEP

TELENCEPHALON
CEREBRAL CORTEX

Limbic cortical
(retrosplenial?) NEP

Neocortical NEP

POSTERIOR
POOL

**TELENCEPHALIC
SUPERVENTRICLE**
(FUTURE LATERAL VENTRICLE)

Amygdaloid NEP

Strionuclear NEP?

BASAL GANGLIA (AMYGDALA)

Lateral

**Hypothalamic
NEP**

Anterior

DIENCEPHALON
(HYPOTHALAMUS)

THALAMIC
POOL

**DIENCEPHALIC
SUPERVENTRICLE
(FUTURE THIRD VENTRICLE)**

SUBTHALAMIC
POOL

HYPO-
THALAMIC
POOL

INFUNDI-
BULAR
RECESS

Migrating epithalamic neurons

Migrating thalamic neurons

Dural outline is external border of
superarachnoid reticulum

Brain surface (heavier line)

*Cortical primordial
plexiform layer*

*Migrating Cajal-
Retzius cells and
subplate neurons*

*Migrating amygdaloid
neurons*

*Migrating bed nucleus of the
stria terminalis neurons?*

*Migrating subthalamic neurons
(zona incerta, Forel's fields)*

Migrating lateral hypothalamic neurons

Migrating anterior hypothalamic neurons

Medial forebrain bundle?

NEP - Neuroepithelium

Arrows indicate the
presumed *direction of
neuron migration* from
neuroepithelial sources.

FONT KEY:
VENTRICULAR DIVISIONS - CAPITALS
Germinal zone - Helvetica bold
Transient structure - Times bold italic
Permanent structure - Times Roman or **Bold**

PLATE 71A
CR 17.5 mm, GW 7.8
M2155
Frontal/Horizontal
Section 325

**Peripheral neural and
non-neural structures labeled**

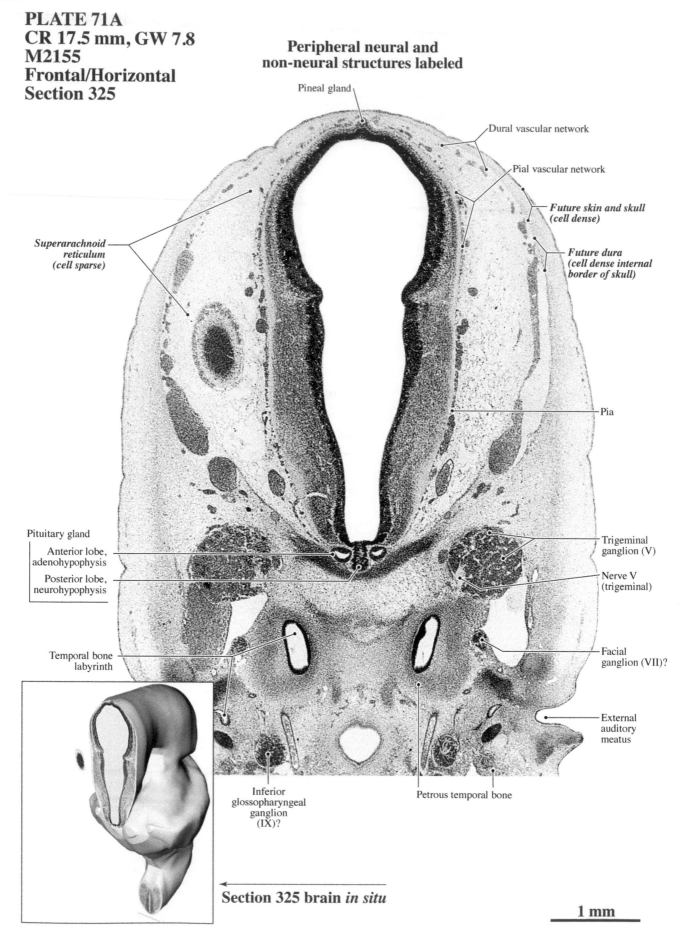

Pineal gland

Dural vascular network

Pial vascular network

*Future skin and skull
(cell dense)*

*Future dura
(cell dense internal
border of skull)*

*Superarachnoid
reticulum
(cell sparse)*

Pia

Pituitary gland
Anterior lobe,
adenohypophysis
Posterior lobe,
neurohypophysis

Trigeminal
ganglion (V)

Nerve V
(trigeminal)

Temporal bone
labyrinth

Facial
ganglion (VII)?

External
auditory
meatus

Inferior
glossopharyngeal
ganglion
(IX)?

Petrous temporal bone

Section 325 brain *in situ*

1 mm

Central neural structures labeled

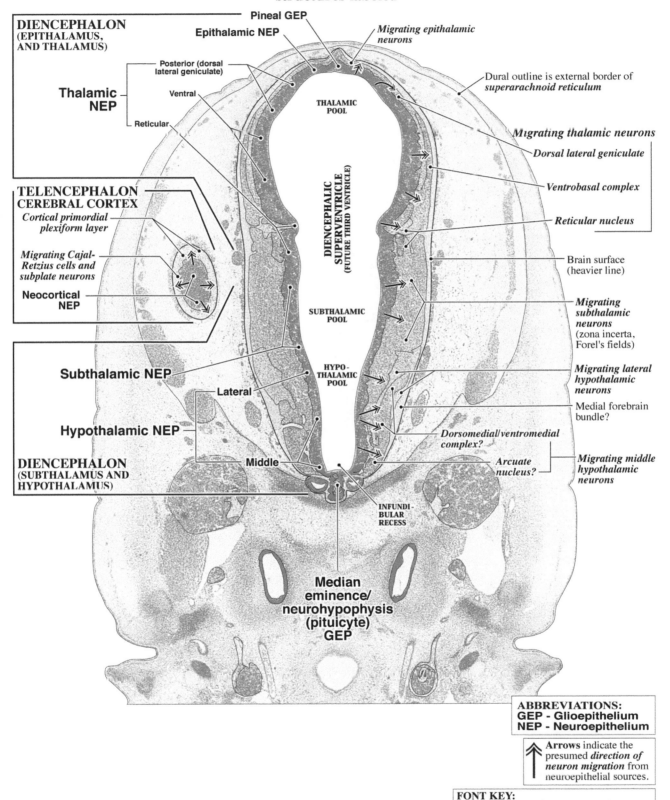

DIENCEPHALON
(EPITHALAMUS,
AND THALAMUS)

Pineal GEP

Epithalamic NEP

*Migrating epithalamic
neurons*

Thalamic
NEP

Posterior (dorsal
lateral geniculate)

Ventral

Reticular

THALAMIC
POOL

*Dural outline is external border of
superarachnoid reticulum*

Migrating thalamic neurons

Dorsal lateral geniculate

Ventrobasal complex

Reticular nucleus

TELENCEPHALON
CEREBRAL CORTEX

*Cortical primordial
plexiform layer*

*Migrating Cajal-
Retzius cells and
subplate neurons*

Neocortical
NEP

DIENCEPHALIC
SUPERVENTRICLE
(FUTURE THIRD VENTRICLE)

Brain surface
(heavier line)

*Migrating
subthalamic
neurons*
(zona incerta,
Forel's fields)

Subthalamic NEP

Hypothalamic NEP

DIENCEPHALON
(SUBTHALAMUS AND
HYPOTHALAMUS)

Lateral

Middle

SUBTHALAMIC
POOL

HYPO-
THALAMIC
POOL

*Migrating lateral
hypothalamic
neurons*

Medial forebrain
bundle?

*Dorsomedial/ventromedial
complex?*

*Arcuate
nucleus?*

*Migrating middle
hypothalamic
neurons*

INFUNDI-
BULAR
RECESS

Median
eminence/
neurohypophysis
(pituicyte)
GEP

ABBREVIATIONS:
GEP - Glioepithelium
NEP - Neuroepithelium

Arrows indicate the
presumed *direction of
neuron migration* from
neuroepithelial sources.

FONT KEY:
VENTRICULAR DIVISIONS - CAPITALS
Germinal zone - Helvetica bold
Transient structure - Times bold italic
Permanent structure - Times Roman or **Bold**

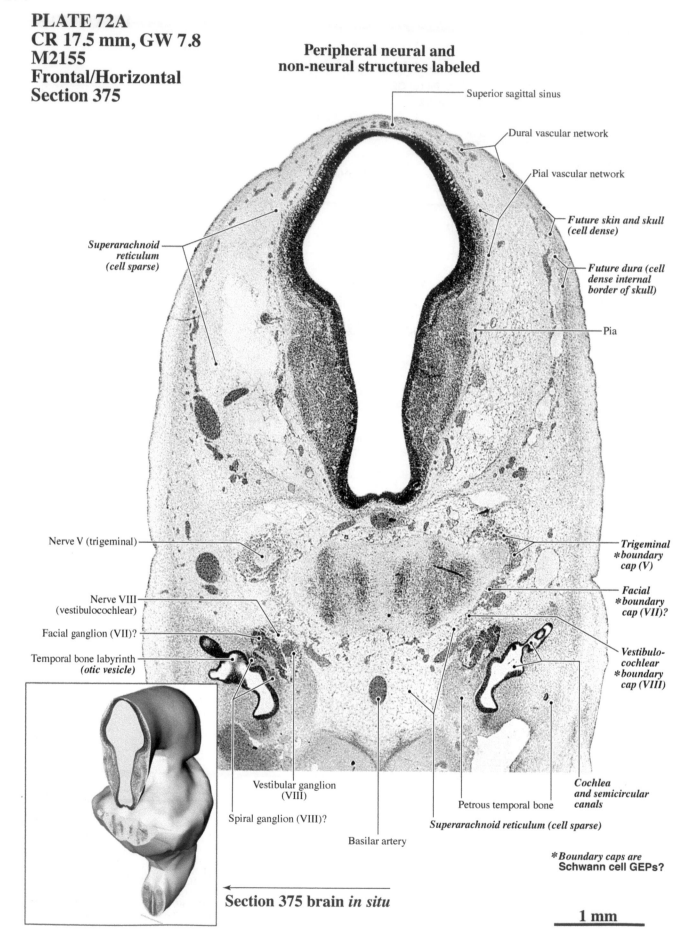

PLATE 72A
CR 17.5 mm, GW 7.8
M2155
Frontal/Horizontal
Section 375

Peripheral neural and
non-neural structures labeled

Superior sagittal sinus

Dural vascular network

Pial vascular network

Future skin and skull
(cell dense)

Future dura (cell
dense internal
border of skull)

Pia

Superarachnoid
reticulum
(cell sparse)

Nerve V (trigeminal)

Trigeminal
*boundary
cap (V)

Facial
*boundary
cap (VII)?

Nerve VIII
(vestibulocochlear)

Facial ganglion (VII)?

Vestibulo-
cochlear
*boundary
cap (VIII)

Temporal bone labyrinth
(otic vesicle)

Vestibular ganglion
(VIII)

Spiral ganglion (VIII)?

Basilar artery

Petrous temporal bone

Superarachnoid reticulum (cell sparse)

Cochlea
and semicircular
canals

**Boundary caps are*
Schwann cell GEPs?

Section 375 brain *in situ*

1 mm

**Central neural
structures labeled**

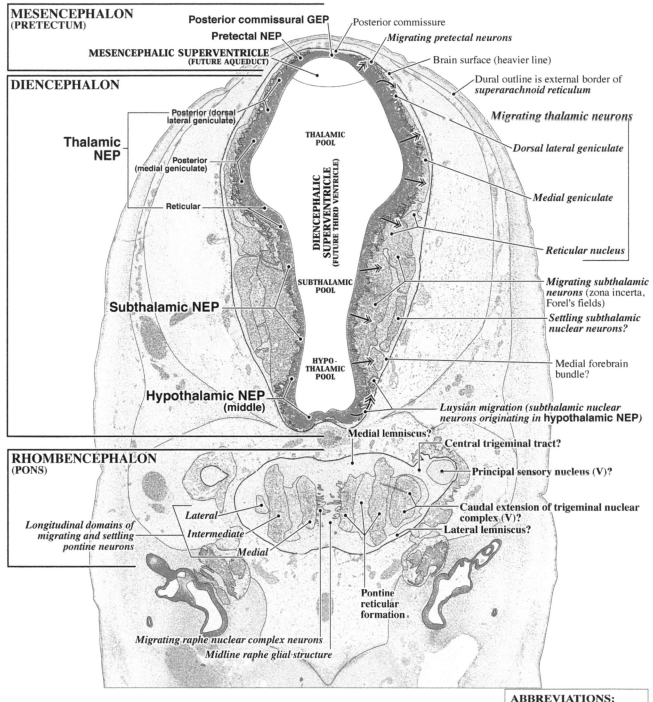

MESENCEPHALON
(PRETECTUM)

Posterior commissural GEP — Posterior commissure

Pretectal NEP — *Migrating pretectal neurons*

MESENCEPHALIC SUPERVENTRICLE
(FUTURE AQUEDUCT)

— Brain surface (heavier line)

DIENCEPHALON

Dural outline is external border of
superarachnoid reticulum

Posterior (dorsal
lateral geniculate)

Migrating thalamic neurons

**Thalamic
NEP**

THALAMIC
POOL

Dorsal lateral geniculate

Posterior
(medial geniculate)

DIENCEPHALIC
SUPERVENTRICLE
(FUTURE THIRD VENTRICLE)

Medial geniculate

Reticular

Reticular nucleus

SUBTHALAMIC
POOL

*Migrating subthalamic
neurons* (zona incerta,
Forel's fields)

Subthalamic NEP

*Settling subthalamic
nuclear neurons?*

HYPO-
THALAMIC
POOL

— Medial forebrain
bundle?

Hypothalamic NEP
(middle)

*Luysian migration (subthalamic nuclear
neurons originating in* **hypothalamic NEP**)

Medial lemniscus? **Central trigeminal tract?**

RHOMBENCEPHALON
(PONS)

Principal sensory nucleus (V)?

Lateral

**Caudal extension of trigeminal nuclear
complex (V)?**

*Longitudinal domains of
migrating and settling
pontine neurons*

Intermediate

Lateral lemniscus?

Medial

Pontine
reticular
formation

Migrating raphe nuclear complex neurons

Midline raphe glial structure

ABBREVIATIONS:
GEP - Glioepithelium
NEP - Neuroepithelium

Arrows indicate the
presumed *direction of
neuron migration* from
neuroepithelial sources.

FONT KEY:
VENTRICULAR DIVISIONS - CAPITALS
Germinal zone - Helvetica bold
Transient structure - Times bold italic
Permanent structure - Times Roman or **Bold**

PLATE 73A
CR 17.5 mm, GW 7.8
M2155
Frontal/Horizontal
Section 410

**Peripheral neural and
non-neural structures labeled**

**See a high-magnification view
of the mesencephalon and
diencephalon from
Section 390 in
Plates 84A/B.**

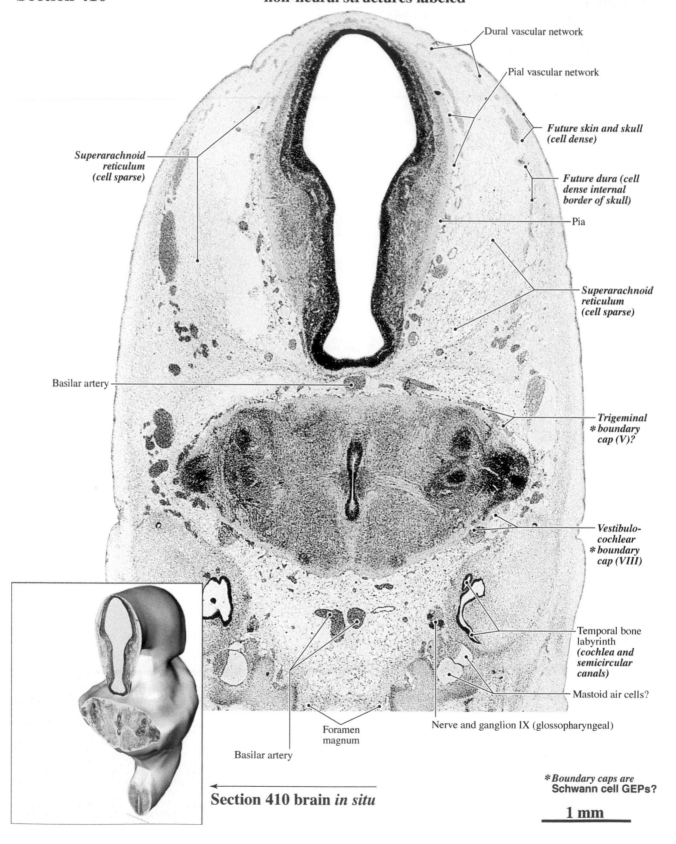

Dural vascular network

Pial vascular network

*Future skin and skull
(cell dense)*

*Future dura (cell
dense internal
border of skull)*

Pia

*Superarachnoid
reticulum
(cell sparse)*

*Superarachnoid
reticulum
(cell sparse)*

Basilar artery

*Trigeminal
* boundary
cap (V)?*

*Vestibulo-
cochlear
* boundary
cap (VIII)*

Temporal bone
labyrinth
*(cochlea and
semicircular
canals)*

Mastoid air cells?

Nerve and ganglion IX (glossopharyngeal)

Foramen
magnum

Basilar artery

Section 410 brain *in situ*

*Boundary caps are
Schwann cell GEPs?

1 mm

Central neural structures labeled

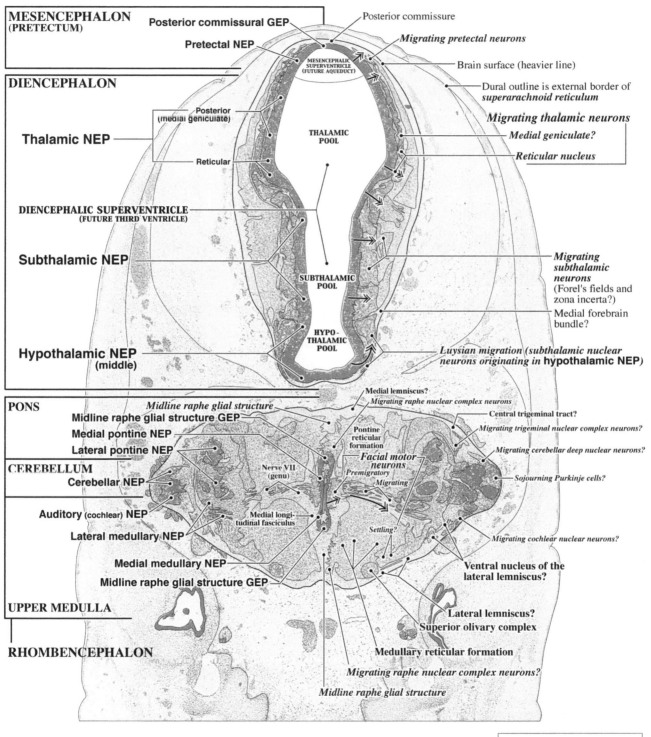

MESENCEPHALON
(PRETECTUM)

Posterior commissural GEP

Pretectal NEP

Posterior commissure

Migrating pretectal neurons

MESENCEPHALIC
SUPERVENTRICLE
(FUTURE AQUEDUCT)

Brain surface (heavier line)

DIENCEPHALON

THALAMIC
POOL

Dural outline is external border of
superarachnoid reticulum

Migrating thalamic neurons

Medial geniculate?

Posterior
(medial geniculate)

Thalamic NEP

Reticular nucleus

Reticular

DIENCEPHALIC SUPERVENTRICLE
(FUTURE THIRD VENTRICLE)

Subthalamic NEP

SUBTHALAMIC
POOL

Migrating
subthalamic
neurons
(Forel's fields and
zona incerta?)

Medial forebrain
bundle?

HYPO-
THALAMIC
POOL

Hypothalamic NEP
(middle)

Luysian migration (subthalamic nuclear
neurons originating in **hypothalamic NEP**)

Medial lemniscus?

Migrating raphe nuclear complex neurons

PONS

Midline raphe glial structure

Midline raphe glial structure GEP

Medial pontine NEP

Lateral pontine NEP

Pontine
reticular
formation

Central trigeminal tract?

Migrating trigeminal nuclear complex neurons?

Migrating cerebellar deep nuclear neurons?

Facial motor
neurons

Premigratory

Migrating

Sojourning Purkinje cells?

CEREBELLUM

Cerebellar NEP

Nerve VII
(genu)

Auditory (cochlear) NEP

Medial longi-
tudinal fasciculus

Settling?

Migrating cochlear nuclear neurons?

Lateral medullary NEP

Medial medullary NEP

Ventral nucleus of the
lateral lemniscus?

Midline raphe glial structure GEP

Lateral lemniscus?

UPPER MEDULLA

Superior olivary complex

Medullary reticular formation

RHOMBENCEPHALON

Migrating raphe nuclear complex neurons?

Midline raphe glial structure

ABBREVIATIONS:
GEP - Glioepithelium
NEP - Neuroepithelium

FONT KEY:
VENTRICULAR DIVISIONS - CAPITALS
Germinal zone - Helvetica bold
Transient structure - Times bold italic
Permanent structure - Times Roman or **Bold**

Arrows indicate the
presumed *direction of*
neuron migration from
neuroepithelial sources.

PLATE 74A
CR 17.5 mm, GW 7.8
M2155
Frontal/Horizontal
Section 424

Peripheral neural and non-neural structures labeled

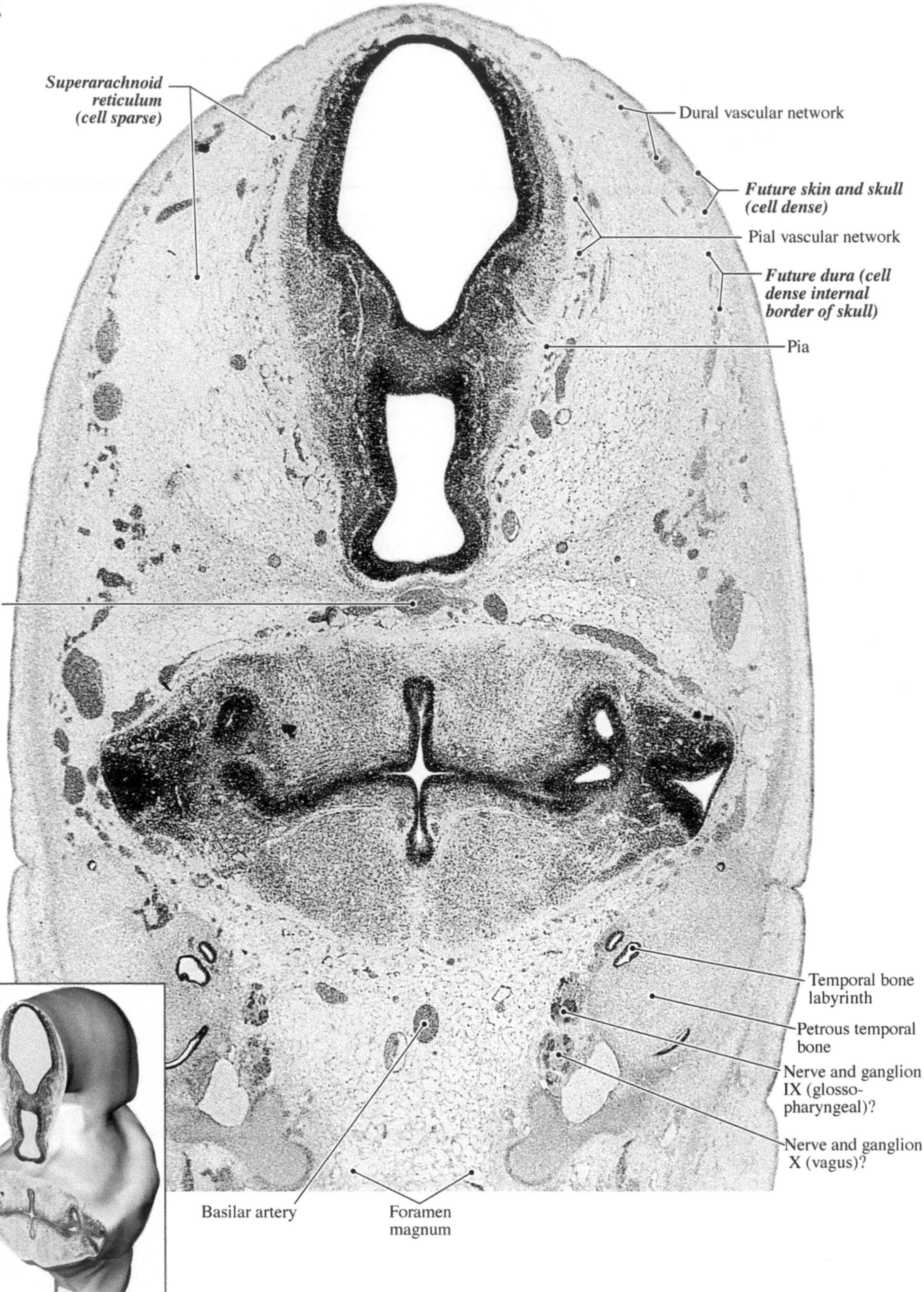

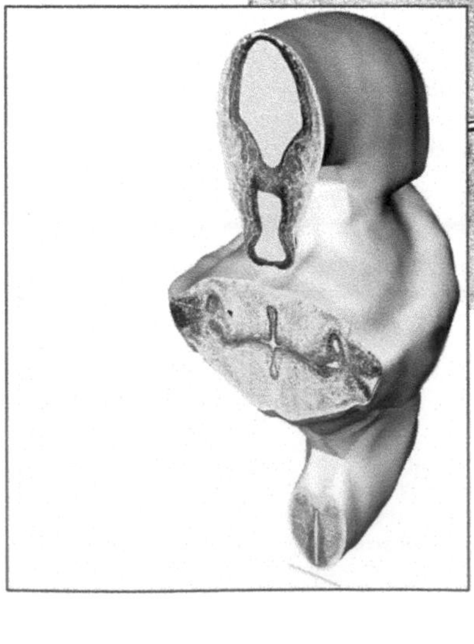

Superarachnoid reticulum (cell sparse)
Dural vascular network
Future skin and skull (cell dense)
Pial vascular network
Future dura (cell dense internal border of skull)
Pia
Basilar artery
Temporal bone labyrinth
Petrous temporal bone
Nerve and ganglion IX (glosso-pharyngeal)?
Nerve and ganglion X (vagus)?
Basilar artery
Foramen magnum
1 mm

Section 424 brain *in situ*

Central neural structures labeled

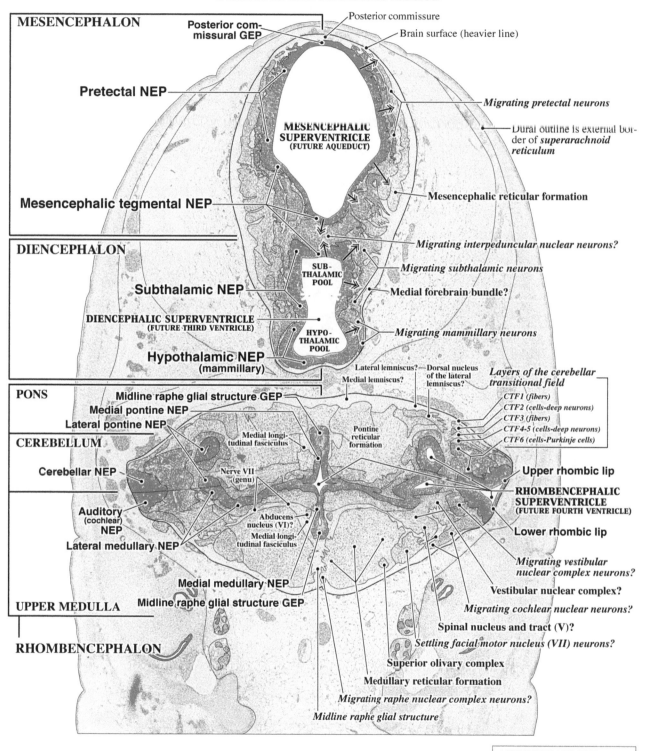

MESENCEPHALON

Posterior com-
missural GEP

Posterior commissure

Brain surface (heavier line)

Pretectal NEP

Migrating pretectal neurons

MESENCEPHALIC
SUPERVENTRICLE
(FUTURE AQUEDUCT)

*Dural outline is external bor-
der of superarachnoid
reticulum*

Mesencephalic tegmental NEP

Mesencephalic reticular formation

Migrating interpeduncular nuclear neurons?

DIENCEPHALON

SUB-
THALAMIC
POOL

Migrating subthalamic neurons

Subthalamic NEP

Medial forebrain bundle?

DIENCEPHALIC SUPERVENTRICLE
(FUTURE THIRD VENTRICLE)

HYPO-
THALAMIC
POOL

Migrating mammillary neurons

Hypothalamic NEP
(mammillary)

Lateral lemniscus?
Medial lemniscus?

Dorsal nucleus
of the lateral
lemniscus?

*Layers of the cerebellar
transitional field*

CTF1 (fibers)
CTF2 (cells-deep neurons)
CTF3 (fibers)
CTF4-5 (cells-deep neurons)
CTF6 (cells-Purkinje cells)

PONS

Midline raphe glial structure GEP
Medial pontine NEP
Lateral pontine NEP

Medial longi-
tudinal fasciculus

Pontine
reticular
formation

CEREBELLUM

Cerebellar NEP

Nerve VII
(genu)

Upper rhombic lip

**RHOMBENCEPHALIC
SUPERVENTRICLE
(FUTURE FOURTH VENTRICLE)**

Auditory
(cochlear)
NEP

Abducens
nucleus (VI)?
Medial longi-
tudinal fasciculus

Lower rhombic lip

Lateral medullary NEP

*Migrating vestibular
nuclear complex neurons?*

Vestibular nuclear complex?

Medial medullary NEP

Migrating cochlear nuclear neurons?

Midline raphe glial structure GEP

UPPER MEDULLA

Spinal nucleus and tract (V)?

Settling facial motor nucleus (VII) neurons?

RHOMBENCEPHALON

Superior olivary complex

Medullary reticular formation

Migrating raphe nuclear complex neurons?

Midline raphe glial structure

ABBREVIATIONS:
GEP - Glioepithelium
NEP - Neuroepithelium

FONT KEY:
VENTRICULAR DIVISIONS - CAPITALS
Germinal zone - Helvetica bold
Transient structure - Times bold italic
Permanent structure - Times Roman or **Bold**

Arrows indicate the
presumed *direction of
neuron migration* from
neuroepithelial sources.

PLATE 75A
CR 17.5 mm, GW 7.8
M2155
Frontal/Horizontal
Section 444

Peripheral neural and
non-neural structures labeled

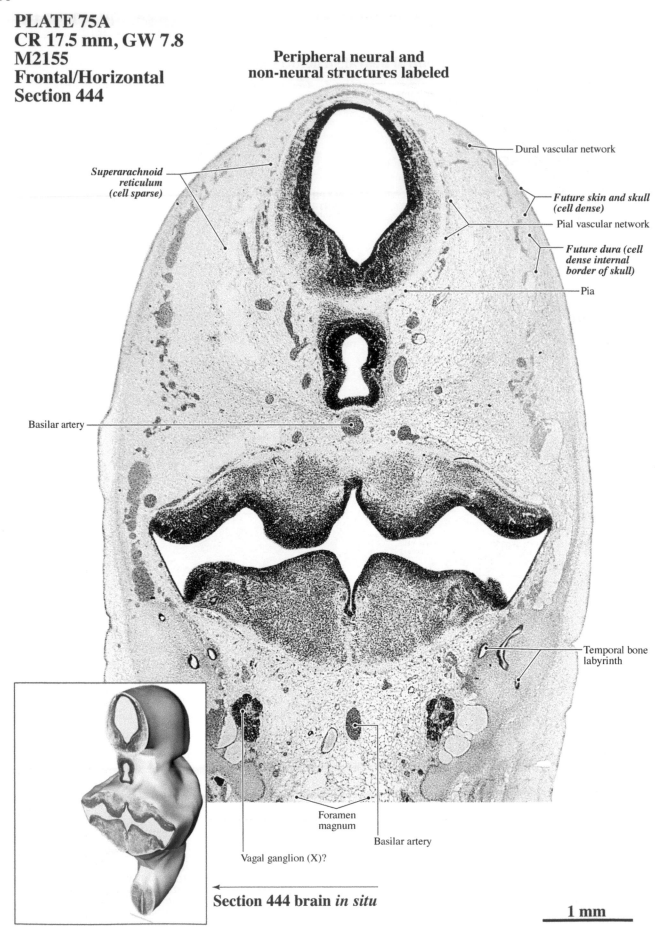

Superarachnoid
reticulum
(cell sparse)

Dural vascular network

Future skin and skull
(cell dense)

Pial vascular network

Future dura (cell
dense internal
border of skull)

Pia

Basilar artery

Temporal bone
labyrinth

Foramen
magnum

Basilar artery

Vagal ganglion (X)?

Section 444 brain *in situ*

1 mm

Central neural structures labeled

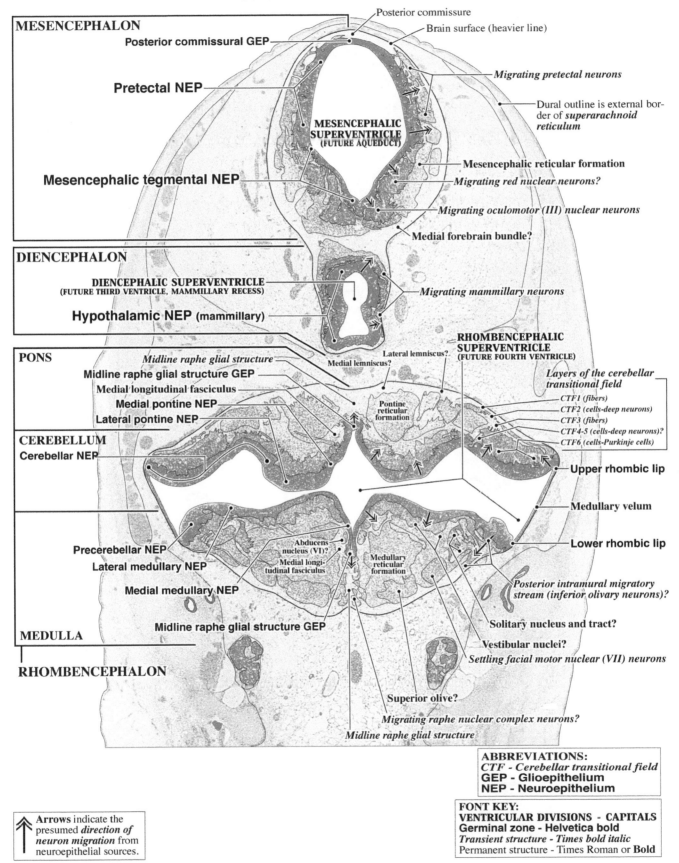

Posterior commissure

Brain surface (heavier line)

MESENCEPHALON

Posterior commissural GEP

Migrating pretectal neurons

Pretectal NEP

Dural outline is external border of *superarachnoid reticulum*

MESENCEPHALIC SUPERVENTRICLE (FUTURE AQUEDUCT)

Mesencephalic reticular formation

Mesencephalic tegmental NEP

Migrating red nuclear neurons?

Migrating oculomotor (III) nuclear neurons

Medial forebrain bundle?

DIENCEPHALON

DIENCEPHALIC SUPERVENTRICLE (FUTURE THIRD VENTRICLE, MAMMILLARY RECESS)

Migrating mammillary neurons

Hypothalamic NEP (mammillary)

RHOMBENCEPHALIC SUPERVENTRICLE (FUTURE FOURTH VENTRICLE)

PONS

Lateral lemniscus?

Medial lemniscus?

Midline raphe glial structure

Midline raphe glial structure GEP

Layers of the cerebellar transitional field

Medial longitudinal fasciculus

Pontine reticular formation

CTF1 (fibers)

Medial pontine NEP

CTF2 (cells-deep neurons)

Lateral pontine NEP

CTF3 (fibers)

CTF4-5 (cells-deep neurons)?

CTF6 (cells-Purkinje cells)

CEREBELLUM

Cerebellar NEP

Upper rhombic lip

Medullary velum

Precerebellar NEP

Abducens nucleus (VI)?

Lower rhombic lip

Medial longitudinal fasciculus

Medullary reticular formation

Lateral medullary NEP

Medial medullary NEP

Posterior intramural migratory stream (inferior olivary neurons)?

Midline raphe glial structure GEP

Solitary nucleus and tract?

Vestibular nuclei?

MEDULLA

Settling facial motor nuclear (VII) neurons

RHOMBENCEPHALON

Superior olive?

Migrating raphe nuclear complex neurons?

Midline raphe glial structure

ABBREVIATIONS:
CTF - Cerebellar transitional field
GEP - Glioepithelium
NEP - Neuroepithelium

Arrows indicate the presumed *direction of neuron migration* from neuroepithelial sources.

FONT KEY:
VENTRICULAR DIVISIONS - CAPITALS
Germinal zone - Helvetica bold
Transient structure - Times bold italic
Permanent structure - Times Roman or **Bold**

182

**Peripheral neural and
non-neural structures labeled**

Dural vascular network

Pial vascular network

Pia

*Superarachnoid
reticulum
(cell sparse)*

*Future skin and skull
(cell dense)*

Nerve III (oculomotor)

*Future dura (cell
dense internal
border of skull)*

*Glossopharyngeal
* boundary cap (IX)?*

*Vagal boundary *
cap (X)?*

Traces of the vagal (X)
and glossopharyngeal (IX) nerve sheaths?

*Boundary caps are
Schwann cell GEPs?

Section 500 brain *in situ*

1 mm

Central neural structures labeled

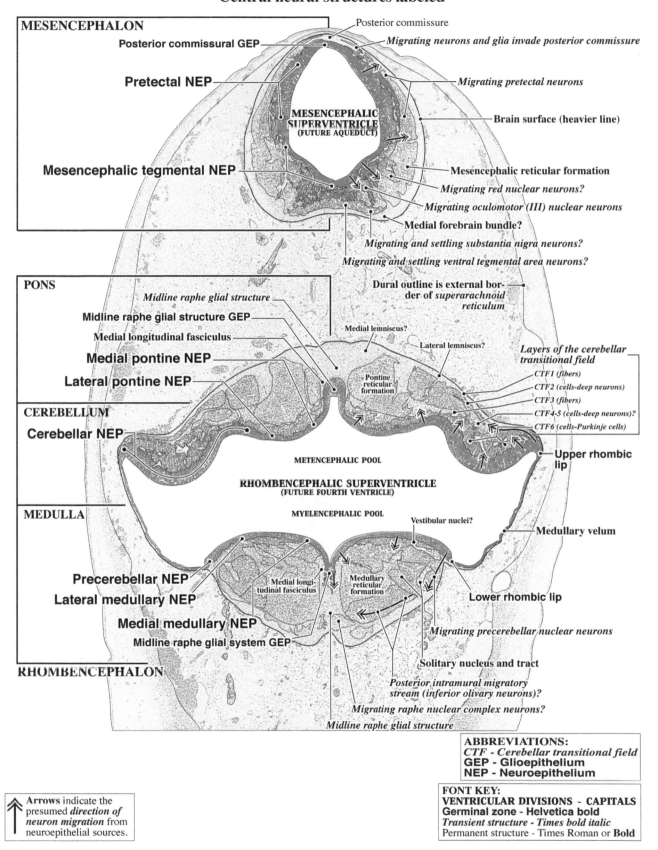

MESENCEPHALON

Posterior commissure

Posterior commissural GEP

Migrating neurons and glia invade posterior commissure

Pretectal NEP

Migrating pretectal neurons

MESENCEPHALIC SUPERVENTRICLE (FUTURE AQUEDUCT)

Brain surface (heavier line)

Mesencephalic tegmental NEP

Mesencephalic reticular formation

Migrating red nuclear neurons?

Migrating oculomotor (III) nuclear neurons

Medial forebrain bundle?

Migrating and settling substantia nigra neurons?

Migrating and settling ventral tegmental area neurons?

PONS

Dural outline is external border of superarachnoid reticulum

Midline raphe glial structure

Midline raphe glial structure GEP

Medial longitudinal fasciculus

Medial lemniscus?

Lateral lemniscus?

Medial pontine NEP

Lateral pontine NEP

Pontine reticular formation

Layers of the cerebellar transitional field

CTF1 (fibers)
CTF2 (cells-deep neurons)
CTF3 (fibers)
CTF4-5 (cells-deep neurons)?
CTF6 (cells-Purkinje cells)

CEREBELLUM

Cerebellar NEP

Upper rhombic lip

METENCEPHALIC POOL

RHOMBENCEPHALIC SUPERVENTRICLE (FUTURE FOURTH VENTRICLE)

MEDULLA

MYELENCEPHALIC POOL

Vestibular nuclei?

Medullary velum

Precerebellar NEP

Medial longitudinal fasciculus

Medullary reticular formation

Lateral medullary NEP

Lower rhombic lip

Medial medullary NEP

Midline raphe glial system GEP

Migrating precerebellar nuclear neurons

Solitary nucleus and tract

RHOMBENCEPHALON

Posterior intramural migratory stream (inferior olivary neurons)?

Migrating raphe nuclear complex neurons?

Midline raphe glial structure

ABBREVIATIONS:
CTF - Cerebellar transitional field
GEP - Glioepithelium
NEP - Neuroepithelium

FONT KEY:
VENTRICULAR DIVISIONS - CAPITALS
Germinal zone - Helvetica bold
Transient structure - Times bold italic
Permanent structure - Times Roman or **Bold**

Arrows indicate the presumed *direction of neuron migration* from neuroepithelial sources.

184

**Peripheral neural and
non-neural structures labeled**

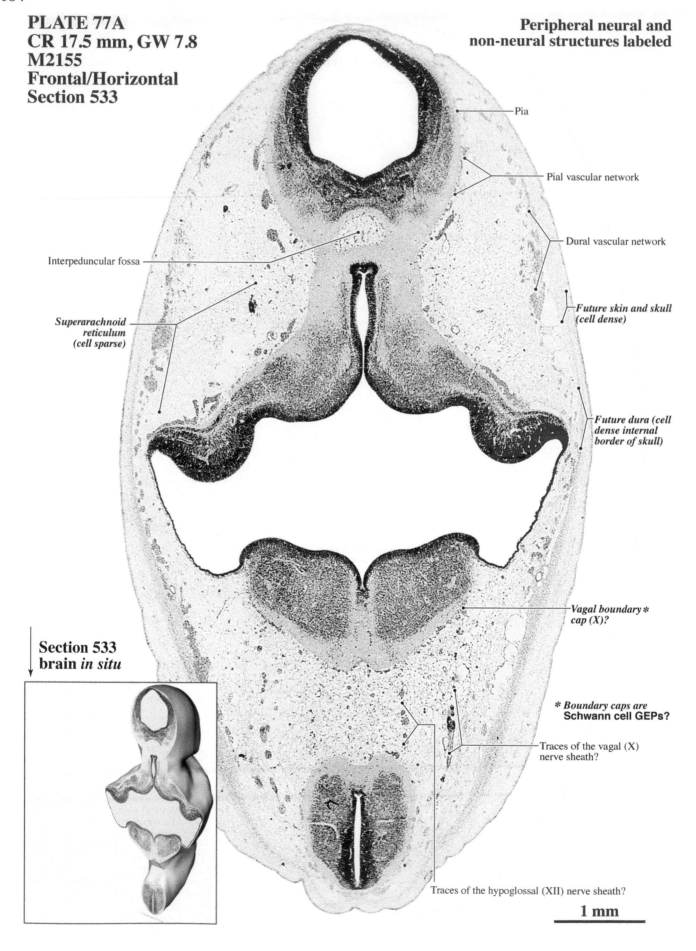

Pia

Pial vascular network

Dural vascular network

Interpeduncular fossa

*Future skin and skull
(cell dense)*

*Superarachnoid
reticulum
(cell sparse)*

*Future dura (cell
dense internal
border of skull)*

*Vagal boundary∗
cap (X)?*

**Section 533
brain *in situ***

∗ *Boundary caps are*
Schwann cell GEPs?

Traces of the vagal (X)
nerve sheath?

Traces of the hypoglossal (XII) nerve sheath?

1 mm

Central neural structures labeled

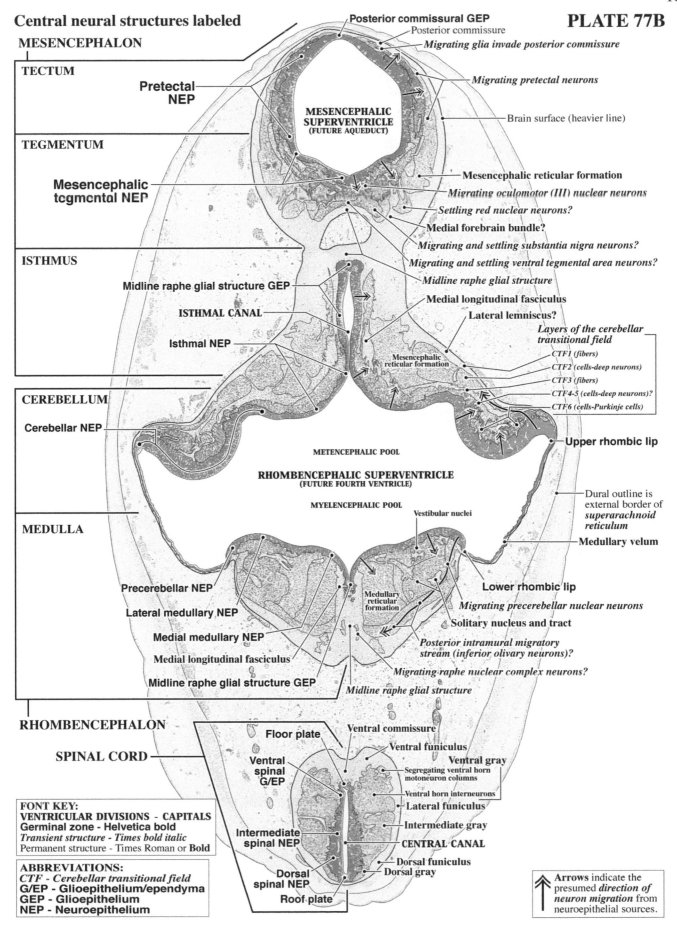

MESENCEPHALON

TECTUM

TEGMENTUM

ISTHMUS

CEREBELLUM

MEDULLA

RHOMBENCEPHALON

SPINAL CORD

Posterior commissural GEP
Posterior commissure
Migrating glia invade posterior commissure

Pretectal NEP

MESENCEPHALIC SUPERVENTRICLE (FUTURE AQUEDUCT)

Migrating pretectal neurons

Brain surface (heavier line)

Mesencephalic tegmental NEP

Mesencephalic reticular formation
Migrating oculomotor (III) nuclear neurons
Settling red nuclear neurons?
Medial forebrain bundle?
Migrating and settling substantia nigra neurons?
Migrating and settling ventral tegmental area neurons?
Midline raphe glial structure

Midline raphe glial structure GEP

Medial longitudinal fasciculus
Lateral lemniscus?

ISTHMAL CANAL

Isthmal NEP

Mesencephalic reticular formation

Layers of the cerebellar transitional field
CTF1 (fibers)
CTF2 (cells-deep neurons)
CTF3 (fibers)
CTF4-5 (cells-deep neurons)?
CTF6 (cells-Purkinje cells)

Cerebellar NEP

Upper rhombic lip

METENCEPHALIC POOL

RHOMBENCEPHALIC SUPERVENTRICLE (FUTURE FOURTH VENTRICLE)

MYELENCEPHALIC POOL

Vestibular nuclei

Dural outline is external border of *superarachnoid reticulum*

Medullary velum

Precerebellar NEP

Lateral medullary NEP

Medial medullary NEP

Medial longitudinal fasciculus

Midline raphe glial structure GEP

Medullary reticular formation

Lower rhombic lip
Migrating precerebellar nuclear neurons
Solitary nucleus and tract
Posterior intramural migratory stream (inferior olivary neurons)?
Migrating raphe nuclear complex neurons?

Midline raphe glial structure

Floor plate
Ventral spinal G/EP

Ventral commissure
Ventral funiculus

Ventral gray
Segregating ventral horn motoneuron columns
Ventral horn interneurons
Lateral funiculus
Intermediate gray
CENTRAL CANAL
Dorsal funiculus
Dorsal gray

Intermediate spinal NEP

Dorsal spinal NEP

Roof plate

FONT KEY:
VENTRICULAR DIVISIONS - CAPITALS
Germinal zone - Helvetica bold
Transient structure - Times bold italic
Permanent structure - Times Roman or **Bold**

ABBREVIATIONS:
CTF - Cerebellar transitional field
G/EP - Glioepithelium/ependyma
GEP - Glioepithelium
NEP - Neuroepithelium

Arrows indicate the presumed *direction of neuron migration* from neuroepithelial sources.

186

PLATE 78A
CR 17.5 mm, GW 7.8
M2155
Frontal/Horizontal
Section 572

Peripheral neural and
non-neural structures labeled

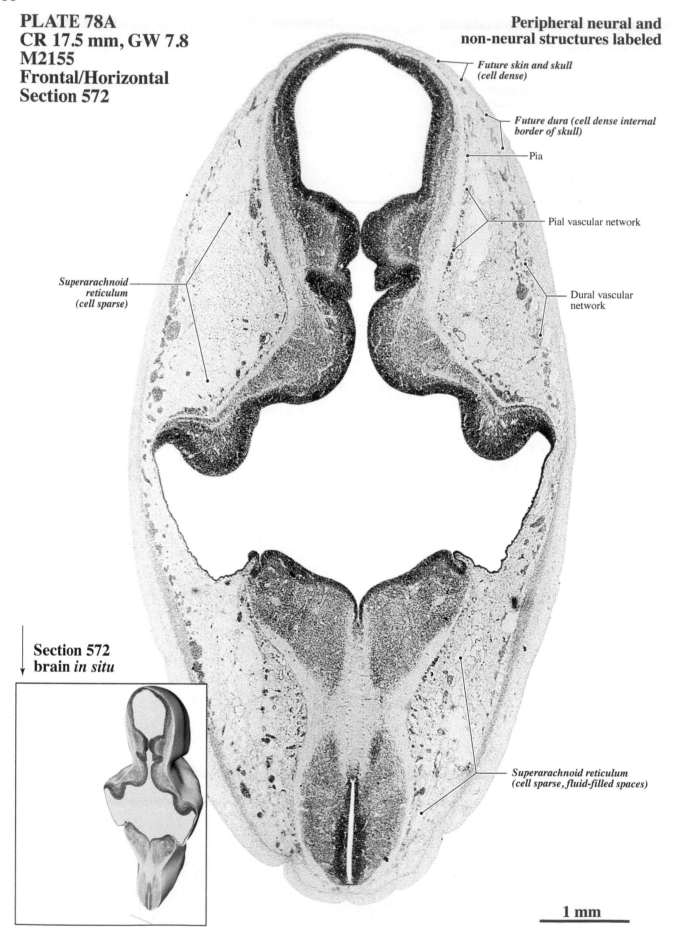

Future skin and skull
(cell dense)

Future dura (cell dense internal
border of skull)

Pia

Pial vascular network

Dural vascular
network

Superarachnoid
reticulum
(cell sparse)

Section 572
brain *in situ*

Superarachnoid reticulum
(cell sparse, fluid-filled spaces)

1 mm

Central neural structures labeled

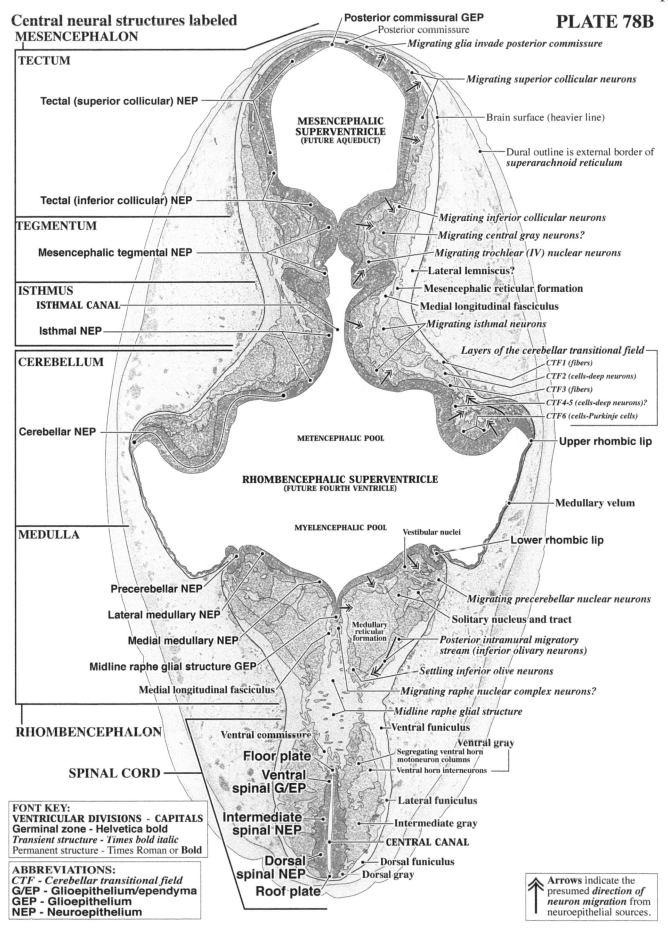

PLATE 78B

Posterior commissural GEP
Posterior commissure
Migrating glia invade posterior commissure

MESENCEPHALON

TECTUM

Tectal (superior collicular) NEP

Migrating superior collicular neurons

MESENCEPHALIC SUPERVENTRICLE (FUTURE AQUEDUCT)

Brain surface (heavier line)

Dural outline is external border of *superarachnoid reticulum*

Tectal (inferior collicular) NEP

Migrating inferior collicular neurons

TEGMENTUM

Migrating central gray neurons?

Migrating trochlear (IV) nuclear neurons

Mesencephalic tegmental NEP

Lateral lemniscus?

Mesencephalic reticular formation

Medial longitudinal fasciculus

ISTHMUS

ISTHMAL CANAL

Migrating isthmal neurons

Isthmal NEP

Layers of the cerebellar transitional field
CTF1 (fibers)
CTF2 (cells-deep neurons)
CTF3 (fibers)
CTF4-5 (cells-deep neurons)?
CTF6 (cells-Purkinje cells)

CEREBELLUM

METENCEPHALIC POOL

Cerebellar NEP

Upper rhombic lip

RHOMBENCEPHALIC SUPERVENTRICLE (FUTURE FOURTH VENTRICLE)

MYELENCEPHALIC POOL

Medullary velum

Vestibular nuclei

MEDULLA

Lower rhombic lip

Migrating precerebellar nuclear neurons

Precerebellar NEP

Solitary nucleus and tract

Lateral medullary NEP

Medullary reticular formation

Posterior intramural migratory stream (inferior olivary neurons)

Medial medullary NEP

Settling inferior olive neurons

Midline raphe glial structure GEP

Migrating raphe nuclear complex neurons?

Medial longitudinal fasciculus

Midline raphe glial structure

Ventral funiculus

RHOMBENCEPHALON

Ventral commissure

ventral gray

Segregating ventral horn motoneuron columns

Floor plate

Ventral horn interneurons

SPINAL CORD

Ventral spinal G/EP

Lateral funiculus

Intermediate spinal NEP

Intermediate gray

CENTRAL CANAL

Dorsal spinal NEP

Dorsal funiculus

Roof plate

Dorsal gray

FONT KEY:
VENTRICULAR DIVISIONS - CAPITALS
Germinal zone - Helvetica bold
Transient structure - Times bold italic
Permanent structure - Times Roman or **Bold**

ABBREVIATIONS:
CTF - Cerebellar transitional field
G/EP - Glioepithelium/ependyma
GEP - Glioepithelium
NEP - Neuroepithelium

Arrows indicate the presumed *direction of neuron migration* from neuroepithelial sources.

**PLATE 79A
CR 17.5 mm, GW 7.8
M2155
Frontal/Horizontal
Section 588**

**Peripheral neural and
non-neural structures labeled**

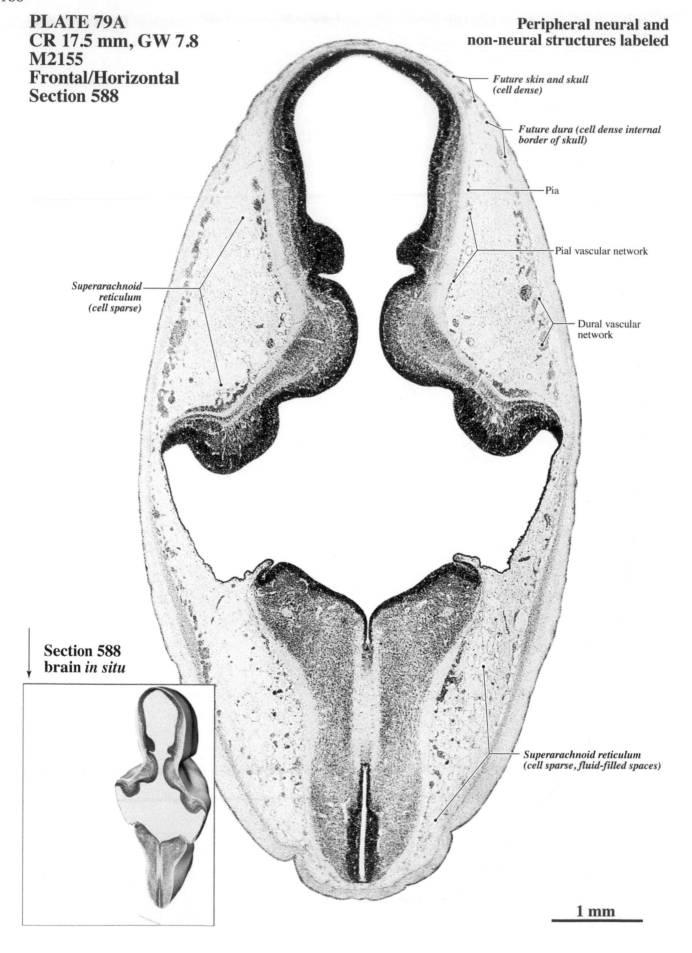

*Future skin and skull
(cell dense)*

*Future dura (cell dense internal
border of skull)*

Pia

Pial vascular network

Dural vascular
network

*Superarachnoid
reticulum
(cell sparse)*

*Superarachnoid reticulum
(cell sparse, fluid-filled spaces)*

**Section 588
brain *in situ***

1 mm

Central neural structures labeled

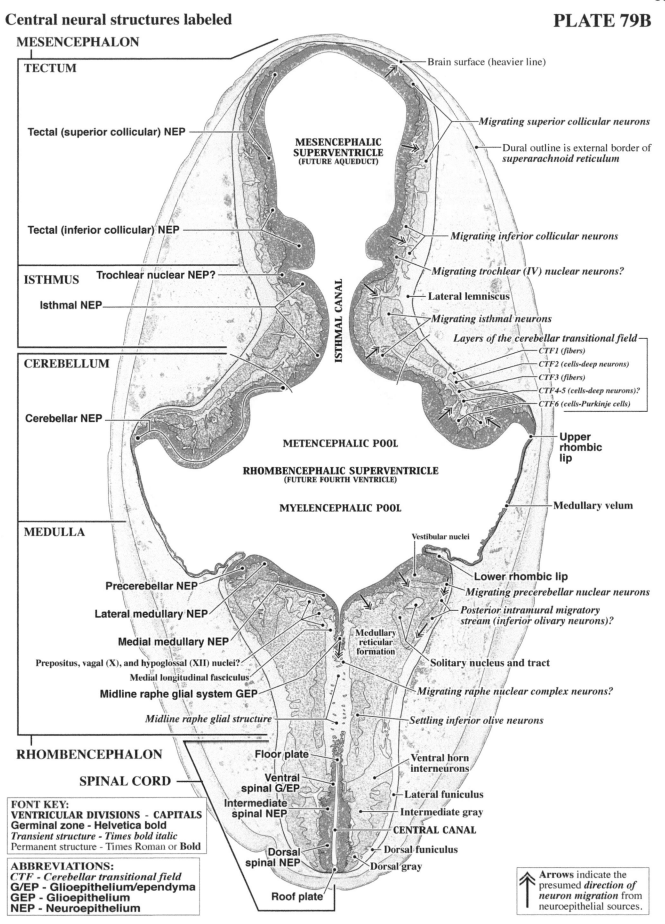

MESENCEPHALON

TECTUM

Tectal (superior collicular) NEP

Tectal (inferior collicular) NEP

ISTHMUS

Trochlear nuclear NEP?

Isthmal NEP

CEREBELLUM

Cerebellar NEP

MEDULLA

Precerebellar NEP

Lateral medullary NEP

Medial medullary NEP

Prepositus, vagal (X), and hypoglossal (XII) nuclei?

Medial longitudinal fasciculus

Midline raphe glial system GEP

Midline raphe glial structure

RHOMBENCEPHALON

SPINAL CORD

Brain surface (heavier line)

Migrating superior collicular neurons

Dural outline is external border of *superarachnoid reticulum*

Migrating inferior collicular neurons

Migrating trochlear (IV) nuclear neurons?

Lateral lemniscus

Migrating isthmal neurons

Layers of the cerebellar transitional field

CTF1 (fibers)
CTF2 (cells-deep neurons)
CTF3 (fibers)
CTF4-5 (cells-deep neurons)?
CTF6 (cells-Purkinje cells)

MESENCEPHALIC SUPERVENTRICLE (FUTURE AQUEDUCT)

ISTHMAL CANAL

METENCEPHALIC POOL

RHOMBENCEPHALIC SUPERVENTRICLE (FUTURE FOURTH VENTRICLE)

MYELENCEPHALIC POOL

Upper rhombic lip

Medullary velum

Vestibular nuclei

Lower rhombic lip

Migrating precerebellar nuclear neurons

Posterior intramural migratory stream (inferior olivary neurons)?

Medullary reticular formation

Solitary nucleus and tract

Migrating raphe nuclear complex neurons?

Settling inferior olive neurons

Floor plate

Ventral spinal G/EP

Intermediate spinal NEP

Dorsal spinal NEP

Roof plate

Ventral horn interneurons

Lateral funiculus

Intermediate gray

CENTRAL CANAL

Dorsal funiculus

Dorsal gray

FONT KEY:
VENTRICULAR DIVISIONS - CAPITALS
Germinal zone - Helvetica bold
Transient structure - Times bold italic
Permanent structure - Times Roman or **Bold**

ABBREVIATIONS:
CTF - Cerebellar transitional field
G/EP - Glioepithelium/ependyma
GEP - Glioepithelium
NEP - Neuroepithelium

Arrows indicate the presumed *direction of neuron migration* from neuroepithelial sources.

190

Peripheral neural and
non-neural structures labeled

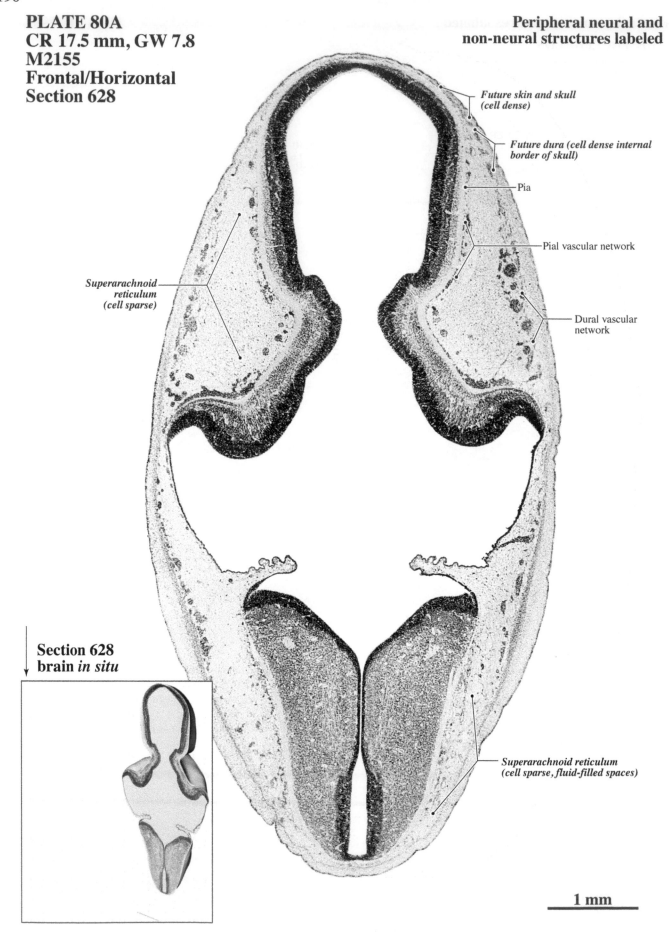

Future skin and skull
(cell dense)

Future dura (cell dense internal
border of skull)

Pia

Pial vascular network

Dural vascular
network

Superarachnoid
reticulum
(cell sparse)

Section 628
brain *in situ*

Superarachnoid reticulum
(cell sparse, fluid-filled spaces)

1 mm

Central neural structures labeled

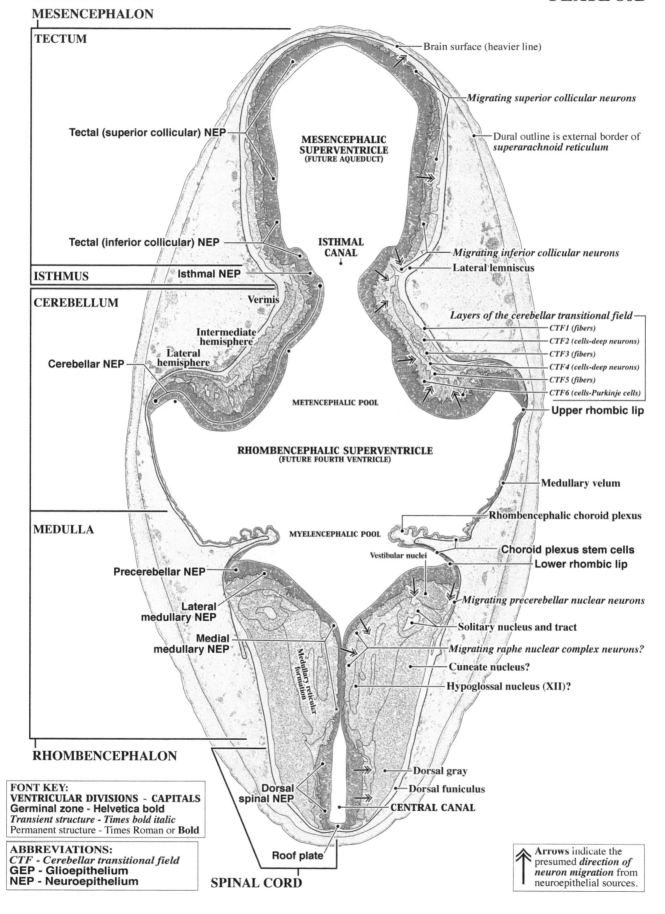

MESENCEPHALON

TECTUM

Brain surface (heavier line)

Migrating superior collicular neurons

Tectal (superior collicular) NEP

MESENCEPHALIC SUPERVENTRICLE (FUTURE AQUEDUCT)

Dural outline is external border of *superarachnoid reticulum*

Tectal (inferior collicular) NEP

ISTHMAL CANAL

Migrating inferior collicular neurons
Lateral lemniscus

ISTHMUS   Isthmal NEP

CEREBELLUM   Vermis

Layers of the cerebellar transitional field
CTF1 *(fibers)*
CTF2 *(cells-deep neurons)*
CTF3 *(fibers)*
CTF4 *(cells-deep neurons)*
CTF5 *(fibers)*
CTF6 *(cells-Purkinje cells)*

Intermediate hemisphere

Lateral hemisphere

Cerebellar NEP

Upper rhombic lip

METENCEPHALIC POOL

RHOMBENCEPHALIC SUPERVENTRICLE (FUTURE FOURTH VENTRICLE)

Medullary velum

Rhombencephalic choroid plexus

MEDULLA

MYELENCEPHALIC POOL

Vestibular nuclei

Choroid plexus stem cells
Lower rhombic lip

Precerebellar NEP

Migrating precerebellar nuclear neurons

Lateral medullary NEP

Medullary reticular formation

Medial medullary NEP

Solitary nucleus and tract

Migrating raphe nuclear complex neurons?
Cuneate nucleus?
Hypoglossal nucleus (XII)?

RHOMBENCEPHALON

Dorsal gray
Dorsal funiculus

Dorsal spinal NEP

CENTRAL CANAL

Roof plate

SPINAL CORD

Arrows indicate the presumed *direction of neuron migration* from neuroepithelial sources.

PLATE 81A
CR 17.5 mm, GW 7.8
M2155
Frontal/Horizontal
Section 677

Peripheral neural and non-neural structures labeled

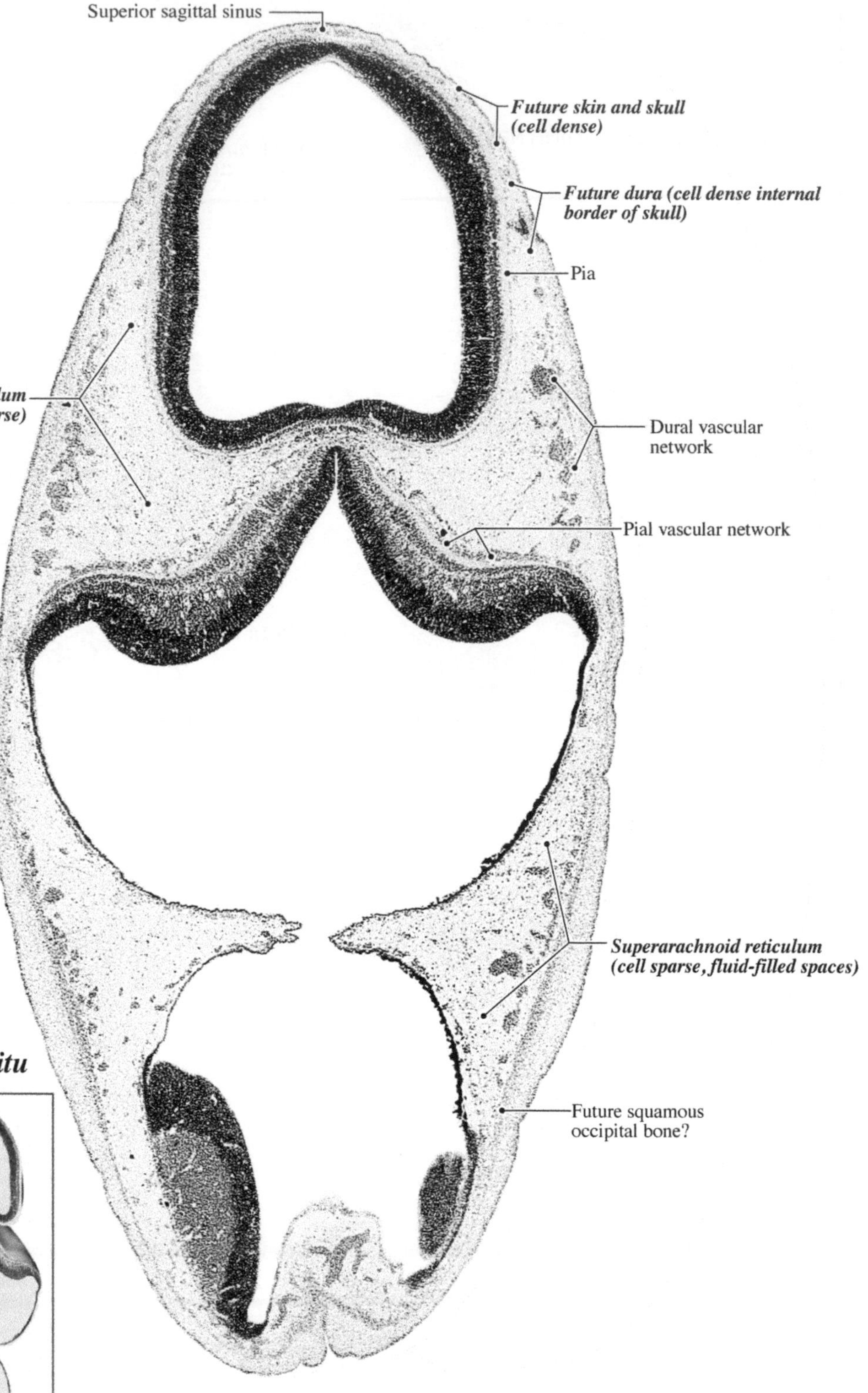

Superior sagittal sinus
Future skin and skull (cell dense)
Future dura (cell dense internal border of skull)
Pia
Superarachnoid reticulum (cell sparse)
Dural vascular network
Pial vascular network
Superarachnoid reticulum (cell sparse, fluid-filled spaces)
Future squamous occipital bone?
1 mm

Section 677 brain *in situ*

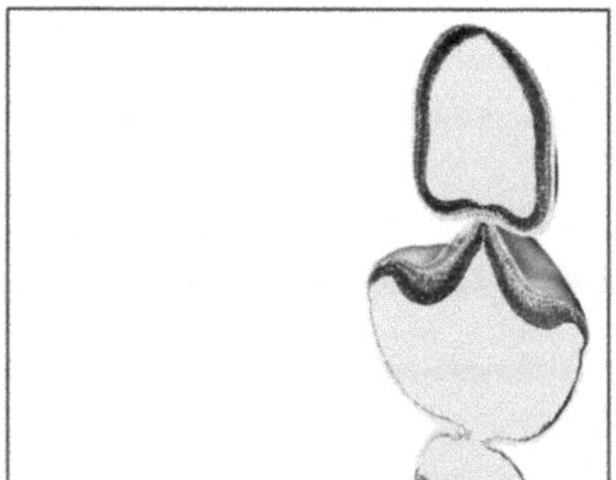

Central neural structures labeled

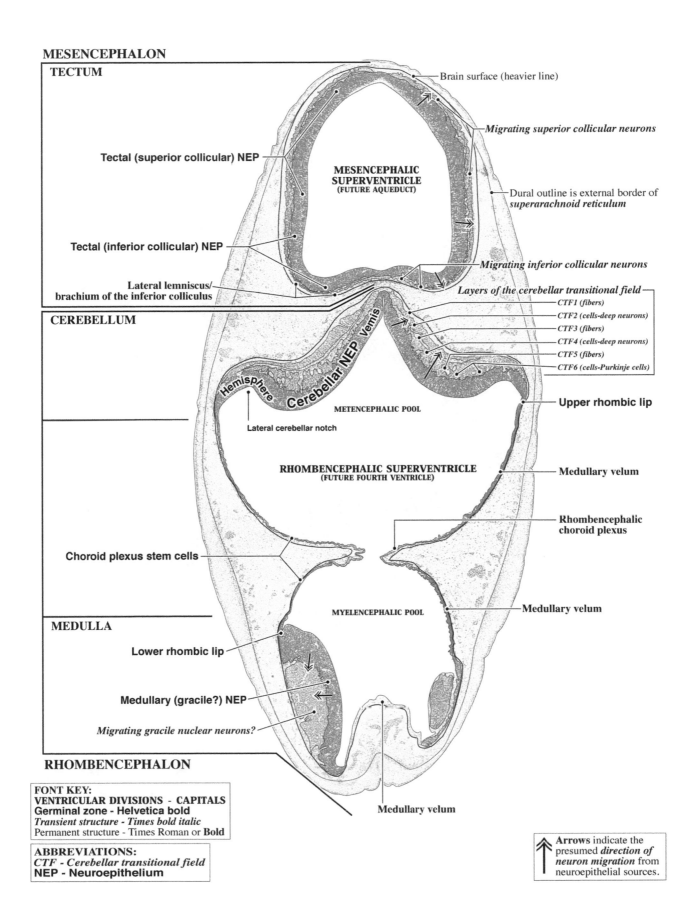

MESENCEPHALON

TECTUM

Brain surface (heavier line)

Migrating superior collicular neurons

Tectal (superior collicular) NEP

MESENCEPHALIC SUPERVENTRICLE
(FUTURE AQUEDUCT)

Dural outline is external border of *superarachnoid reticulum*

Tectal (inferior collicular) NEP

Migrating inferior collicular neurons

Lateral lemniscus/ brachium of the inferior colliculus

Layers of the cerebellar transitional field

CTF1 (fibers)
CTF2 (cells-deep neurons)
CTF3 (fibers)
CTF4 (cells-deep neurons)
CTF5 (fibers)
CTF6 (cells-Purkinje cells)

CEREBELLUM

Hemisphere

Cerebellar NEP Vemis

Upper rhombic lip

Lateral cerebellar notch

METENCEPHALIC POOL

RHOMBENCEPHALIC SUPERVENTRICLE
(FUTURE FOURTH VENTRICLE)

Medullary velum

Rhombencephalic choroid plexus

Choroid plexus stem cells

Medullary velum

MYELENCEPHALIC POOL

MEDULLA

Lower rhombic lip

Medullary (gracile?) NEP

Migrating gracile nuclear neurons?

Medullary velum

RHOMBENCEPHALON

FONT KEY:
VENTRICULAR DIVISIONS - CAPITALS
Germinal zone - Helvetica bold
Transient structure - Times bold italic
Permanent structure - Times Roman or **Bold**

ABBREVIATIONS:
CTF - Cerebellar transitional field
NEP - Neuroepithelium

Arrows indicate the presumed *direction of neuron migration* from neuroepithelial sources.

194

CEREBRAL CORTEX AND THALAMUS

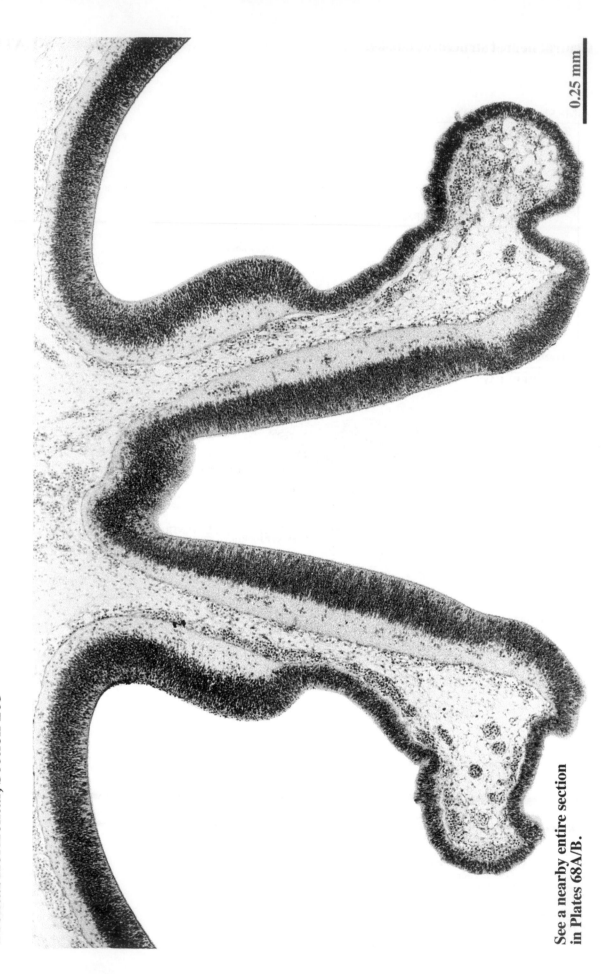

PLATE 82A
CR 17.5 mm, GW 7.8, M2155
Frontal/Horizontal, Section 203

See a nearby entire section
in Plates 68A/B.

0.25 mm

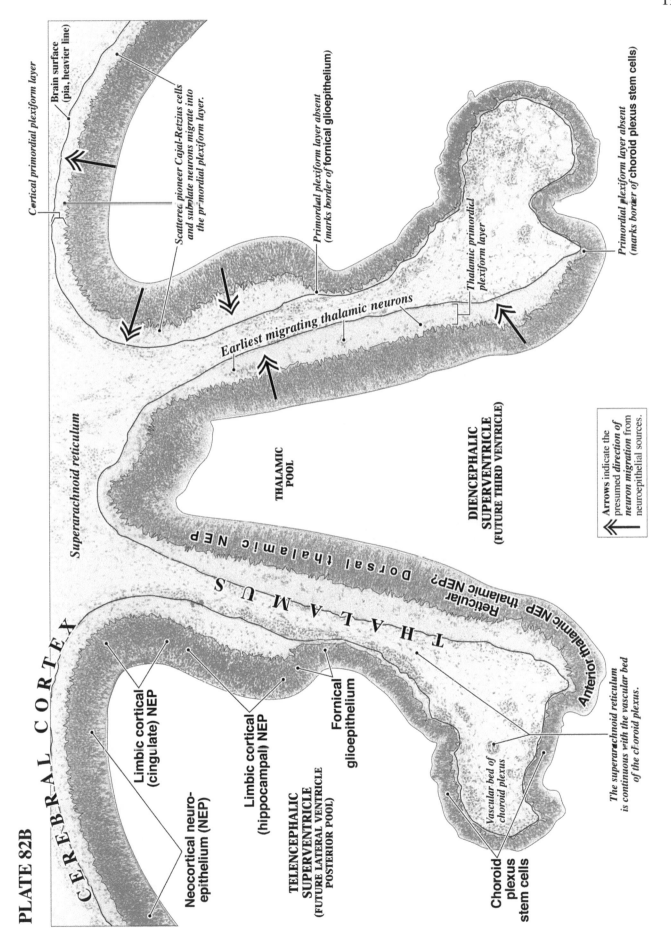

PLATE 82B

CEREBRAL CORTEX

Cortical primordial plexiform layer

Brain surface (pia, heavier line)

Scattered pioneer Cajal-Retzius cells and subplate neurons migrate into the primordial plexiform layer.

Primordial plexiform layer absent (marks border of fornical glioepithelium)

Thalamic primordial plexiform layer

Primordial plexiform layer absent (marks border of choroid plexus stem cells)

Superarachnoid reticulum

Earliest migrating thalamic neurons

THALAMIC POOL

DIENCEPHALIC SUPERVENTRICLE (FUTURE THIRD VENTRICLE)

Arrows indicate the presumed direction of neuron migration from neuroepithelial sources.

THALAMUS

Dorsal thalamic NEP

Reticular thalamic NEP?

Anterior thalamic NEP?

Neocortical neuro-epithelium (NEP)

Limbic cortical (cingulate) NEP

Limbic cortical (hippocampal) NEP

TELENCEPHALIC SUPERVENTRICLE (FUTURE LATERAL VENTRICLE POSTERIOR POOL)

Fornical glioepithelium

Vascular bed of choroid plexus

Choroid plexus stem cells

The superarachnoid reticulum is continuous with the vascular bed of the choroid plexus.

CEREBRAL CORTEX AND THALAMUS

PLATE 83A
CR 17.5 mm, GW 7.8
M2155
Frontal/Horizontal
Section 236

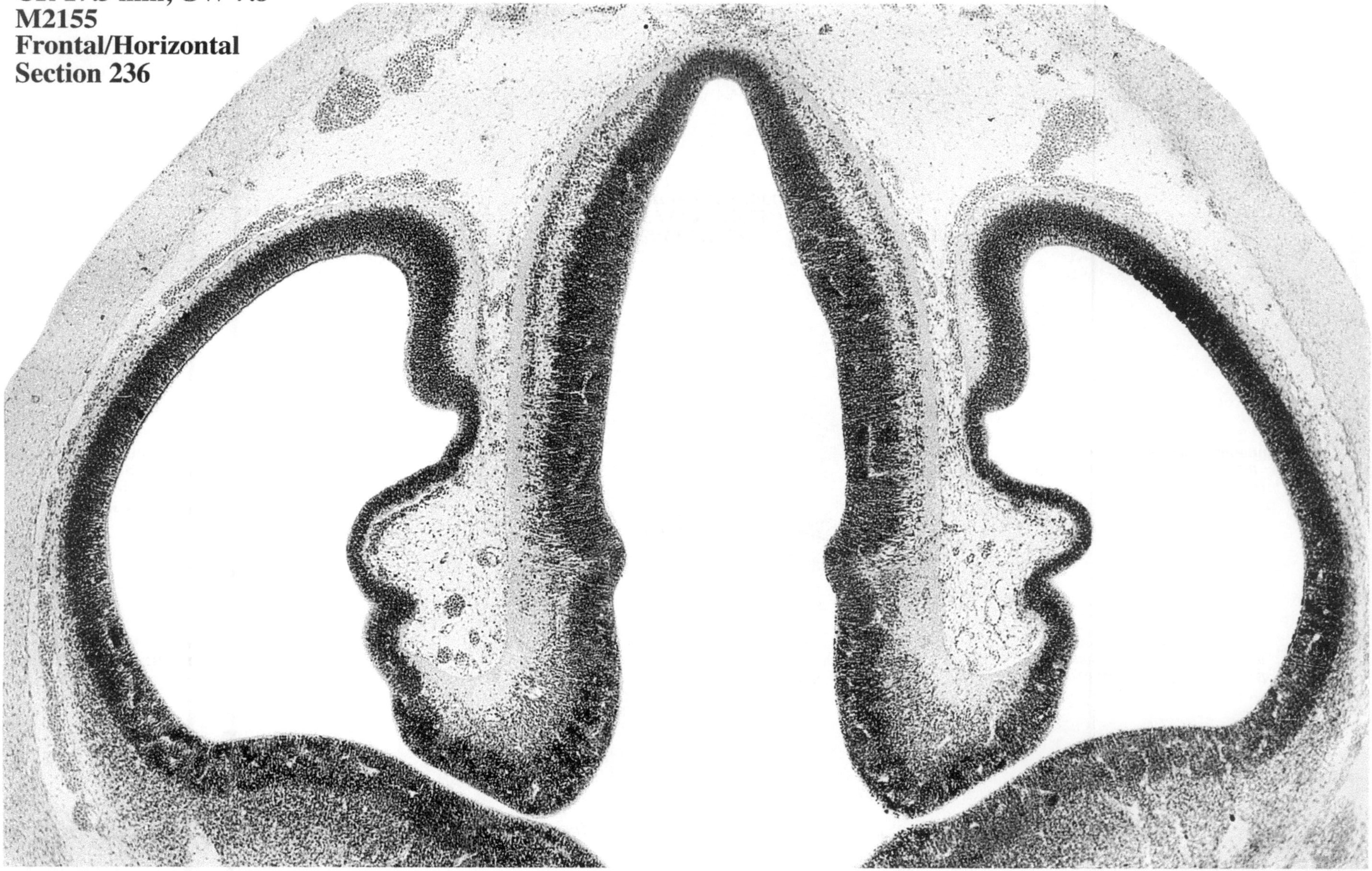

See a nearby entire section in Plates 69A/B.

0.25 mm

PLATE 83B

Arrows indicate the presumed *direction of neuron migration* from neuroepithelial sources.

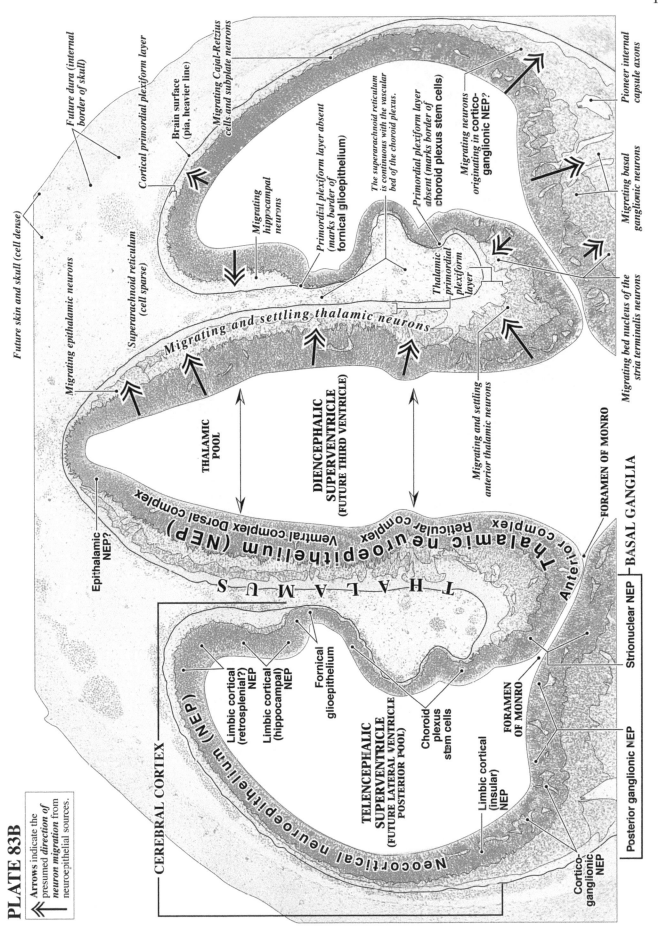

Future skin and skull (cell dense)

Future dura (internal border of skull)

Cortical primordial plexiform layer

Brain surface (pia, heavier line)

Migrating Cajal-Retzius cells and subplate neurons

Pioneer internal capsule axons

Migrating epithalamic neurons

Superarachnoid reticulum (cell sparse)

Migrating hippocampal neurons

Primordial plexiform layer absent (marks border of fornical glioepithelium)

The superarachnoid reticulum is continuous with the vascular bed of the choroid plexus.

Primordial plexiform layer absent (marks border of choroid plexus stem cells)

Migrating neurons originating in cortico-ganglionic NEP?

Migrating basal ganglionic neurons

Thalamic primordial plexiform layer

Migrating bed nucleus of the stria terminalis neurons

Migrating and settling thalamic neurons

Migrating and settling anterior thalamic neurons

Epithalamic NEP?

THALAMIC POOL

DIENCEPHALIC SUPERVENTRICLE (FUTURE THIRD VENTRICLE)

Thalamic neuroepithelium (NEP) Ventral complex Reticular complex Dorsal complex

T H A L A M U S

Anterior complex

FORAMEN OF MONRO

BASAL GANGLIA

CEREBRAL CORTEX

Neocortical neuroepithelium (NEP)

Limbic cortical (retrosplenial?) NEP

Limbic cortical (hippocampal) NEP

Fornical glioepithelium

TELENCEPHALIC SUPERVENTRICLE (FUTURE LATERAL VENTRICLE POSTERIOR POOL)

Choroid plexus stem cells

Limbic cortical (insular) NEP

FORAMEN OF MONRO

Strionuclear NEP

Posterior ganglionic NEP

Cortico-ganglionic NEP

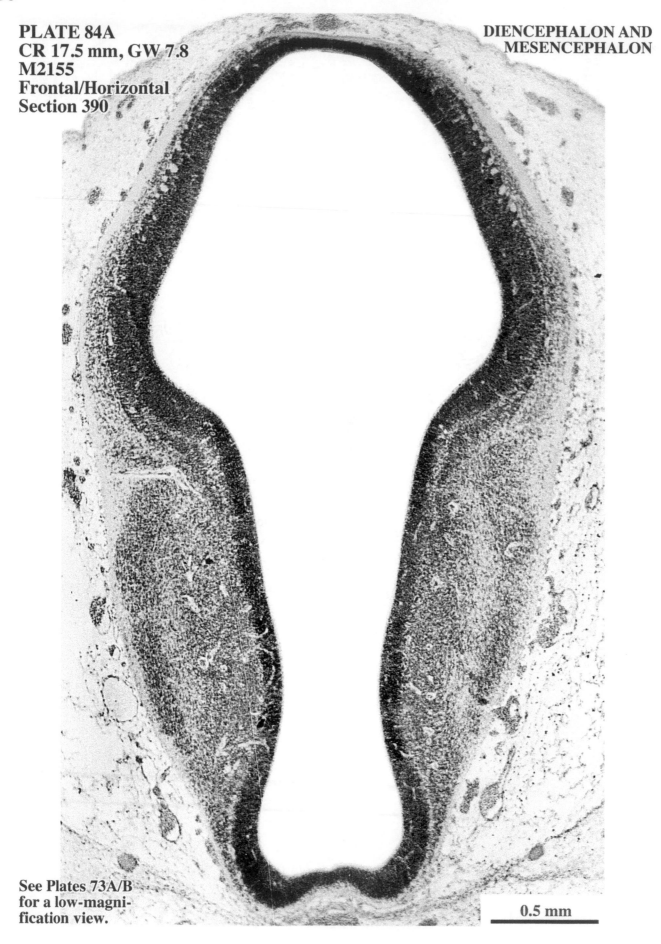

PLATE 84A
CR 17.5 mm, GW 7.8
M2155
Frontal/Horizontal
Section 390

DIENCEPHALON AND
MESENCEPHALON

See Plates 73A/B
for a low-magni-
fication view.

0.5 mm

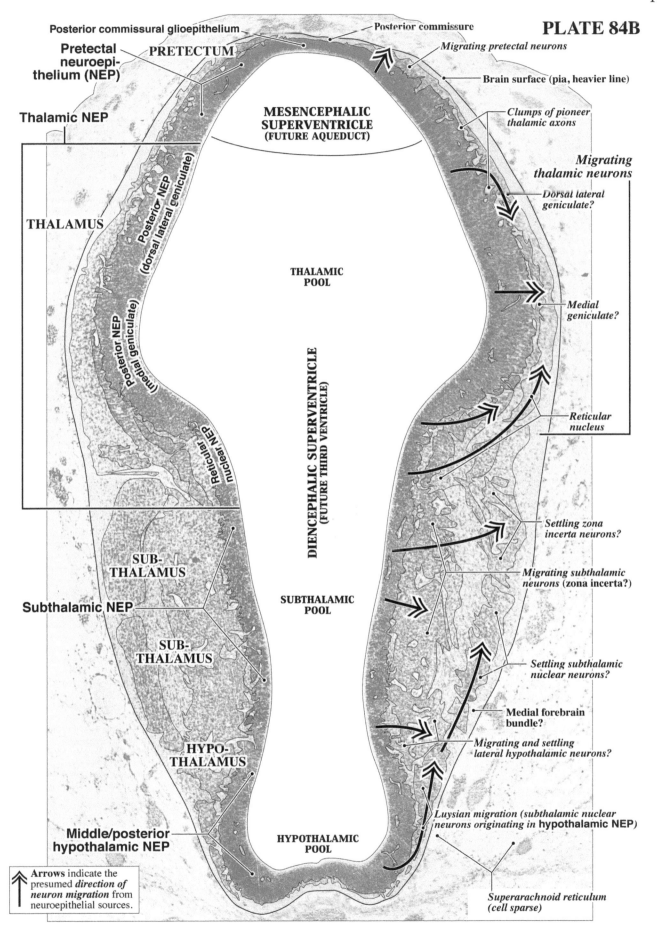

Posterior commissural glioepithelium

Posterior commissure

PRETECTUM

Migrating pretectal neurons

Pretectal neuroepi-thelium (NEP)

- Brain surface (pia, heavier line)

Thalamic NEP

Clumps of pioneer thalamic axons

MESENCEPHALIC SUPERVENTRICLE (FUTURE AQUEDUCT)

Migrating thalamic neurons

THALAMUS

Posterior NEP (dorsal lateral geniculate)

Dorsal lateral geniculate?

Posterior NEP (medial geniculate)

THALAMIC POOL

Medial geniculate?

Reticular nuclear NEP

DIENCEPHALIC SUPERVENTRICLE (FUTURE THIRD VENTRICLE)

Reticular nucleus

Settling zona incerta neurons?

SUB-THALAMUS

Migrating subthalamic neurons (zona incerta?)

Subthalamic NEP

SUBTHALAMIC POOL

SUB-THALAMUS

Settling subthalamic nuclear neurons?

Medial forebrain bundle?

HYPO-THALAMUS

Migrating and settling lateral hypothalamic neurons?

Middle/posterior hypothalamic NEP

HYPOTHALAMIC POOL

Luysian migration (subthalamic nuclear neurons originating in **hypothalamic NEP**)

Arrows indicate the presumed *direction of neuron migration* from neuroepithelial sources.

Superarachnoid reticulum (cell sparse)

9 781032 219288